D1499828

ANIMAL HOMING

Chapman & Hall Animal Behaviour Series

SERIES EDITOR

D.M. Broom

Colleen Macleod Professor of Animal Welfare, Department of Clinical Veterinary Medicine, University of Cambridge, UK

Detailed studies of behaviour are important in many areas of physiology, psychology, zoology and agriculture. Each volume in this series provides a concise and readable account of a topic of fundamental importance and current interest in animal behaviour.

The major areas covered range from behavioural ecology and socio-biology to general behavioural mechanisms and physiological psychology. Many facets of the study of animal behaviour are explored and the topics included reflect the broad scope of the subject. Each volume provides a rigorous and balanced view of the subject at a level appropriate for senior undergraduates and research workers.

Books for the series are currently being commissioned and those interested to contribute, or who wish to know more about the series, are invited to contact one of the editors.

Other books in the series

Animal Conflict

Felicity A. Huntingford and Angela K. Turner

Animal Motivation

Patrick Colgan

Searching Behaviour
The behavioural ecology of finding resources

W.J. Bell

ANIMAL HOMING

Edited by

F. Papi

Dipartimento di Scienze del Comportamento Animale e dell 'Uomo
Università degli Studi di Pisa
Pisa, Italy

CHAPMAN & HALL
London · New York · Tokyo · Melbourne · Madras

Published by Chapman & Hall, 2–6 Boundary Row, London SE1 8HN

Chapman & Hall, 2–6 Boundary Row, London SE1 8HN, UK

Blackie Academic & Professional, Wester Cleddens Road, Bishopbriggs, Glasgow G64 2NZ, UK

Chapman & Hall, 29 West 35th Street, New York NY10001, USA

Chapman & Hall Japan, Thomson Publishing Japan, Hirakawacho Nemoto Building, 6F, 1-7-11 Hirakawa-cho, Chiyoda-ku, Tokyo 102, Japan

Chapman & Hall Australia, Thomas Nelson Australia, 102 Dodds Street, South Melbourne, Victoria 3205, Australia

Chapman & Hall India, R. Seshadri, 32 Second Main Road, CIT East, Madras 600 035, India

First edition 1992
© 1992 Chapman & Hall

Typeset in 11/12 pt Bembo by Best-set Typesetter Ltd, Hong Kong
Printed in Great Britain by Cambridge University Press, Cambridge, England

ISBN 0 412 36390 9

A catalogue record for this book is available from the British Library

Library of Congress Cataloging-in-Publication data
Animal homing/edited by F. Papi.
    p. cm. – (Chapman and Hall animal behaviour series)
    Includes bibliographical references and index.
    ISBN 0–412–36390–9
    1. Animal homing. I. Papi, F. (Floriano), 1926– . II. Series.
QL782.5.A53 1992   92-8431
591.51 – dc20   CIP

∞ Printed on permanent acid-free text paper, manufactured in accordance with the proposed ANSI/NISO Z 39.48-199X and ANSI Z 39.48-1984

Contents

Contributors

Jacques Bovet
Université Laval
Faculté des Sciences et de Génie
Department de Biologie
Cité Universitaire
Québec
Canada GlK 7P4

Guido Chelazzi
Università degli Studi di Firenze
Dipartimento di Biologia Animale
e Genetica
Via Romana, 17
I-50125 Firenze
Italy

Andrew Dittman
University of Washington
Fisheries Research Institute
Seattle, Washington 98195
USA

Floriano Papi
Università degli Studi di Pisa
Dipartimento di Scienze del
Comportamento Animale e dell'Uomo
Via A. Volta, 6
I-56126 Pisa
Italy

Thomas P. Quinn
University of Washington
Fisheries Research Institute
Seattle, Washington 98195
USA

Ulrich Sinsch
Institut für Bioloqie
Universität Koblenz-Landau
Rheinau 3–4
D-5400 Koblenz
Germany

Hans Georg Wallraff
Max-Planck-Institut für Verhaltensphysiologie
D-8130 Seewiesen über Starnberg
Germany

Rüdiger Wehner
Zoologisches Institut der
Universität Zürich
Winterthurerstrasse 190
CH-8057 Zürich
Switzerland

Acknowledgements

The editor is particularly indebted to the Accademia Nazionale dei Lincei and the Centro per la Faunistica ed Ecologia Tropicali del Consiglio Nazionale delle Ricerche for supporting the workshop on animal homing held in Rome in May 1989 which enabled the authors to meet and draw up the plan of this book. He also wishes to express his gratitude to the authors for their wonderful spirit of co-operation; their criticisms and ideas were also most valuable in the preparation of the introductory chapter. Special thanks are due to the Series Editor, Professor D. Broom, and to Dr R.C.J. Carling and Dr Clem Earle of Chapman & Hall, for their precious help and suggestions. Lastly, he is most grateful to Dr Resi Mencacci for having assisted him with great care and skill in all the phases of this volume's production.

Acknowledgements

Preface

Homing phenomena must be considered an important aspect of animal behaviour on account of their frequent occurrence, their survival value, and the variety of the mechanisms involved. Many species regularly rely on their ability to home or reach other familiar sites, but how they manage to do this is often uncertain. In many cases the goal is attained in the absence of any sensory contact, by mechanisms of indirect orientation whose complexity and sophistication have for a long time challenged the skill and patience of many researchers. A series of problems of increasing difficulty have to be overcome; researchers have to discover the nature of orienting cues, the sensory windows involved, the role of inherited and acquired information, and, eventually, how the central mechanisms process information and control motory responses. Naturally, this book emphasizes targets achieved rather than areas unexplored and mysteries unsolved. Even so, the reader will quickly realize that our knowledge of phenomena and mechanisms has progressed to different degrees in different animal groups, ranging from the mere description of homing behaviour to a satisfactory insight into some underlying mechanisms.

In the last few dacades there have been promising developments in the study of animal homing, since new approaches have been tried out, and new species and groups have been investigated. Despite this, homing phenomena have not recently been the object of exhaustive reviews and there is a tendency for them to be neglected in general treatises on animal behaviour. While taking part in a workshop on animal homing held in Rome in May 1989, the authors of the present book agreed to meet the need for a textbook on homing, which would combine the criteria of exhaustive scientific information and readability.

As explained at the beginning of the introductory chapter, the present book not only reviews animal homing in the narrow sense, but includes all phenomena of goal orientation which share mechanisms with homing. Since animal groups have different habits, each posing a variety of problems of major interest, a systematic presentation of the subject can be adopted without this creating a risk of serious overlapping. Each group has been discussed by one or two authors with plenty of research experience in that group, but every reasonable effort has been made to reduce dissimilarities between chapters. Each chapter devoted to a group first reports the natural history of homing in the animals considered, and then reviews our knowledge of their mechanisms of orientation and navigation; in some cases, where this has been possible, the neurobiological machinery that underlies homing has been illustrated and current phylogenetic views are discussed.

F.P.
Pisa

Chapter 1
General aspects

F. Papi

1.1 WHAT IS HOMING?

In many animals, locomotion may be directed to any place where suitable environmental conditions, food or sexual partners can be found. Species whose behaviour is complex, however, usually move towards specific sites where they feed, spawn, mate or find shelter. Such sites can be personal or shared with conspecifics, but their users can tell each one from all others, even if they are similar. To perform this kind of spatial activity, animals have to rely on orientational mechanisms which guarantee homing.

The term **homing**, as it is ordinarily understood, indicates a return to the place where an animal lives and has its shelter, den or nest. However, since the same mechanisms may operate for movements to and from the sites that the animal visits regularly, the study of homing must include consideration of any movement undertaken to reach a spatially restricted area which is known to an animal. All such movements are referred to in the chapters of this book, so that its scope is enlarged beyond that of the narrower meanings of homing. This book, in fact, reviews both homing and related behaviour patterns, and, therefore, all phenomena of **goal orientation** in which the goal is a specific, spatially restricted area.

Some special cases are discussed with respect to this definition of true homing and related phenomena, hereafter called homing. Animals living near the boundary between two different environments (e.g. the sea-shore) display so-called **zonal orientation** (Jander, 1975) or **y-axis orientation** (Ferguson, 1967). This phenomenon has been the object of many studies on animals from sea, lake, or river shores (Herrnkind, 1972). After an active or passive movement away from the shore, they orientate perpendicu-

Animal Homing. Edited by Floriano Papi. Published in 1992 by Chapman & Hall, 2–6 Boundary Row, London SE1 8HN. ISBN 0 412 36390 9.

larly to the shore, until they reach it. Unless the animals further manage to reach the spot from which they moved or were displaced, zonal orientation cannot be considered homing.

Let us now examine the case of a foraging bee, which, when informed by the dance of a companion about the distance and direction of a new food source, flies to the goal. By widening the definition of homing a little, it may be granted that the bee's flight belongs to homing phenomena, though the goal is only known through information given by a conspecific, not by direct experience.

A third case is even more puzzling. In some bird species, the young can perform their first migration by themselves and reach a wintering area far away. The birds have no experience of it, but they possess information about distance and direction from the native quarters. In this case, information derives from inheritance (Berthold, 1990): should a first migratory flight of this kind be considered a form of homing behaviour?

Even if this last case may raise doubt, it is clear that all migratory journeys after the first constitute homing phenomena. Animals often aim at the very site they had lived in during a previous stay in their reproductive, wintering, or feeding area. In birds this phenomenon has been known since ancient times; in the thirteenth century it was reported by the Emperor Frederick II of Germany in his famous treatise on the art of hunting with birds (*De Arte Venandi cum Avibus*), yet the first experimental evidence was only obtained at the end of the Age of Enlightenment. Lazzaro Spallanzani put red bands on the legs of some swallows and observed their coming back to nest on the same house the following year.

Man has always been intrigued by the animal capabilities that he does not possess as a natural faculty. Birds orientating from one continent to another, displaced cats, dogs and pigeons which find their way home, salmon swimming back to their native stream, and also insects which can explore and forage over areas which are enormous for such small animals, all strike the imagination of the layman, and prod the zoologist into investigating these phenomena.

At first sight, other animals look less attractive as regards their spatial activities. Some have a definite, central shelter from which they leave and return to for feeding purposes (**radial cropping**, radial looping, or central place foraging). Typical central place foragers are social insects, but limpets and chitons display the same behaviour (Fig. 2.2b, p. 22). Many vertebrates, and invertebrates, regularly visit a specific area, the **home range** that some authors (e.g. Baker, 1978) distinguish from a larger, surrounding **familiar area**. It may appear obvious that the user masters its home range and skilfully moves in it, but we often actually ignore many fundamental facts, such as the previous experience necessary for this performance, the nature of the cues which the animal relies on, and how the sensory information is centrally processed. Answers to these questions are, of course, different for animals of different groups or living in different environments, and this increases the number of 'black boxes' we should open and investigate. Recently, due attention has been devoted to these less exciting aspects of animal homing, so yielding rewarding discoveries and achievements, and opening up new prospects for the future. If many of the puzzles that have emerged from the study of both spectacular and 'humble' spatial performances have not yet been resolved, the main reason lies in the need to combine a variety of behavioural, physiological and computational approaches to finally understand how animals find their way home.

1.2 METHODS

In studying homing behaviour in any species, the first step obviously consists of recording movements of identifiable individuals in their natural environment. Unless single individuals can be recognized on the basis of distinctive natural or accidental markings, such as pigmentation or scars (e.g. in whales), distinctive tags, rings, collars or similar devices must be attached. The amputation of one or more phalanges has been used to mark amphibians, and fin clipping for fishes. Insects are often marked with colour spots, more seldom with numbered labels.

Recordings of successive positions of marked animals can be entrusted to casual recoveries or to one or more recapture or resighting efforts (**capture–marking–recapture method, CMR**). This method proved to be indispensable in the study of migration. Mass ringing studies in birds produced the bulk of our knowledge on the location of their reproduction and wintering areas, both at the species and the population levels. Migration of several fish species was also successfully investigated using the CMR method.

The CMR method is also applied in the study of the size of the home range, the routes followed inside and the excursions outside it. Two locations of the same individual at different sites are evidence that this animal moved from the first to the second site, the maximum time interval being that between the two captures or sightings. However, this conveys nothing about the actual route travelled, stops and possible detours made between the two sites.

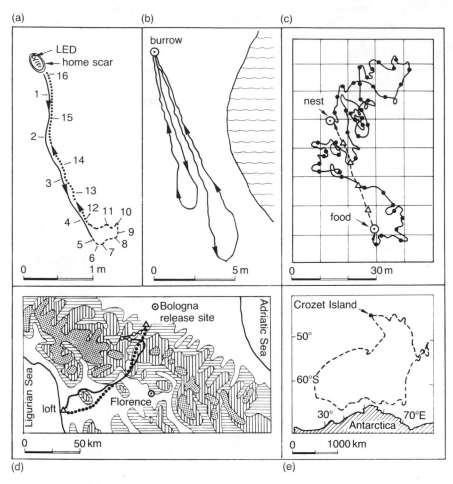

Figure 1.1 Homing paths of animals recorded with different methods. (a) Foraging roundtrip of the intertidal chiton *Acanthopleura gemmata*, which moved from its home scar to a feeding ground near the low tide limit. The chiton was equipped with a light emitting diode (LED), which enabled time-lapse photographic recordings to be made. Time marks (1–16) are given at 10-min intervals. Original: courtesy of P. Della Santina. (b) Two return excursions of a specimen of the tortoise, *Testudo hermanni*, from its burrow in a coastal dune. The tracks were obtained by thread trailing: the tortoise left behind an Ariadne's thread unwinding from a reel attached on its back. After Chelazzi and Francisci (1979), modified. (c) Starting from its nest, the desert ant, *Cataglyphis albicans*, follows a winding path until it finds food, then it runs back on a straight course (dashed). The path was reconstructed by visual observation made easier by a grid painted on the terrain. Time marks (filled circles and open triangles) are given every 10 s.

The paths of walking animals can be recorded by different methods according to the animal's size and speed. In fast running insects, track reconstruction can be supported by a grid painted on the soil (Fig. 1.1c); in tortoises and toads, by thread trailing (Fig. 1.1b). The construction of radio and ultrasonic transmitters suitable for use in **tracking experiments** has been one of the major technological advances affecting the investigation of animal homing (Amlaners and Macdonald, 1980). This made it possible to reconstruct animal movements either by successive locations (or 'fixes'), or by continuous monitoring. The position of animals equipped with suitable radiotransmitters can be recorded several times in a day by means of satellites. This technique, initially applicable to large animals only, is bound to become more widely used thanks to the progressive miniaturization of transmitters. It has even been applied recently to birds with impressive results (Jouventin and Weimerskirch, 1990; Fig. 1.1e). The cost of satellite tracking is surprisingly low (Fancy *et al.*, 1989). A new animal-borne device which detects and stores directions of movement with respect to the earth's magnetic field allows route reconstruction (Fig. 1.1d) provided that both the animal and the device are recovered (Bramanti *et al.*, 1988).

The repeated field recordings of movements in an animal species are enough by themselves to shed light on mechanisms underlying homing, particularly if recordings are made when a range of orientational cues are available, as for instance under clear or overcast skies. In most cases, however, experimental interference is still necessary. It consists in any purposeful alteration either of the course of events, which the homing animal is involved in, or of the related phenomena of sensory reception, information processing, and motor output.

The most simple and frequent type of experiment (thousands have been reported in the literature) is the **displacement experiment**,

---

From Wehner (1982), modified. (d) Two flights of pigeons crossing the Appennine chain. The tracks were obtained by a direction recorder carried by the birds. Note the course correction of a bird and its frequent stops (open dots) in the first leg. Original: courtesy of P. Dall'Antonia. (e) A foraging roundtrip flight of a male wandering albatross, *Diomedea exulans*, reconstructed by satellite tracking. The bird homed at Crozet Island after a 33-day flight over open ocean and along the Antarctic coast. From Jouventin and Weimerskirch (1990), modified.

which is performed by releasing a marked animal far away from the capture site. This appears to be a meaningful variation on the CMR experiment. In most cases, animals are removed from their permanent or temporary quarters, but displacement experiments have also been performed while migration is in progress.

The percentage of animals which home – known as **homing success** – and their **homing speed** are both important, but the fate and behaviour of individuals which completely fail to home are of interest too. The most important datum, however, is the route followed by homers. It can be costly, in terms of time and money, to record this in releases of large groups of specimens from great distances; observations are therefore often limited to the movements that immediately follow release. The outcome is often disappointing, since many animals do no more than look for shelter as soon as they become free. Others, on the contrary, move away until they vanish from sight, sometimes heading more or less precisely towards their goal. In such cases the orientation after release, or **initial orientation**, is one of the most valuable findings, and is evaluated from the distribution of single bearings (the so-called vanishing or initial bearings, or take-off directions), which are usually pooled in circular diagrams.

The aim of experimental interference in sensory information is usually that of preventing or restricting the reception of possible orientational cues. The effects of preventing animals from seeing, hearing or smelling have been tested with a variety of methods. Permanent magnets or varying magnetic fields are used when reliance on magnetic cues is suspected, while damage done to the labyrinth or transportation in rotating drums should disturb homing based on inertial navigation (see later). To investigate the role of information gathered *en route*, treatment can be limited to transportation to the release site. Even deep anaesthesia or continually bombarding animals with a variety of noisy stimuli (Wallraff, 1980) have been used as a radical procedure. Negative outcomes from such experiments must be evaluated cautiously. Natural selection not only promoted the acquisition of homing mechanisms which are complex and refined, but also provided animals with the ability to use more than one mechanism and to rely on alternative or auxilary information coming from different sources.

Intensive field and laboratory experiments on some species have paved the way to the study of the neural hardware of the homing mechanism. In insects, the celestial compass sits in an anatomically and physiologically specialized part of the eye (Wehner, 1987). In pigeons, field tests on animals subjected to ablation of some cerebral areas were successful in determining their role in homing (Bingman *et al.*, 1989; Papi and Casini, 1990).

Circular statistics are a necessary foundation for the evaluation of the results obtained. Vector analysis yields a mean vector, whose length is inversely proportional to the scattering of bearings. The Rayleigh and V-test are routinely applied to test bearing sets for randomness, while U^2, Mardia or other tests are used to compare bearing distributions. The Hotelling test and other bivariate methods are used when mean vector sets have to be tested or when each sample is bivariate to account for direction and distance. The excellent monograph of Batschelet (1981) on circular statistics in biology can be recommended to any beginner in these studies.

1.3 COMPASS ORIENTATION

The first aim of any basic research on homing is to determine the source of the spatial information allowing proper orientation. When it derives from sensory information relative to physical or chemical parameters in the environment, orientation is called **extrinsic** (Jander, 1975) or, more frequently, **allothetic** (Mittelstaedt and Mittelstaedt, 1973). Having determined the nature of orienting cues, the successive steps arise from the investigation of the physiological machinery which acquires information, then processes it, and eventually produces the appropriate movement. In this connection, a formidable breakthrough was made when it was discovered that animals can use the position of celestial bodies and related astronomical phenomena to select and maintain specific directions. A star compass, for birds, and a sun compass, for a great many animals, play an important role in homing, whereas a moon compass is used by crustacean amphipods for their zonal orientation (Papi and Pardi, 1953).

The reliance on the sun's position for orientation was first observed by Santschi (1911) in ants, but the possession of a true **sun compass**, i.e. an animal's ability to compensate for the sun's daily movement, was not reported until the 1950s when astronomical orientation was investigated in bees, shore-dwelling arthropods, and birds (see p. 91, for references). This time-compensated sun orientation mechanism turned out to be widespread in the animal kingdom and is apparently based on the same principles. The animals only take into account the sun's azimuth, i.e. its geographical direction, and compensate for the azimuth change during the day by varying their angle of orientation with the sun. The process is regulated by an endogenous rhythm or **biological clock**. The best evidence that an animal relies on a sun compass is obtained by subjecting it to a shifted light–dark cycle. As a result, the internal clock is also shifted and the orientation under the sun is deflected in a predictable way (Hoffmann, 1953). The rate of change of the sun azimuth is on average 15°/h, but it varies during the day and the season. If an animal compensated for the sun's movement at a constant rate, this would lead to considerable errors in temperate and tropical zones. To keep a constant bearing, the rate of change of the sun's angle of orientation must vary according to the local sun movement; this has actually been demonstrated in some cases (Papi et al., 1957; Braemer and Schwassmann, 1963).

Customarily, one distinguishes orientation with the sun from orientation with polarized light coming from the blue sky. The complex pattern made up of polarized light, with its differences in direction of vibration and intensity of polarization determined by the sun's position (pp. 87–8), is probably perceived by animals as a single picture comprising the sun's position and the differences in light brightness and spectrum in different sky areas. The picture is redundant enough to allow orientation when only part of it is visible.

A view of the sky with the setting sun appears to play a role in night-migrating birds, since orientation is significantly worse when birds can only see the stars after dark (Alerstam, 1990). Also, bats are able to use the evening twilight to orientate (p. 348).

Reliance on stars in passerine birds migrating at night was first reported in the late 1950s (Sauer, 1957). In some species at least, orientation by a **star compass** does not require compensation for the rotation of the starry sky (Emlen, 1967). At present it seems that a star compass is restricted to birds.

The old idea that animals may use the earth's magnetic field to determine compass directions was not supported by experimental evidence until the 1960s, when Merkel and Wiltschko (1965) showed that migrating robins deprived of visual cues rely on the magnetic field for orientation. Since then intensive research has been performed with the result that a **magnetic compass** has been attributed to a variety of animals belonging to all main phyla (Kirschvink et al., 1985). However, the state of the art is unsatisfactory for several reasons. Excepting sharks and rays, which may use their electroreceptors for magnetic detection (Kalmijn, 1971), no evidence of the receptor or transduction mechanism of magnetic stimuli has so far been discovered in other groups. In addition, evidence of magnetic orientation is often difficult to obtain, as orientation based on only the magnetic field is often inaccurate, showing very scattered bearings. Besides this, attempts to replicate the results of other authors or even of authors' own previous results have often proved inconclusive (Griffin, 1987, 1990). Many claims in favour of magnetic orientation are based on results which show magnetic sensitivity only, without demonstrating the existence of a working magnetic compass. A convincing demonstration of magnetic orientation should be based on a shift in orientation in the predicted way following a corresponding shift in the surrounding field.

1.4 CLASSIFICATION OF HOMING PHENOMENA

Homing may be based on cues of different kinds. It involves various orientation mechanisms and depends to a greater or lesser extent on genetic information or individual experience. Thus the task of classifying homing phenomena is a difficult one. In discussing the homing of pigeons, Griffin (1952) proposed a classification based on three levels of ability, which has often been adopted to classify homing phenomena in a variety of animals. The first type is 'reliance on visual landmarks within familiar territory and the use of exploration or some form of undirected wandering when released in unfamiliar territory' (this type therefore comprises two different homing mechanisms, which will be referred to below as 'pilotage' and 'random or systematic search'). The second type occurs when an animal roughly maintains a fixed compass direction, usually that adhered to in previous journeys, whether its bearing is related to the actual goal or not. This type does not imply establishing home bearing. Conversely, the third type is that allowing goal orientation even from an unfamiliar area and an unusual direction. Griffin's scheme had the advantage of permitting classification on the basis of overt behaviour only, but, after 40 years of continuous research work in the field, it has become obsolete and is no longer suitable for the arrangement of the many different homing phenomena that have been described (Schmidt-Koenig, 1975; Able, 1980; Merkel, 1980). In the meantime, new terms and classifications of orientation phenomena have been proposed (Jander, 1975; Schöne, 1980; Beugnon, 1986), which represent a substantial improvement with respect to the classical arrangement of Fraenkel and Gunn (1940).

The classification used in this book (Table 1.1) has been adopted from a recent proposal (Papi, 1990) and is only based on the origin of the information that is used to home. Broadly speaking, there are only three sources of information: genetic (or innate), sensory, and memory dependent (Jander, 1975). Since, however, they can be combined in various ways, and information from the three sources

Table 1.1 Classification of homing phenomena

Adopted term	Source of information (and homing mechanism)
Random or systematic search	Not available
Genetically based orientation	Genetic
Trail following	Trail left during a previous journey
Route-based orientation	Outward journey
Route reversal	Outward journey (following a landmark series)
Course reversal	Outward journey (reversing compass direction without integration)
Path integration	Outward journey (integrating distance and direction of each leg)
Pilotage	Acquired (topographic or cognitive) map (without using compass)
True navigation	Location-specific stimuli (deduced direction is selected by a compass)
Map-based navigation	Directly or indirectly acquired mosaic of local cues
Grid-based navigation	Physical or chemical gradients

can be subdivided into several items, the real number of categories is much larger.

As in any biological classification, difficulties arise from intermediate situations and from the many cases in which the animal appears to use two or more homing mechanisms, sometimes integrating them. Further problems derive from homing strategies which are still poorly understood. All the phenomena taken into account here concern homing by means of indirect orientation, i.e. without sensory contact with the goal. The following presentation of the single categories (see also Table 1.1) will allow further discussion of homing mechanisms and terminology.

1.4.1 Random or systematic search

Accidents occurring in nature or experimental displacements can deprive animals of the information necessary to home. They may then exhibit **transecting** behaviour by going on straight trips, which can be radially oriented excursions followed by a return to the starting point, as observed, for instance, in red squirrels (pp. 341–3). When the animal is informed that the goal is not far away, it may perform a convoluted search, which consists of loops of increasing size with regular returns to the starting point, as in the ant *Cataglyphis* (pp. 104–6).

1.4.2 Genetically-based orientation

Homing is often based on both inherited and acquired information, but the former appears to be prevalent in some cases. Some young birds reared in the laboratory are able to select the correct migratory direction and display migratory restlessness for the time needed to carry out their journey (Berthold, 1990). A similar ability was shown by young starlings (*Sturnus vulgaris*) displaced by Perdeck (1958) during their autumn migration. As mentioned above, it is doubtful whether this behaviour can be considered true homing or not. However, when hooded crows (*Corvus coronae*) were displaced during the period of their spring migration towards familiar quarters, they behaved like the young starlings just referred to (Rüppel, 1944), so one can conclude that this kind of orientation is a homing mechanism. Since both distance and direction are incorporated in it, this type of homing, along with others, has been labelled **vectorial orientation** or **vectorial navigation**. A more specific indication is, however, given by the term genetically-based orientation.

1.4.3 Trail following

Some animals find their way home by following a trail left during the outward journey. It is well known that molluscs (Fig. 2.5) and ants leave a continuous mucous and scent trail, respectively, while other animals mark their route with an intermittent series of odorous marks (for instance *Trigona* apids; Lindauer and Kerr, 1958). This mechanism of trail following does not involve memory information of the outward journey, whereas all the mechanisms classified as route-based orientation do.

1.4.4 Route-based orientation

This comprises three different mechanisms: route reversal, course reversal and path integration.

In a **route reversal** the animal retraces the outward path thanks to a series of reference points or landmarks memorized during the outward journey. Some mammals do this while moving from and to the home range (pp. 330–3). Insects rely on chains of memory images (pp. 106–15) to orientate during these routine trips; one can therefore suppose that, when they first perform a new return excursion, they find home by a route reversal based on visual images.

Course reversal requires that the animal records the compass direction of the outward journey. Plain orientation in the opposite direction will then lead towards the goal; if the distance has been recorded, finding home will be made easier. An example of course reversal is given by bees trained to fly to a feeding station. Evidence of their homing mechanisms derives from the simple experiment of displacing the bee from the feeding site at a right angle to the hive direction. The bee then takes off in the compass direction opposite to that of the outward flight (Otto, 1959). Course reversal after a passive outward journey has been reported in wasps (Ugolini, 1987).

Path integration is the third case of route-based homing and allows animals a straight return after a more or less winding outward trip (Fig. 1.1c). By taking into account the direction and length of each leg of the route covered, the animal continuously updates its position with respect to the starting point. The position is calculated in terms of distance and direction; for this reason, this mechanism has sometimes been known as **vectorial navigation** or **vectorial integration**, whereas the only term used here will be path integration.

Hymenopterans relying on path integration use external references to calculate the directions followed on an outward journey (p. 87) and geese do the same even during passive transportation (Saint-Paul, 1982). Orientation mechanisms based on external references are called **allothetic**; some animals on the contrary rely on **idiothetic** mechanisms based on centrally stored recordings of their own movements or programmes of movement. Idiothetic path integration (p. 347) is often contrasted with the allothetic path integration. However, both internal and external references probably play a role in many cases. Different terms have been used for cases of reliance on exclusively internal sources of reference; of these, **kinaesthetic orientation** (Görner, 1958) has a broader meaning than **endokinetic** or **intrinsic orientation** (Jander, 1975), while currently **idiothetic orientation** (Mittelstaedt and Mittelstaedt, 1973) is the term most commonly used. **Inertial navigation** is a hypothetical form of idiothetic orientation, which requires that an animal could record all the linear and angular accelerations during the outward journey and double integrate them to compute the direct route home (Barlow, 1964).

It is interesting to recall that Darwin (1873) was the first to suppose that man and animals can fix their position with respect to the goal (or the starting point) by path integration. He spoke about the possible role of visual reference(s) and of 'the sense of muscular movement', and called the mechanism by a nautical term, **dead reckoning** (probably a reminiscence of his travels on the *Beagle*). There is a fascinating correspondence between Darwin and Fabre (Fabre, 1924), in which the former, hypothesizing path integration in homing pigeons, proposed to disturb their computation of home position first by carrying them in the opposite direction to the release site and then by rotating them in a kind of centrifuge. Neither Darwin nor Fabre had the opportunity to perform this experiment on pigeons (it was performed by Matthews (1951) 70 years later!), but Fabre performed it on mason bees and cats, in both cases with negative results.

1.4.5 Pilotage and true navigation

The mechanisms which we now discuss, pilotage and true navigation, work on a principle different from that of path integration, and do not require any route-based information. They have been referred to as **geocentric systems** to be set against the **egocentric system** constituted by path integration (Wehner and Wehner, 1990, see also pp. 84–5). We speak of **pilotage** when an animal is familiar with an array of landmarks, visual or of other types, and is able to switch from one to the other in the appropriate order to reach any known site without the use of a compass. The mechanism is used by many vertebrates over familiar areas and presupposes an acquired **topographic map** that psychologists, following Tolman (1948), would recognize as being an example of a **cognitive map** or **cognitive mapping** (Thinus-Blanc, 1987). **Mental map** is another synonym for it.

Whereas piloting does not require the use of a compass, **true navigation**, in the sense we attribute to it here, does. True navigation occurs when an animal has to rely on local cues to calculate the goal direction, which is selected by means of a compass, no matter whether it is over a familiar area or not. This definition of true navigation is very close to Kramer's (1961) **map and compass** concept, which he expressed in relation to pigeon homing. Since Kramer was referring to homing from unfamiliar sites and had no idea how pigeons could fix their position, his 'map' is an indefinite notion. The present definition is wider than that of Baker (1984), which restricts its meaning to finding one's goal across unfamiliar terrain.

Two different types of true navigation can be distinguished. In the first, that of **map-based navigation**, the map consists of a mosaic of familiar landmarks, or, more generally, a system of local cues used to calculate home bearing. This map may have been acquired by direct experience, each landmark being associated with its direction with respect to home or other goals. Some distant landmarks and local cues can be indirectly acquired, so contributing to extend the map: mountain outlines and wind-borne olfactory information (pp. 295–9) are examples. One can imagine that an animal relies on map-based navigation to the extent to which the landmarks are too distant from each other to allow pilotage, but switches to pilotage as soon as they become close enough.

The second type of true navigation, that of **grid-based navigation**, has so far only been hypothesized, but never demonstrated. In this hypothesis, animals rely on physical or chemical factors whose scalar values rise or fall steadily in given directions; the relative sketches are often called **grid maps** or **gradient maps**. In an often presented scheme (e.g. Matthews, 1968), two different factors give rise to two gradients, the isolines of which form a grid similar to that of parallels and meridians. The grid is expected to be world-wide or, at least, to cover a huge area. Since each point in the grid would be determined by two coordinates, navigation based on it has also been called **bicoordinate navigation**. Even short trips around home might inform an animal about how the scalar values of the two gradients vary in different directions. When away from home, the comparison of local values of the gradients (the mechanism is also called the **comparative system**) with the home values would allow the animal to fix its position. According to Yeagley's (1947, 1951) hypothesis, the values of the vertical component in the earth's magnetic field and those of Corioli's force might form a reliable grid for navigational purposes. This idea, however, has not been supported by experimental evidence.

Concluding this short survey of homing phenomena, it must be stressed that the present arrangement has a provisional character, which mirrors our unsatisfactory knowledge of the many different mechanisms that animals rely on to control their spatial activity.

1.5 SUMMARY AND CONCLUSIONS

Homing comprises any form of movement whose aim is to reach a specific, familiar site. Trips to and from the sites that the animal regularly visits, return journeys away from the home range, and most migratory journeys are typical homing phenomena. Research work aims to analyse the orientation mechanisms underlying homing. It starts with recording homing tracks in natural conditions and proceeds with displacement experiments inside and outside the familiar area. Subjecting animals to sensory limitations during passive transportation and/or at the moment of release have often revealed what and when orienting cues play a role in homing.

The orientational behaviour consists of both innate and acquired components. Compass orientation based on sun position, stars or the earth's magnetic field is reported to be a widespread occurrence. Flexible components of homing behaviour include reliance on familiar cues and maps, as well as information gathered during active – and, sometimes, passive – outward journeys. To explain homing ability from distant, unfamiliar areas, reliance on indirectly acquired maps or geophysical or chemical grids is often considered and tested.

Homing phenomena can be arranged according to the origin of the information which allows an animal to travel towards its goal. It must, however, be borne in mind that animals often rely on two or more sources of information, sometimes switching from one to the other, and sometimes taking all of them into account. Thus, the classification presented in Table 1.1 is suitable for the arrangement of single mechanisms rather than integrated homing behaviour.

Homing phenomena represent an immense field of research, whose potential has so far only been achieved to a small extent. Every species apparently possesses redundant ways of homing, but only a few species have been the object of exhaustive investigations. Experience shows that it is unlikely that the findings obtained with a species or a group will offer any unequivocal clue to understanding the homing strategies of other species or groups.

Many neural mechanisms related to orientation, such as those involved in magnetoreception, information processing and the central control of homing, are promising topics for future research. It is therefore likely that, over the next 10–20 years, research teams will tend to adopt a wide variety of approaches, and to extend their investigations to a larger number of animal groups.

REFERENCES

Able, K.P. (1980) Mechanisms of orientation, navigation, and homing, in *Animal Migration, Orientation, and Navigation* (ed. S.A. Gautreaux), Academic Press, London, pp. 283–373.

Alerstam, T. (1990) *Bird Migration*, Cambridge University Press, Cambridge.

Amlaner, C.J. and Macdonald, D.W. (eds) (1980) *A Handbook on Biotelemetry and Radio Tracking*, Pergamon Press, Oxford.

Baker, R.R. (1978) *The Evolutionary Ecology of Animal Migration*, Hodder and Stoughton, London.

Baker, R.R. (1984) *Bird Navigation: the solution of a mystery?*, Hodder and Stoughton, London.

Barlow, J.S. (1964) Inertial navigation as a basis for animal navigation. *J. Theor. Biol.*, **6**, 76–117.

Batschelet, E. (1981) *Circular Statistics in Biology*, Academic Press, New York.

Berthold, P. (1990) Spatiotemporal programs and genetics of orientation. *Experientia*, **46**, 363–71.

Beugnon, G. (1986) Spatial orientation memories, in *Orientation in Space* (ed. G. Beugnon), Privat, I.E.C., Toulouse, pp. 9–19.

Bingman, V.P., Bagnoli, P., Ioalè, P. and Casini, G. (1989) Behavioural and anatomical studies of the avian hippocampus, in *The Hippocampus – New Vistas, Neurology and Neurobiology Series* (eds V. Chan-Palay and C. Koeler), A. Liss, New York, pp. 379–94.

Braemer, W. and Schwassmann, H.O. (1963) Von Rhythmus der Sonnenorientierung am Äquator (bei Fischen). *Ergbn. Biol.*, **26**, 182–201.

Bramanti, M., Dall'Antonia, L. and Papi, F. (1988) A new technique to monitor the flight paths of birds. *J. Exp. Biol.*, **134**, 467–72.

Chelazzi, G. and Francisci, F. (1979) Movement patterns and homing behaviour of *Testudo hermanni* Gmelin (Reptilia Testudinidae). *Monitore Zool. Ital.*, **13**, 105–27.

Darwin, C. (1873) Origin of certain instincts. *Nature*, **7**, 417–18.

Emlen, S.T. (1967) Migration: orientation and navigation, in *Avian Biology*, Vol. 5, Academic Press, New York, pp. 129–219.

Fabre, J.H. (1924) *Souvenirs Entomologiques*, Deuxième Série, Libraire Delagrave, Paris.

Fancy, S.G., Pank, L.F., Whitten, K.R. and Regelin, W.L. (1989) Seasonal movements of caribou in arctic Alaska as determined by satellite. *Can. J. Zool.*, **67**, 644–50.

Ferguson, D.E. (1967) Sun-compass orientation in anurans, in *Animal Orientation and Navigation* (ed. R.M. Storm), Oregon State University Press, Corvallis, pp. 21–32.

Fraenkel, G. and Gunn, D.L. (1940) *The Orientation of Animals*, 2nd edn, Dover Publ., New York.

Frederick II of Germany (XIIIth century) *De Arte Venandi cum Avibus*, *Italian translation* (ed. University Library, Bologna), G. Mondadori, Milano, Ms 717.

Görner, P. (1958) Die optische und kinästhetische Orientierung der Trichterspinnen *Agelena labyrinthica*. *Z. Vergl. Physiol.*, **41**, 111–53.

Griffin, D.R. (1952) Bird navigation. *Biol. Rev.*, **27**, 359–400.

Griffin, D.R. (1987) Foreword to papers on magnetic sensitivity in birds. *Anim. Learn. Behav.*, **15**, 108–9.

Griffin, D.R. (1990) Orientation in birds: a foreword. *Experientia*, **46**, 335–6.

Herrnkind, W.F. (1972) Orientation in the shore living Arthropods, especially the sand fiddler crab, in *Behavior of Marine Animals*, Vol. 1, *Invertebrates* (eds H.E. Winn and E.L. Olla), Plenum Press, New York, pp. 1–55.

Hoffmann, K. (1953) Experimentelle Änderung des Richtungsfindens beim Star durch Beeinflussung der "inneren Uhr". *Naturwiss*, **40**, 608–9.

Jander, R. (1975) Ecological aspects of spatial orientation, in *Annual Review of Ecology and Systematics* (eds R.F. Johnston, P.W. Frank and C.D. Michener), Annual Rev. Inc., Palo Alto, pp. 171–88.

Jouventin, P. and Weimerskirch, H. (1990) Satellite tracking of Wandering Albatrosses. *Nature*, **343**, 746–8.

Kalmijn, A.J. (1971) The electric sense of sharks and rays. *J. Exp. Biol.*, **55**, 371–83.

Kirschvink, J.L., Jones, D.S. and MacFadden, B.J. (eds) (1985) *Magnetite Biomineralization and Magnetoreception in Organisms*, Plenum Press, New York.

Kramer, G. (1961) Long-distance orientation, in *Biology and Comparative Physiology in Birds* (ed. A.J. Marshall), Academic Press, New York, pp. 341–71.

Lindauer, M. and Kerr, W.E. (1958) Die gegenseitige Verständigung bei den stachellosen Bienen. *Z. Vergl. Physiol.*, **41**, 405–34.

Matthews, G.V.T. (1951) The experimental investigation of navigation in homing pigeons. *J. Exp. Biol.*, **28**, 508–36.

Matthews, G.V.T. (1968) *Bird Navigation*, 2nd edn, Cambridge University Press, Cambridge.

Merkel, F.W. (1980) *Orientierung im Tierreich*, Fischer, Stuttgart.

Merkel, F.W. and Wiltschko, W. (1965) Magnetismus und Richtungsfinden zugunruhiger Rotkehlchen (*Erithacus rubecola*). *Vogelwarte*, **23**, 71–7.

Mittelstaedt, H. and Mittelstaedt, M.L. (1973) Mechanismen der Orientierung ohne richtende Aussenreize. *Fortschr. Zool.*, **21**, 46–58.

Otto, F. (1959) Die Bedeutung des Rückfluges für die Richtungs- und Entgernungsangabe der Bienen. *Z. Vergl. Physiol.*, **42**, 303–33.

Papi, F. (1990) Homing phenomena: mechanisms and classifications. *Ethol. Ecol. Evol.*, **2**, 3–10.

Papi, F. and Casini, G. (1990) Pigeons with ablated pyriform cortex home from familiar but not from unfamiliar sites. *Proc. Natl. Acad. Sci.*, *U.S.A.*, **87**, 3783–7.

Papi, F. and Pardi, L. (1953) Ricerche sull'orientamento di *Talitrus saltator* (Montagu) (Crustacea Amphipoda). II. Sui fattori che regolano la variazione dell'angolo di orientamento nel corso del giorno. L'orientamento di notte. L'orientamento diurno di altre popolazioni. *Z. Vergl. Physiol.*, **35**, 459–89.

Papi, F., Serretti, L. and Parrini, S. (1957) Nuove ricerche sull'orientamento e il senso del tempo di *Arctosa perita* (Latr.) (Araneae Lycosidae). *Z. Vergl. Physiol.*, **39**, 531–61.

Perdeck, A.C. (1958) Two types of orientation in migrating starlings, *Sturnus vulgaris* L., and chaffinches, *Fringilla coelebs* L., as revealed by displacement experiments. *Ardea*, **46**, 1–37.

Rüppel, W. (1944) Versuche ueber das Heimfinden ziehender Nebelkraehen nach Verfrachtung. *J. Ornithol., Lpz.*, **92**, 106–32.

Saint-Paul, U. von (1982) Do geese use path integration for walking home?, in *Avian Navigation* (eds F. Papi and H.G. Wallraff), Springer, Berlin, pp. 298–307.

Santschi, F. (1911) Observations et remarques critiques sur le mécanisme de l'orientation chez les fourmis. *Rev. Suisse Zool.*, **19**, 305–38.

Sauer, F. (1957) Die Sternenorientierung nächlich ziehender Grasmücken (*Sylvia atricapilla, borin* und *curruca*). *Z. Tierpsychol.*, **14**, 29–70.

Schmidt-Koenig, K. (1975) *Migration and Homing in Animals*, Springer, Berlin.

Schöne, H. (1980) *Orientierung im Raum*, Wissenschaftliche Verlagsgesellschaft mbH, Stuttgart.

Spallanzani, L. (1934) Opuscoli cinque sopra diverse specie di rondini, in *Le opere di Lazzaro Spallanzani*, Vol. 3, U. Hoepli, Milano, pp. 383–844.

Thinus-Blanc, C. (1987) The cognitive map concept and its consequences, in *Cognitive Processes and Spatial Orientation in Animal and Man*, Vol. 1 (eds P. Ellen and C. Thinus-Blanc), Nijhoff Publishers, Dordrecht, pp. 1–19.

Tolman, E.C. (1948) Cognitive maps in rats and men. *Psychol. Rev.*, **55**, 189–208.

Ugolini, A. (1987) Visual information acquired during displacement and initial orientation in *Polistes gallicus* L. (Hymenoptera, Vespidae). *Anim. Behav.*, **35**, 590–5.

Wallraff, H.G. (1980) Does pigeon homing depend on stimuli perceived during displacement? I. Experiments in Germany. *J. Comp. Physiol.*, **139**, 193–201.

Wehner, R. (1982) Himmelsnavigation bei Insekten. Neurophysiologie und Verhalten. *Neujahrsbl. Naturforsch. Ges. Zürich*, **184**, 1–132.

Wehner, R. (1987) 'Matched filters' – neural models of the external world. *J. Comp. Physiol.*, **161**, 511–31.

Wehner, R. and Wehner, S. (1990) Insect navigation: use of maps or Aradne's thread? *Ethol. Ecol. Evol.*, **2**, 27–48.

Yeagley, H.L. (1947) A preliminary study of a physical basis of bird navigation. *J. Appl. Phys.*, **18**, 1035–63.

Yeagley, H.L. (1951) A preliminary study of a physical basis of bird navigation. II. *J. Appl. Phys.*, **22**, 746–60.

Chapter 2
Invertebrates (excluding Arthropods)
G. Chelazzi

2.1 INTRODUCTION

Among invertebrates the general evolutionary trend toward more efficient locomotory systems, coupled with the evolution of a sophisticated central nervous system and receptors, make the possibility of active displacement an increasingly important factor in the relationship of animals to the ecological (biotic and abiotic) elements characterizing their specific environment. Even if some forms of all the major invertebrate phyla are sedentary, for example barnacles among crustaceans and bivalves among molluscs, none the less, an increase in locomotor efficiency can be seen not only as a critical factor in reducing the risk of exposure to sources of stress but also as a means of improving resource exploitation and reducing intra- and interspecific competition.

The many different movement patterns exhibited by invertebrates vary between groups and also according to the onthological phase and functional state of each species. Just as apparently random movements can appear in the behavioural repertoire of highly mobile and neurally sophisticated animals, so can movements oriented with respect to single cues or arrays of stimuli be incorporated in the displacements of such relatively simple invertebrates as some flatworms, nematodes and annelids (Schöne, 1984). The evolution of basic oriented responses, such as phototaxis, into more complex movement patterns is obviously constrained by the morpho-functional equipment of a given species but is also closely linked to its ecology. For instance, living in a gradient environment such as the sea–land border favours the evolution of directional movements by which intertidal molluscs, crustaceans, spiders and insects can maintain their zonation or rhythmically vary their

Animal Homing. Edited by Floriano Papi. Published in 1992 by Chapman & Hall, 2–6 Boundary Row, London SE1 8HN. ISBN 0 412 36390 9.

position in a specific zone. This orientation pattern, which is normal to the shore-line, is usually termed γ-axis orientation after Ferguson (1967), or zonal orientation after Jander (1975).

Homing behaviour usually appears in two different contexts (Chapter 1). The capacity to return to a specific area where resource disposition is known may be extremely adaptive when external factors, such as winds, waves or predators, displace the animal outside the boundaries of its home range (external homing). However, homing is also regularly performed by animals repeatedly exploiting different resources distributed around their rest sites (internal homing).

While more or less complex migrations have been described in members of almost all the main invertebrate phyla, homing behaviour – both internal (or spontaneous) and external (or induced) – has been analysed most frequently in arthropods (Chapter 3). Among the other invertebrates, information is becoming less episodic and anecdotal only for molluscs and echinoderms thanks to the collection of quantitative data based on good experimental techniques.

2.2 OCCURRENCE OF HOMING BEHAVIOUR IN MOLLUSCS

The locomotion most typical of molluscs, that of crawling on the substratum by pedal contraction, is found in three classes of this phylum: Monoplacophora (a few deep-sea species, resembling limpets), Polyplacophora (chitons) and Gastropoda (limpets, snails and slugs). Their locomotion is based on waves of muscular contraction moving forward or backward along the basal part of the animal, the foot. However, a few species of gastropods and many species of Cephalopoda have lost contact with the substratum, or benthic habitat, and have become pelagic, actively moving in the water. This highly mobile group has evolved an original jet propulsion system based on the expulsion of water from mantle cavity.

2.2.1 Cephalopods

That movement patterns, particularly homing behaviour, are preferentially studied in groups showing a higher behavioural complexity is not completely true in the case of molluscs. Cephalopods, one of the most sophisticated invertebrate classes both in terms of nervous organization and behavioural richness, are much sought after for laboratory studies on neurophysiology, neuroethology and learning capacity, but their natural behaviour has rarely been studied in the field. The few accounts on their homing behaviour that have been published mostly concern the benthic group of octopuses.

The adults of many species of *Octopus* occupy constant shelters during certain periods of the year, to which they return after feeding excursions (Hartwick *et al.*, 1984). This is true of *O. bimaculatus*, which inhabits the same home for more than one month, and of *O. cyanea* and *O. vulgaris*, which home to a constant shelter for 3–5 weeks. The shift to different dens in the same general area, together with their seasonal migrations, make the homing behaviour of these molluscs less rigid than that observed in other molluscs. However, a recent study conducted by Mather (1991) on *O. vulgaris* and *O. rubescens* in Bermuda, showed that these species perform foraging trips lasting up to about one hour, precisely centred to a personal home. Field observations and laboratory experiments showed that octopuses home from hunting trips using a long-term memory of visual landmarks. No evidence was obtained supporting the use of mucous trails.

2.2.2 Intertidal chitons and gastropods

Movements play an important role in the ecology of chitons and gastropods despite their relatively low mobility in comparison with such fast-moving animals as cephalopods, crustaceans and insects. In fact, the appropriate organization of movement patterns greatly

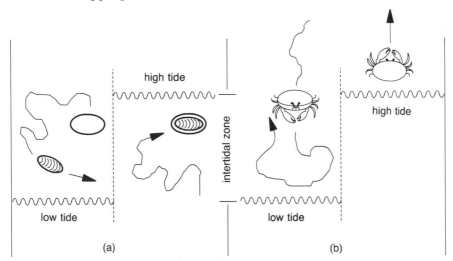

Figure 2.1 The two movement patterns generally adopted by intertidal animals. The isozonal pattern (a) is typical of animals which rest and feed in the same zone. The isophasic pattern (b) is typical of mobile animals which enter the intertidal zone only when it is exposed to air (terrestrial migrators) or submerged by water (marine migrators).

improves the morphophysiological adaptation of these molluscs in two ecological contexts: on rocky shores, where gastropods and chitons can face strong physical and biological constraints (Underwood, 1977, 1979), and on land, where water loss is one of the major threats to the many pulmonate gastropods living there. In both instances these groups tend to adopt movement patterns limited to a definite area or centred around a refuge where both physical and biotic constraints can be momentarily dodged.

Intertidal chitons and gastropods often show an isozonal movement strategy (Fig. 2.1a): they occupy a constant zone of the shore where they are alternately exposed to air and sprayed, splashed, or submerged by water according to the tidal phase. This is an alternative to the isophasic model (Fig. 2.1b) adopted by several arthropods and vertebrates which enter the intertidal zone only when it is exposed to air (terrestrial isophasics) or submerged by water (marine isophasics). The isozonal model may simply comprise the capacity to maintain a specific zonation, but not a constant site along the shore (Fig. 2.2a). This pattern, described in some periwinkles and top-shells, requires only simple orientation mechanisms to limit long-term diffusion along the sea–land axis (y-axis) and to locate suitable shelters, but the short-term movements look like random walks and long-term diffusion occurs parallel to the coastline (x-axis).

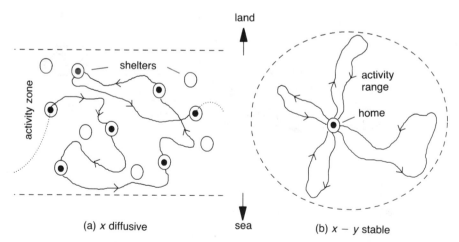

Figure 2.2 Different complexity of the isozonal pattern. (a) Some intertidal animals maintain a constant zonation, but range parallel to the coastline (x-axis), occupying natural shelters. (b) Homing to a constant rest site and radial looping for foraging inside a definite activity range allow the maintenance of a stable position both along the x- and y-axis.

Other isozonal molluscs remain within a subunit of the shore defined along both x- and y-axes, and return to a more or less permanent shelter after each excursion (Fig. 2.2b). Despite the many scattered reports on the homing capacity of chitons (Boyle, 1977), only a few species of this class have been studied in detail: among these are three species of the intertropical genus *Acanthopleura*. A quantitative analysis of their movement patterns (Fig. 2.3) shows a definite interspecific scaling of homing capacity (Chelazzi *et al.*, 1983a,b, 1987; Chelazzi and Parpagnoli, 1987; Focardi and Chelazzi, 1990). *A. gemmata* (Indian and Pacific Ocean) is able to relocate an almost permanent home scar after foraging excursions averaging more than 2 m. *A. granulata* (Caribbean Sea) performs foraging excursions averaging less than 1 m, and usually within a roughly defined activity range, rarely returning to the starting site. *A. brevispinosa* (western Indian Ocean), whose foraging excursions are of intermediate length, shows a mixed strategy, alternating homing and ranging behaviour.

Such interspecific variation in the homing behaviour of these *Acanthopleura* species is strictly related to the ecological differences between them. Due to different exposure to waves and tidal regime, the predation pressure decreases from *A. gemmata* to *A. brevispinosa* to *A. granulata* which is the basis of the observed

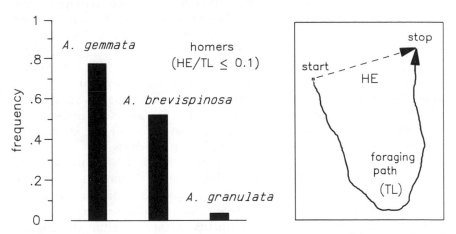

Figure 2.3 Different precision in the homing behaviour of three species of intertidal tropical chitons of the genus *Acanthopleura*. The homing performance was computed using the ratio between start–stop distance (HE) and the total length of foraging path (TL). By assuming a good homing performance when HE/TL ≤ 0.1, the frequency of homer chitons decreases from *A. gemmata* to *A. brevispinosa* to *A. granulata*. (After Focardi and Chelazzi, 1990.)

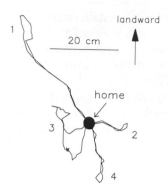

Figure 2.4 Four consecutive foraging paths of the Mediterranean high shore limpet *Patella rustica* recorded in the field. (After Chelazzi *et al.*, 1990.)

differences in spatial strategy. The better homing performance of *A. gemmata* depends on the presence of a tetraodontid fish which preys heavily on chitons not snugly ensconced in deep shelters at high tide.

Well-known examples of homing behaviour concern limpets (Fig. 2.4). Homing to a scar protects them from dehydration when exposed to air (Cook, 1976) due to shell–substrate matching, but defence against predators has also been suggested (Garrity and Levings, 1983). However, there is great variability in the homing capacity of different species of Patellidae and Acmaeidae (Galbraith, 1965; Jessee, 1968; Branch, 1981), which is evidently not due to any morphophysiological constraints but rather to the physical and biotic configuration of their environment. Upper-shore species are usually good homers, while lower-shore species are less so. Intraspecific size-related variations in homing are also frequent, as shown by species which range as juveniles and home as adults.

However, homing must not be regarded simply as a mechanism to relocate a constant shelter. The homing behaviour of these isozonal molluscs is clearly part of a complex central place foraging and has evolved as a multi-adaptive behavioural device to minimize stress and improve access to resources. Other possible ethological elements of this strategy are the active defence of a feeding territory or shelter against conspecifics or heterospecifics, or even the gardening of a specific algal patch (Branch, 1975).

Moreover, some chitons and gastropods perform 'collective homing', sheltering in clusters to improve protection from desiccation, overheating or predation (Magnus and Haacker, 1968; Willoughby, 1973). Such rhythmical clustering at each rest phase is common in limpets, neritids, cerithids and onchidians, and must not be confused with occasional aggregations for breeding. That communal refuging differs from independent convergence toward the same shelter, such as that observed in some littorinids, is evident from two characteristics: (1) the animals obtain mutual benefits from collective sheltering and sometimes congregate in conventional places no different from many others left empty (Chelazzi et al., 1984); (2) collective homing relies on interindividual communication mechanisms (Chelazzi et al., 1985; Focardi et al., 1985).

2.2.3 Land pulmonates

Homing under natural conditions has seldom been investigated in terrestrial gastropods; in slugs this is partly due to the difficulty of using individual marking and tracking techniques to record their movements. Nevertheless, some early qualitative reports are available on different Helicidae and Limacidae, and recently the behaviour of a few species has been analysed on a quantitative basis in the field or under semi-natural conditions (Cook, 1980; Rollo and Wellington, 1981).

The first detailed report on pulmonate homing is that of Edelstam and Palmer (1950) on *Helix pomatia*. Marked individuals released at increasing distances from their shelters show a good homing capacity up to 40 m but none whatsoever between 150 and 2000 m. A distinct variation in homing was observed in relation to weather and season, with the best performances occurring in the summer. A similar homing performance was observed by Southwick and Southwick (1969) in the giant land snail, *Achatina fulica*, which is able to home when displaced 5–30 m from its shelter.

More evidence is available on the homing of slugs in which the loss of an external shell makes the availability of daytime resting places a critical factor. This is particularly true for large slugs which tend to home to permanent shelters after night-time foraging or mating activity. But the homing habit varies both intra- and interspecifically. Some species such as *Limax pseudoflavus* home to a permanent shelter (Cook, 1979), while others such as *Ariolimax reticulatum* are more diffusive. In *Ariolimax columbianus*, Rollo and Wellington (1981) observed a clearly climate-related variation in homing capacity according to the season. Also slugs show the communal sheltering mentioned in intertidal gastropods and chitons. Cook (1981) described this behaviour in four sympatric slug species which form both homo- and heterospecific aggregations. Collective shelters are preferred by *A. columbianus* (Rollo and

Wellington, 1981) and frequently preferred by the common slug, *Limacus flavus* (syn. *Limax f.*) (Chelazzi *et al.*, 1988). Experiments on slugs demonstrate that collective homing and huddling significantly reduce their water loss (Cook, 1981), an advantage so important that non-aggressive slug species may clump with the highly aggressive *Limax maximus*. The land snails *Achatina fulica* (Chase *et al.*, 1980) and *Helicella virgata* (Pomeroy, 1968) sometimes adopt communal refuging and collective homing as well.

2.3 ORIENTATION MECHANISMS IN THE HOMING OF CHITONS AND GASTROPODS

It is evident that chemoreception is deeply involved in the homing capacity of both intertidal and land molluscs. Visual orientation has been described in some intertidal gastropods (Hamilton, 1977; Chelazzi and Vannini, 1980), but the importance of both distance and contact chemoreception in this and other functional contexts of gastropod biology has been widely recognized (Kohn, 1961; Croll, 1983). This is also true in chitons (Boyle, 1977).

2.3.1 Trail following

The orientation mechanisms on which intertidal chitons and gastropods rely for homing have recently been reviewed (Chelazzi *et al.*, 1988a). The mechanisms considered are idiothetic orientation, path integration based on external cues, and different forms of piloting based on cues from the substratum on which the animal moves. Experiments made on *Siphonaria* (Cook, 1969) and *Patella* (Cook *et al.*, 1969) reduced the number of possible homing mechanisms to the memorization of micro-landmarks, and to the trail-following mechanism, i.e. the active deposition of information on the substratum and its subsequent recognition. A complex series of tests were performed by Funke (1968) on the homing and home recognition mechanisms of *Patella vulgata*. His results support the importance of trail following and home chemical marking. Since then the evidence on trail following in intertidal gastropods has accumulated (Stirling and Hamilton, 1986) while no direct proof of microtopographic memorization has been obtained.

Similar results have been obtained for chitons. The return and outgoing paths of the strong homer *Acanthopleura gemmata* coincide greatly, with only a small divergence in the grazing area (Chelazzi *et al.*, 1987, 1990). Trail interruption using different methods decreases homing performance, but there remains a residual homing capacity inversely related to the amount of trail removed. Tracking

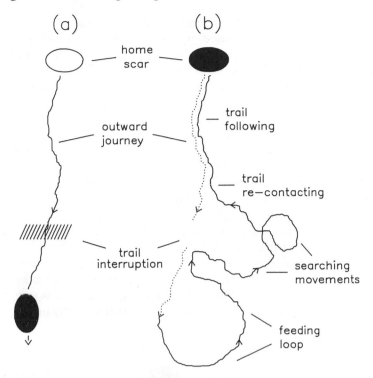

Figure 2.5 Schematic representation of an experiment conducted on the intertidal chiton *Acanthopleura gemmata*. (After Chelazzi *et al.*, 1987.) (a) When the animal reaches the feeding area the trail is interrupted using chemical or physical methods. (b) The detailed tracking of the chiton during foraging and return shows searching movements at the trail interruption point and trail following after trail recontact.

chitons after trail breaking shows that the mechanism responsible for their success in re-contacting the outward trail on the homeward side of the interruption is exploration (Fig. 2.5). Less trail coincidence was recorded in *A. brevispinosa* and *A. granulata*, but when homing performance was high it invariably coincided with a high degree of trail overlap. These findings demonstrate the unique importance of trail following in the natural homing of these chitons.

However, the following aspects of homing behaviour in inter-tidal molluscs seem to disprove the trail-following hypothesis: in some species the outward and return trail do not always coincide; some species seem able to home after being displaced from home; and trail interruption using chemical or physical techniques does not always decrease the homing performance. The first two observations can be explained by the fact that trail information is long-lasting, as has been proved in siphonarid limpets (Cook, 1969, 1971). The third finding can be explained by the explorative behaviour of homers at the interruption point. The trail recontact discussed in *A. gemmata* has also been observed in *Onchidium verruculatum* by McFarlane (1980).

Collective intertidal homers use trail following as well (cf. Chelazzi *et al.*, 1988a). The gastropod *Nerita textilis* spreads a web of trails from the communal shelter which collects both previous members of the cluster and scattered animals. Long-lasting substrate marking of the clustering area is also evident from displacement experiments (Chelazzi *et al.*, 1985).

An obvious difference between collective and solitary homers concerns the specificity of trail information, which must be species-specific in the former to allow interindividual following, and somehow individual in the latter in order to avoid mistakes at the intersection of two individual trails. The capacity to follow conspecific trails has been demonstrated in such communal homers

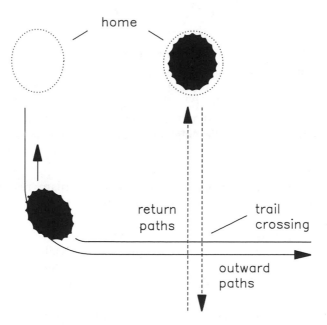

Figure 2.6 Simplified representation of an experiment made by Funke (1968) on *Patella vulgata* in the laboratory. Limpets crossing the trail of a conspecific continue to head homeward on their own outgoing trails.

as *Ilyanassa obsoleta* (Trott and Dimock, 1978) and *Nerita textilis* (Chelazzi *et al.*, 1985), but the degree of individual specificity has seldom been investigated directly. In some solitary limpets discrimination between personal and allotrails seems to be lacking (Cook, 1971) and if confirmed may suggest that the choice of a personal trail is based on cues external to the trail. On the contrary, Funke's (1968) experiments show that *Patella vulgata* is able to discriminate between personal and conspecific trails (Fig. 2.6).

A detailed study on this aspect of solitary homing was conducted on the chiton *Acanthopleura gemmata* (Chelazzi *et al.*, 1987): cross experiments demonstrated a decrease in homing performance when animals were displaced on conspecific trails (Fig. 2.7). Deneubourg *et al.* (1988) used a mathematical model to show that the trails of

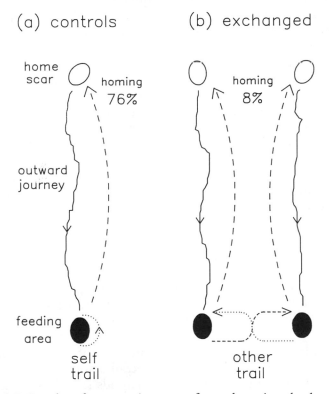

Figure 2.7 Results of an experiment performed on *Acanthopleura gemmata*. (After Chelazzi *et al.*, 1987.) Chitons dislodged from the rock while feeding and replaced on their own trail after head–tail rotation show a high homing performance (a). Chitons displaced from the rock and placed on a conspecific trail do not return to their homes, but show a very low return to the home of the conspecific who released the trail (b).

these chitons are polymorphic, which permits quasi-personal trail following. The estimated number of trail types is about 12 in the studied population of this species, as compared to 4–5 in *A. granulata* and *A. brevispinosa*, which agrees with their lower attachment to an individual home scar.

A particular problem connected to the use of trail following is that of the binary choice which faces an animal encountering its previous trail: which is the correct homeward direction? Some experiments (Cook and Cook, 1975) ruled out the possibility that limpets rely upon such external cues as light or gravity to solve this task, favouring the hypothesis that the trail of intertidal molluscs is intrinsically polarized. Cook (1971) considers three main mechanisms of trail polarization in siphonarid limpets: chemical macro-gradient along the trail; polarized sequence of different chemicals; and polarized physical structure of the trail and consequent differential friction upon retracing the trail in the two opposite directions. To these Stirling and Hamilton (1986) added three more possible mechanisms: chemical micro-gradient; bilateral asymmetry in chemical or physical properties of the trail; and optical detection of the different light reflection by the trail in the two opposite directions. The experimental evidence does not conclusively support any of these but *Littorina irrorata* and *Ilyanassa obsoleta* seem to detect the microstructural polarization of the trail, while *Littorina sitkana* and *L. littorea* probably choose the right direction by detecting a chemical macro-gradient (Gilly and Swenson, 1978).

Strong evidence supporting the importance of trail following has also been obtained for freshwater gastropods (Wells and Buckley, 1972; Townsend, 1974), but there are only a few demonstrations of its importance in the homing of terrestrial species. The slug *Limax grossui* (Cook, 1977) can discriminate between the trails of conspecifics and of *L. flavus* (Fig. 2.8), and the snail *Achatina fulica* between conspecific and heterospecific trails (Chase *et al.*, 1980), but these studies reveal that the degree of precision in homing-related trail following is lower in pulmonates than intertidal chitons and limpets.

Surprisingly, despite the large number of studies on trail following in molluscs, the nature of trail-associated information utilized for homing in chitons and gastropods is still uncertain. For intertidal gastropods it is sometimes debated whether it has a chemical or physical basis. Cook (1971) considers radular scraping of the substrate, physical discontinuities provided by the mucus, and attached or diffusible chemicals as factors. Trott and Dimock (1978) argue for chemical recognition in *Ilyanassa obsoleta*; Peters (1964) and Hall (1973) claim recognition of chemical cues in *Littorina*

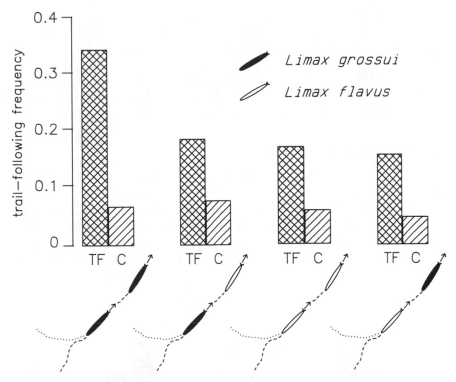

Figure 2.8 Results of intra- and interspecific trail following experiments on *Limax* species in the laboratory. (After Cook, 1977.) Cross-hatched bars indicate the frequency of coincidence (TF) between the path of the follower slug and that of a marker animal of the same or another species. Hatched bars indicate the chance coincidence (C) between the path of the test animal and that of a slug moving on a different substratum.

spp. as does Raftery (1983) who also stresses the probable importance of physical cues. The physical nature of orienting cues involved in trail following is supported by the findings of Bretz and Dimock (1983) on *I. obsoleta* which demonstrate that changes in trail-following responses are associated with changes in the structural elements (parallel filaments) of the trails.

In freshwater and land pulmonates the most accredited hypothesis is the chemical one (Townsend, 1974; Wells and Buckley, 1972; Cook, 1979; Rollo and Wellington, 1981; Croll, 1983). There are also data on the receptors and detection mechanisms involved in the trail following of these animals. Chase and Croll (1981) found that bilateral lesion of the posterior cephalic tentacles does not impair trail-following efficiency, while resection of both anteriors has a marked effect. Owing to their closeness to the substratum the anterior tentacles can detect trail-associated chemical information even though contact chemoreception does not seem to be involved. In *Limax pseudoflavus* both pairs of tentacles are involved in trail following (Cook, 1985): the anteriors are used for trailing once the animal is on the trail while the posteriors locate the trail from outside, allowing the slug to turn onto it.

2.3.2 Distant chemoreception

Whether the generally low accuracy of trail following in land snails and slugs is real or a laboratory artifact is not presently obvious (Cook, 1985), but it is in apparent contrast to the high homing performance shown by some land gastropods. This is due to the existence of an alternative or complementary mechanism – distance chemoreception (Croll, 1983). In some slugs the distant detection of

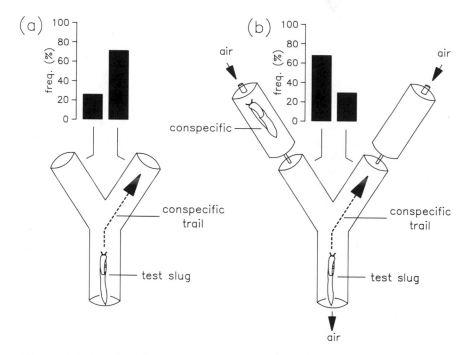

Figure 2.9 Results of two experiments conducted on *Limacus flavus* using a binary choice chamber in the laboratory. (After Chelazzi *et al.*, 1988.) The slugs move mostly into the maze arm where a conspecific had passed previously (a). However, if a conspecific is caged in the other arm, the slugs prefer to head toward this (b).

home cues seems to be the only mechanism involved in homing (Gelperin, 1974), while other species mingle this with trail following (Cook, 1980). Recent experiments on *L. flavus* (Chelazzi *et al.*, 1988b) show that distant chemoreception predominates over trail following when the two mechanisms are simultaneously tested by using a binary choice apparatus (Fig. 2.9), but they may work together when homing occurs under natural conditions.

Snails and slugs frequently home to a communal shelter, as mentioned above, and Rollo and Wellington (1981) found that *Ariolimax columbianus* homes more frequently to an occupied rather than to an unoccupied shelter. *Limacus flavus* heads toward a communal refuge when offered this as an alternative to a single one, even if both are empty (Chelazzi *et al.*, 1988). The orientation toward conspecifics or chemicals extracted from their pedal glands has been investigated in *Achatina fulica* (Chase and Boulanger, 1978). Moreover, Chase *et al.* (1980) discovered that in *A. fulica* the chemical cue is family-specific, which has interesting sociobiological and operational implications. All these observations support the hypothesis that one or more pheromones are emitted by the resting animals and used as 'homing beacons' by conspecifics or relatives. Nevertheless, the involved molecules have yet to be isolated.

There are basically two mechanisms used by land pulmonates for distant chemoreception: klinotaxis to an airborne odour and tropotaxis in a diffusion gradient. The first is evident when slugs and snails home upwind; the latter when animals are able to locate the home or conspecifics in still air. The two posterior tentacles are primarily responsible for distant chemoreception, while the anterior ones are mainly contact, or short range, detectors and thus fundamental in trail following (Chase, 1986).

However, in land pulmonates the mechanisms of trail following and distance chemoreception may be closely linked: some results question 'whether the distant olfaction and trail-following behaviour of *Achatina fulica* might not represent a single phenomenon' (Chase *et al.*, 1978). Cook (1979) favours a dual chemosensory mechanism in *Limax grossui*, probably involving two different pheromones, but later concludes that in *Limax pseudoflavus* trail following does not need to involve contact chemoreception. Recent tests on *Limacus flavus* (Chelazzi *et al.*, 1988b) indicate distant detection of the trail and support the simpler hypothesis.

2.4 HOMING BEHAVIOUR IN ECHINODERMS

An original locomotor device, probably born as a food-collecting device, has evolved in echinoderms (sea-stars, sea-urchins and sea-cucumbers). Part of their complex inner cavity, the hydrocoele, has developed into a system of canals, ampullae and externally projecting appendages called pedicelli which allow crawling on the substratum.

Several sea–urchins (Echinoidea) are algal grazers on hard coastal substrata and hide in crevices during rest phases. Different species show varying degrees of fidelity to a constant shelter under natural circumstances, based on different homing behaviour. In general, such species as *Diadema antillarum* and *D. setosum* which live on highly irregular substrata providing many shelters show less precise homing (Smith, 1969; Fricke, 1974). On the contrary, where natural shelters are scarce or must be actively dug in the rock by the urchins, homing behaviour is more precise. The Australian species *Centrostephanus rodgersii* rests in rock crevices during the day and moves up to 1 m away during the night when it forages. At the end of each activity phase it relocates the original shelter or settles in a new one (Sinclair, 1959).

On the Californian coast, Nelson and Vance (1979) observed precise homing in the diadematid urchin *Centrostephanus coronatus*, whose site fidelity seems related to the daytime predation exerted by sheephead fish. On the contrary, the juveniles of the same species, which use smaller shelters frequently occurring throughout their range, show no homing behaviour. Such rock-boring species as the Caribbean *Echinometra lucunter* and the eastern Pacific *Strongylocentrotus purpuratus* have an even stronger attachment to a personal shelter, which is defended from intruders (Grünbaum *et al.*, 1978; Neill, 1988).

However, homing behaviour also varies intraspecifically, both with age and in relation to the ecological situation. A detailed study on the dependence of homing behaviour upon such biological factors as predation and population density was conducted by Carpenter (1984) on the Caribbean echinoid *Diadema antillarum*. About 80% of the animals return to a specific crevice, but where predation is potentially greater the urchins are more faithful to a constant shelter. The effect of predation may also be indirect, as it reduces population density which in turn increases the abundance of algae. More abundant algal grounds possibly favour the exclusive use of a single foraging range and return to a constant shelter.

By considering the feeding excursions of each urchin of the same species on different nights (Fig. 2.10) it is evident that individuals do not graze the same areas during successive excursions but rather avoid regrazing the same ground. This regular shifting of the foraging areas around the home also seems to occur in *Centrostephanus coronatus* (Nelson and Vance, 1979).

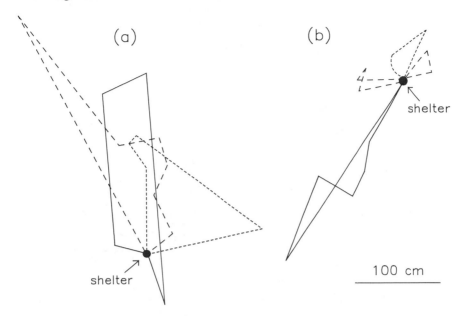

Figure 2.10 Examples of foraging paths recorded in two specimens (a–b) of the sea-urchin *Diadema antillarum* on three consecutive nights. (After Carpenter, 1984.)

A common phenomenon in echinoids is the aggregation between conspecifics, on which a large amount of research has been conducted (Warner, 1979). However, lumping behaviour rarely takes the form of regular homing to a communal shelter such as that described in intertidal molluscs. It can best be seen as a response to predation or as a feeding strategy which appears under specific ecological conditions.

In conclusion, the few quantitative analyses of the movements performed by sea-urchins in the field show a 'central place foraging' strategy based on various degrees of homing capacity, collective sheltering, home digging and shelter defence resembling that of intertidal molluscs. There is scanty evidence on the natural occurrence of homing in the other echinoderm classes. Sea-stars show different degrees of homing capacity after experimental displacement. In the Caribbean, Scheibling (1980) translocated individual *Oreaster reticulatus* from their home sand-patch to sites up to 20 m away in the surrounding seagrass bed. The sea-stars oriented toward the sand-patch containing groups of conspecifics (Fig. 2.11) which supports the conclusion that asteroids are probably able to 'navigate by chemotaxis' to chemical stimuli released by foraging conspecifics.

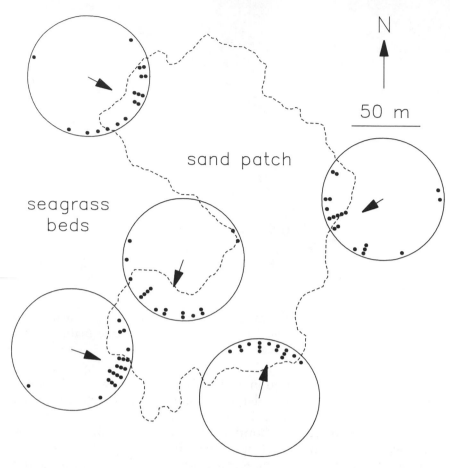

Figure 2.11 Results of an experiment performed on the sea-star *Oreaster reticulatus*. (After Scheibling, 1980.) Animals displaced from their habitat (sand–patch) to seagrass beds head toward sites inhabited by groups of conspecifics. Black dots represent individual bearings after displacement. Arrows represent average directions.

A different behaviour was reported by Pabst and Vicentini (1978) in *Astropecten jonstoni*. Undisturbed sea-stars shift seasonally from shallow to deeper water and back across a Mediterranean sand beach. After landward and seaward experimental displacement, the stars move toward deeper and shallower water, respectively (Fig. 2.12), but animals displaced parallel to the coastline show no difference with respect to undisturbed animals. These findings suggest that the responses of *A. jonstoni* must be included in the

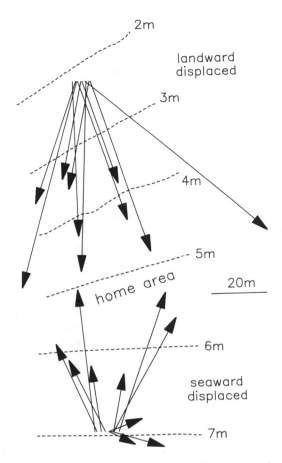

Figure 2.12 Zonal recovery of the sea-star *Astropecten jonstoni*. (After Pabst and Vicentini, 1978.) Arrows indicate the movement of animals after seaward and landward displacements. Dashed lines indicate isobaths.

class of *γ*-axis orientation or zonal adjustment capacity shown by several littoral animals, and is not a true homing behaviour.

Sea-stars, as well as the brittle stars or ophiuroids, often show gregarious lumping related to feeding and reproduction. Such goal orientation is apparently based on chemical orientation toward conspecifics (Broom, 1975; Keegan *et al.*, 1985).

2.5 CONCLUSIONS

Only a few general conclusions can be reached when considering the most significant contributions on homing in invertebrates, if we exclude arthropods. The first obvious consideration is that the

occurrence of homing behaviour under natural conditions, and
the adaptive significance of this movement pattern, is still poorly
documented on a quantitative basis. Homing behaviour is apparently
adopted by the different species whenever a central place strategy
becomes adaptive due to the spatial segregation of resources
with respect to a constant shelter. The appearance of this spatial
strategy is largely independent of the morphophysiological complexity of a group but homing has not yet been described in the
lower invertebrates.

Among the non-arthropods, the ecological significance of
homing is now being assessed in general terms only in gastropods
and chitons. Nevertheless, some evident analogies can be observed
in the central place patterns of echinoids and intertidal molluscs.

The difference in quantity and quality of experimental analysis
between all the other invertebrates and arthropods (Chapter 3) is
even more evident when considering the orientation mechanisms
and the cues used for homing. The only firm point in the operational study of homing in non-arthropods is the importance of
chemical cues.

Convergences and diversity in the operational aspects of homing
behaviour in intertidal and terrestrial molluscs may be better
understood by taking into account their different organization and
ecology. First, the specific environmental conditions in which they
operate may differentiate the kind of orienting cues available;
second, the morphophysiological organization of chitons (and
Archaeogastropods) differs greatly from that of the sensorially and
neurally sophisticated Pulmonates; third, varying locomotion
efficiency influences the spatial scale of their excursions which range
from a few centimetres up to a few metres in the first group, and
may reach several metres in land snails and slugs.

However, despite the generally recognized importance of trail
following in the homing of intertidal molluscs, the nature of trail-
associated orienting cues is still obscure. The complexity of the
trail-following mechanism, including individual specificity and trail
directionality, makes the chemical hypothesis probable in intertidal
chitons and gastropods.

Even if the available figures on the foraging paths of echinoids do
not show an evident overlap in the outgoing and return branches of
each excursion, the general resemblance of the homing patterns of
some echinoids to those of intertidal molluscs – though not directly
supporting the importance of trail-following in the homing of the
latter – suggest that future research should include the analysis of
trail following in this group as well.

Despite   the   abundance   of   laboratory   studies   on   the chemoorientation cues emanating from conspecifics, predators or food items, the nature of orientating cues involved in true homing of the echinoderms is another problem to be solved in the biology of this intriguing group of invertebrates.

REFERENCES

Boyle, P.R. (1977) The physiology and behaviour of chitons (Mollusca: Polyplacophora). *Oceanogr. Mar. Biol. A. Rev.*, **15**, 461–509.

Branch, G.M. (1975) Mechanisms reducing intraspecific competition in *Patella* species: migration, differentiation and territorial behaviour. *J. Anim. Ecol.*, **44**, 575–600.

Branch, G.M. (1981) The biology of limpets: physical factors, energy flow and ecological interactions. *Oceanogr. Mar. Biol. A. Rev.*, **19**, 235–380.

Bretz, D.D. and Dimock, R.V. (1983) Behaviorally important characteristics of the mucous trail of the marine gastropod *Ilyanassa obsoleta* (Say). *J. Exp. Mar. Biol. Ecol.*, **71**, 181–91.

Broom, D.M. (1975) Aggregation behaviour of the brittle-star *Ophiothrix fragilis*. *J. Mar. Biol. Ass. U.K.*, **55**, 191–7.

Carpenter, R.C. (1984) Predator and population density control of homing behaviour in the Caribbean echinoid *Diadema antillarum*. *Mar. Biol.*, **82**, 101–8.

Chase, R. (1986) Lessons from snail tentacles. *Chem. Sens.*, **11**, 411–26.

Chase, R. and Boulanger, C.M. (1978) Attraction of the snail *Achatina fulica* to extracts of conspecific pedal glands. *Behav. Biol.*, **23**, 107–11.

Chase, R. and Croll, R.P. (1981) Tentacular function in snail olfactory orientation. *J. Comp. Physiol.*, **143**, 357–62.

Chase, R., Croll, R.P. and Zeichner, L.L. (1980) Aggregation in snails, *Achatina fulica*. *Behav. Neural Biol.*, **30**, 218–30.

Chase, R., Pryer, K., Baker, R. and Madison, D. (1978) Responses to conspecific chemical stimuli in the terrestrial snail *Achatina fulica* (Pulmonata: Sigmurethra). *Behav. Biol.*, **22**, 302–15.

Chelazzi, G., Della Santina, P. and Parpagnoli, D. (1987) Trail following in the chiton *Acanthopleura gemmata*: operational and ecological problems. *Mar. Biol.*, **95**, 539–45.

Chelazzi, G., Della Santina, P. and Parpagnoli, D. (1990) The role of trail following in the homing in intertidal chitons: a comparison between three Acanthopleura spp., *Mar. Biol.*, **105**, 445–50.

Chelazzi, G., Della Santina, P. and Vannini, M. (1985) Long-lasting substrate marking in the collective homing of the gastropod *Nerita textilis*. *Biol. Bull.*, **168**, 214–21.

Chelazzi, G., Focardi, S. and Deneubourg, J.L. (1983a) A comparative study on the movement patterns of two sympatric tropical chitons (Mollusca: Polyplacophora). *Mar. Biol.*, **74**, 115–25.

Chelazzi, G., Focardi, S., Deneubourg, J.L. and Innocenti, R. (1983b) Competition for the home and aggressive behaviour in the chiton *Acanthopleura gemmata* Blainville (Mollusca: Polyplacophora). *Behav. Ecol. Sociobiol.*, **14**, 15–20.

Chelazzi, G., Focardi, S. and Deneubourg, J.L. (1984) Cooperative interactions and environmental control in the intertidal clustering of *Nerita textilis* (Gastropoda; Prosobranchia). *Behaviour*, **90**, 151–66.

Chelazzi, G., Focardi, S. and Deneubourg, J.L. (1988a) Analysis of movement patterns and orientation mechanisms in intertidal chitons and gastropods, in *Behavioral Adaptation to Intertidal Life* (eds G. Chelazzi and M. Vannini), Plenum Press, New York, pp. 173–84.

Chelazzi, G., Le Voci, G. and Parpagnoli, D. (1988b) Relative importance of airborne odours and trails in the group homing of *Limacus flavus* (Linnaeus) (Gastropoda, Pulmonata). *J. Moll. Stud.*, **54**, 173–80.

Chelazzi, G. and Parpagnoli, D. (1987) Behavioural responses to crowding modification and home intrusion in *Acanthopleura gemmata* (Mollusca, Polyplacophora). *Ethology*, **75**, 109–18.

Chelazzi, G., Terranova, G. and Della Santina, P. (1990) A technique for recording the activity of limpets. *J. Moll. Stud.*, **56**, 595–600.

Chelazzi, G. and Vannini, M. (1980) Zonal orientation based on local visual cues in *Nerita plicata* L. (Mollusca: Gastropoda) at Aldabra Atoll. *J. Exp. Mar. Biol. Ecol.*, **46**, 147–56.

Cook, A. (1977) Mucus trail following by the slug *Limax grossui* Lupu. *Anim. Behav.*, **25**, 774–81.

Cook, A. (1979) Homing by the slug *Limax pseudoflavus. Anim. Behav.*, **27**, 545–52.

Cook, A. (1980) Field studies of homing in the pulmonate slug *Limax pseudoflavus* (Evans). *J. Moll. Stud.*, **46**, 100–5.

Cook, A. (1981) A comparative study of aggregation in pulmonate slugs (genus *Limax*). *J. Anim. Ecol.*, **50**, 703–13.

Cook, A. (1985) Tentacular function in trail following by the pulmonate slug *Limax pseudoflavus* Evans. *J. Moll. Stud.*, **51**, 240–7.

Cook, A., Bamford, O.S., Freeman, J.B. and Teidman, D.J. (1969) A study on the homing habit of the limpet. *Anim. Behav.*, **17**, 330–9.

Cook, S.B. (1969) Experiments on homing in the limpet *Siphonaria normalis. Anim. Behav.*, **17**, 679–82.

Cook, S.B. (1971) A study on homing behaviour in the limpet *Siphonaria alternata. Biol. Bull. Mar. Biol. Lab. Woods Hole*, **141**, 449–57.

Cook, S.B. (1976) The role of the 'home scar' in pulmonate limpets. *Bull. Am. Mal. Union Inc.*, **1976**, 34–7.

Cook, S.B. and Cook, C.B. (1975) Directionality in the trail following response of the pulmonate limpet *Siphonaria alternata. Mar. Behav. Physiol.*, **3**, 147–55.

Croll, R.P. (1983) Gastropod chemioreception. *Biol. Rev.*, **58**, 293–319.

Deneubourg, J.L., Focardi, S. and Chelazzi, G. (1988) Homing mechanisms of intertidal chitons: field evidence and the hypothesis of trail polymorphism, in *Behavioural Adaptation to Intertidal Life* (eds G. Chelazzi and M. Vannini), Plenum Press, New York, pp. 185–95.

Edelstam, C. and Palmer, C. (1950) Homing behaviour in gastropods. *Oikos*, **2**, 259–70.

Feare, C.J. (1971) The adaptive significance of aggregation behaviour in the dogwhelk *Nucella lapillus (L.). Oecologia, Berl.*, **7**, 117–26.

Ferguson, D.E. (1967) Sun-compass orientation in anurans, in *Animal Orientation and Navigation* (ed. R.M. Storm), Oregon State University Press, Carvallis, pp. 21–34.

Focardi, S. and Chelazzi, G. (1990) Ecological determinants of bioeconomics in three intertidal chitons (*Acanthopleura* spp.). *J. Anim. Ecol.*, **59**, 347–62.

Focardi, S., Deneubourg, J.L. and Chelazzi, G. (1985) How shore morphology and orientation mechanisms can affect spatial organization of intertidal molluscs. *J. Theor. Biol.*, **112**, 771–82.

Fricke, H.W. (1974) Möglicher einfluss von Feinden auf das Verhalten von *Diadema*-Seeigeln. *Mar. Biol.*, **27**, 59–62.

Funke, W. (1968) Heimfindevermogen und ortstreue bei *Patella* L. (Gastropoda, Prosobranchia). *Oecologia, Berl.*, **2**, 19–142.

Galbraith, R.T. (1965) Homing behaviour in the limpets *Acmaea digitalis* and *Lottia gigantea*. *Am. Midl. Nat.*, **74**, 245–6.

Garrity, S.D. and Levings, S.C. (1983) Homing to scars as a defense against predators in the pulmonate limpet *Siphonaria gigas* (Gastropoda). *Mar. Biol.*, **72**, 319–24.

Gelperin, A. (1974) Olfactory basis of homing behavior in the giant garden slug, *Limax maximus*. *Proc. Natl. Acad. Sci., U.S.A.*, **71**, 966–70.

Gilly, W.F. and Swenson, R.P. (1978) Trail following by *Littorina*: washout of polarized information and the point of paradox test. *Biol. Bull.*, **155**, 439.

Grünbaum, H., Bergman, G., Abbott, D.P. and Ogden, J.C. (1978) Intraspecific agonistic behaviour in the rock-boring sea urchin *Echinometra lucunter* (L.) (Echinodermata: Echinoidea). *Bull. Mar. Sci.*, **28**, 181–8.

Hall, J.R. (1973) Intraspecific trail-following in the marsh periwinkle *Littorina irrorata* (Say). *Veliger*, **16**, 72–5.

Hamilton, P.V. (1977) Daily movements and visual location of plant stems by *Littorina irrorata* (Mollusca: Gastropoda). *Mar. Behav. Physiol.*, **4**, 293–304.

Hartwick, E.B., Ambrose, R.F. and Robinson, S.M.C. (1984) Den utilization and the movements of tagged *Octopus dofleini*. *Mar. Behav. Physiol.*, **11**, 95–110.

Jander, R. (1975) Ecological aspects of spatial orientation. *A. Rev. Ecol. Syst.*, **6**, 171–88.

Jessee, W.F. (1968) Studies on homing behaviour in the limpet *Acmaea scabra*. *Veliger* (Suppl.), **11**, 52–5.

Keegan, B.F., O'Connor, B.D.S. and Konnecker, G.F. (1985) Littoral and benthic investigations on the west coast of Ireland – XX. Echinoderm aggregations. *Proc. R. Ir. Acad.*, **85b**, 91–9.

Kohn, A.J. (1961) Chemoreception in gastropod molluscs. *Am. Zool.*, **1**, 291–308.

Magnus, D.B.E. and Haacker, U. (1968) Zum phanomen der ortsunsteten ruheversammlungen der strandschnecke *Planaxis sulcatus* (Born) (Mollusca, Prosobranchia). *Sarsia*, **34**, 137–48.

McFarlane, I.D. (1980) Trail following and trail searching behaviour in homing of the intertidal gastropod mollusc, *Onchidium verruculatum*. *Mar. Behav. Physiol.*, **7**, 95–108.

Mather, J.A. (1991) Navigation by spatial memory and use of visual landmarks in octopuses. *J. Comp. Physiol. A*, **168**, 491–97.

Neill, J.B. (1988) Experimental analysis of burrow defense in *Echinometra mathaei* (de Blainville) on Indo-West Pacific reef flat. *J. Exp. Mar. Biol. Ecol.*, **115**, 127–36.

Nelson, B.V. and Vance, R.R. (1979) Diel foraging patterns of the sea urchin *Centrostephanus coronatus* as a predator avoidance strategy. *Mar. Biol.*, **51**, 251–8.

Pabst, B. and Vicentini, H. (1978) Dislocation experiments in the migrating sea star *Astropecten jonstoni*. *Mar. Biol.*, **48**, 271–8.

Peters, R.L. (1964) Function of the cephalic tentacles in *Littorina planaxis* Philippi (Gastropoda: Prosobranchiata). *Veliger*, **7**, 143–8.

Pomeroy, D.E. (1968) Dormancy in the land snail, *Helicella virgata* (Pulmonata: Helicidae). *Austr. J. Zool.*, **16**, 857–69.

Raftery, R.E. (1983) *Littorina* trail following: sexual preference, loss of polarized information, and trail alterations. *Veliger*, **25**, 378–82.

Rollo, C.D. and Wellington, W.G. (1981) Environmental orientation by terrestrial Mollusca with particular reference to homing behaviour. *Can. J. Zool.*, **59**, 225–39.

Scheibling, R.E. (1980) Homing movements of *Oreaster reticulatus* L. (Echinodermata: Asteroidea) when experimentally translocated from a sand patch habitat. *Mar. Behav. Physiol.*, **7**, 213–23.

Schöne, H. (1984) *Spatial Orientation: the spatial control of behavior in animals and man*, Princeton University Press, Princeton, New Jersey.

Sinclair, A.N. (1959) Observations on the behavior of sea urchins. *Aust. Mus. Mag.*, **13**, 3–8.

Smith, D.P. (1969) Daily migrations of tropical sea urchins. *Am. Zool.*, **9**, 1075.

Southwick, C.H. and Southwick, H.M. (1969) Population density and preferential return in the giant African snail *Achatina fulica*. *Am. Zool.*, **9**, 566.

Stirling, D. and Hamilton, P.V. (1986) Observations on the mechanisms of detecting mucous trail polarity in the snail *Littorina irrorata*. *Veliger*, **29**, 31–7.

Townsend, C.R. (1974) Mucus trail following by the snail *Biomphalaria glabrata* (Say). *Anim. Behav.*, **22**, 170–7.

Trott, T.J. and Dimock, R.V. (1978) Intraspecific trail following by the mud snail *Ilyanassa obsoleta*. *Mar. Behav. Physiol.*, **5**, 91–101.

Underwood, A.J. (1977) Movement of intertidal gastropods. *J. Exp. Mar. Biol. Ecol.*, **26**, 191–201.

Underwood, A.J. (1979) The ecology of intertidal gastropods. *Adv. Mar. Biol.*, **16**, 111–210.

Warner, G.F. (1979) Aggregation in echinoderms, in *Biology and Systematics of Colonial Organisms*, Systematic Association Special

Volume No. 11 (eds G. Larwood and B.R. Rosen), Academic Press, London, pp. 375–96.

Wells, M.J. and Buckley, K.L. (1972) Snails and trails. *Anim. Behav.*, **20**, 345–55.

Willoughby, J.W. (1973) A field study on the clustering and movement behavior of the limpet *Acmaea digitalis*. *Veliger*, **15**, 223–30.

Chapter 3
Arthropods

Rüdiger Wehner

3.1 INTRODUCTION

More than 200 years ago, Lazzaro Spallanzani (Accademia d'Italia, 1934) wondered how birds – sand martins in his case – found their way back home after they had been displaced inadvertently to unknown territory, but the first to marvel at the amazing homing abilities of insects was Jean Henri Fabre (1879, 1882). He released some megachilid bees and sphecid wasps up to 4 km away from home and was surprised to find that many of them returned to their nesting sites the very same day. Even though he performed a number of experiments including the attachment of tiny magnets to the homing bees, he finally was left with the conclusion that his experimental animals possessed some enigmatic sense of directionality. Nevertheless he started what can be called the first period of research on homing in insects. In this period, which culminated in the discovery of the insect's celestial compass (Santschi, 1911, 1923; see Wehner, 1990a), most investigators focused on the sensory basis of insect navigation. John Lubbock, better known as Lord Avebury, discovered that ants could perceive ultraviolet light and use it for homeward navigation (Lubbock, 1889), and Felix Santschi was the first to demonstrate that walking ants deposited secretions from their anal glands on the ground and later used these trail substances as chemical signposts (Santschi, 1913).

The second period of intensive research on insect homing, now accompanied by studies on other arthropods as well, spans the 1950s and early 1960s. The discoveries of Karl von Frisch and Martin Lindauer on homing in honey bees and of Leo Pardi and

Animal Homing. Edited by Floriano Papi. Published in 1992 by Chapman & Hall, 2–6 Boundary Row, London SEI 8HN. ISBN 0 412 36390 9.

Floriano Papi on navigation in talitrid crustaceans and lycosid spiders mark the highlights of this time. Again, emphasis was placed largely on sensory mechanisms rather than navigational strategies. This is borne out by the topic studied most extensively: sensory guidance by sunlight, light patterns in the sky and moonlight (e.g. Frisch, 1949, 1950; Pardi and Papi, 1953; Papi and Pardi, 1953; Frisch and Lindauer, 1954; Papi, 1959; Pardi, 1960; Lindauer, 1963).

It might come as somewhat of a surprise that with the advent of the neurosciences in the early 1970s research on how small-brain navigators performed their necessary feats of computation did not gather momentum. Most research projects such as those on spiders (Görner and Claas, 1985), isopods (Hoffmann, 1983a and b), beetles (Frantsevich et al., 1977) or bees (Brines and Gould, 1979) remained completely within the behavioural realm. No attempt was made to research the neural hardware of the navigational mechanisms deduced from behavioural studies, and the conceptual potential offered by the neurosciences was not exploited in shaping the questions to be asked. Instead, even very recently, researchers have been led astray by inappropriate formalism as expressed, for instance, in the simplistic taxis concept. 'Pharotaxis' and 'mnemotaxis' (Rosengren and Fortelius, 1986), 'tropo-phototaxis' (Fourcassie and Beugnon, 1988) and 'menotaxis' (Aron et al., 1988) are just some of those monstrous terms still used to 'explain' certain types of navigation.

Given this state of the art, the present chapter on homing in arthropods is intended to serve a two-fold function. First, various homing phenomena are described as they occur in the main groups of arthropods: spiders, crustaceans and insects. These descriptions, which are accompanied by discussions of some crucial experiments performed in that context, will familiarize the reader with the animal's actual navigational performances as well as with the ecological framework within which they are displayed, and within which the animal must finally solve its computational tasks. In focusing on the animal's real life and on the minute details of its behavioural repertoire I try to disconnect the behaviour of the animal from whatever conceptual context it was embedded in by the original investigator. Furthermore, by discussing these particular examples in some detail, I hope to stimulate future research which certainly will yield still further surprises.

The second part of this chapter deals with the mechanisms potentially underlying the animal's homing performances. In deriving and categorizing such mechanisms I shall be as parsimonious as possible, and hence try to keep Ockham's razor sharp. In the present account navigational mechanisms used in homing are discussed in neurobiological and computational terms. This might be more of a programme than a finished treatise, for it was only relatively recently, in studies on desert ants (Fig. 3.18), that both the neurobiological (Wehner, 1982) and the computational approach (Wehner and Srinivasan, 1981) were entertained. Since then the latter has been forcefully promoted by work on landmark-based homing behaviour in honey bees (Cartwright and Collett, 1983; Collett and Kelber, 1988). By emphasizing the need for linking behavioural studies to neurobiological and computational investigations, the second part of this chapter is intended to be a prolegomena – a preliminary attempt to pave the way for a deeper understanding of the mechanisms mediating homing behaviour.

It is not surprising that this attempt is made in small-sized navigators. Insects, crabs and spiders are equipped with relatively small and experimentally accessible neural networks responsible for rather stereotyped modes of behaviour. Furthermore, in many cases the navigational performances of a homing arthropod can be monitored precisely and analysed in detail, so that the minute subtleties of a particular trait of behaviour will become apparent. These subtleties might cast some light on the underlying selection pressures and, subsequently, on the ways in which the animal copes with these pressures. During the vicissitudes of their evolutionary history animals were forced to respond to particular pressures rather than to develop first-principle solutions to general problems. Seen in this light, the various case studies described in the first part of this chapter are not mere examples illustrating some general principles outlined in the second part. Careful explications of them should guard against simplistic all-inclusive models of homing behaviour and might help us to 'think small enough', that is to see the particular problems the animal rather than the human investigator must solve.

3.2 PHENOMENA

Homing is the apogee of spatial behaviour. Even small-brain navigators like many arthropods may exhibit such amazing powers of homeward orientation that some investigators felt entitled to attribute 'intelligence' (Pricer, 1908), 'insight' (Herrnkind, 1965) and even 'consciousness' (Wigglesworth, 1987) to the crabs, ants and bees they studied. The following account on how homing arthropods actually behave might give a more analytical view.

Figure 3.1 Wandering spider, *Cupiennius salei* (female). Body length (excluding legs): 3 cm. Photograph: courtesy of E.A. Seyfarth.

3.2.1 Araneae

Near the edge of the neotropical rainforest, in the coffee plantations of the highlands of central Guatemala, Costa Rica and Panama, banana plants are cultivated as protective shade trees for the young coffee plants. It is behind the leaves of these banana plants, or at similar localities of agaves and bromeliads, that wandering spiders of the genus *Cupiennius* (Fig. 3.1) spend their daytime hours in the ready-made shelters formed by the crevices between the bases of the fleshy leaves and the stem. These hiding places, which are often supported or even closed up by a strong sheath of silk, provide the spider not only with a humid and shady and hence favourable microclimate but also with mechanical and optical protection (Barth *et al.*, 1988). At nightfall the spiders leave their retreats, walk up the leaves and hunt for prey. After about 3–4 h they return to their home base (Seyfarth, 1980), which they will occupy for several days (Melchers, 1967; Barth and Seyfarth, 1979).

Besides the remarkable homing ability exhibited by *Cupiennius* in full darkness, there is yet another situation in which the spider must depend on memorized information about its own previous movements. Similar to other hunting spiders living on foliage, e.g. lycosid spiders (Rovner and Knost, 1974), *Cupiennius* restrains large prey by wrapping it in silk. Should the spider get disturbed while handling its prey, for example by the approach of another nocturnal predator, it will immediately leave the partially wrapped prey and later return to it along a straight course. In navigating back to its prey, and finally to its home, the spider relies largely on non-visual information to find its way.

A situation analogous to this natural setting can easily be created under laboratory conditions. When *Cupiennius* spiders are transferred to a circular arena, they are easily attracted by a humming housefly presented at the tip of two electrodes. As soon as the spider attacks and bites, the fly is electrically charged via the electrodes. The spider immediately releases its prey and is chased off with a brush from the capture site. After a period of several minutes it will turn around and walk back to the original site from which the fly has meanwhile been removed by the investigator (Fig. 3.2a).

How does *Cupiennius* find its way back to its former capture site? Olfactory cues are not used, and neither are visual ones. The former can be ruled out by placing the fly in such a position that the spider, after having been chased away from its prey, is closer to the fly's actual position than to the former capture site. Nevertheless, the spider returns to where the fly has been (the capture site) rather than to where it is now. Visual orientation is excluded by blinding the spiders with black paint and later, after the following moult, examining the exuviae for opacity. The most likely hypothesis that remains is that *Cupiennius*, while wandering back to its start, relies on proprioceptive cues, i.e. on memorized information about its own previous movements (idiothetic orientation: Mittelstaedt and Mittelstaedt, 1973; endokinetic orientation: Jander, 1975; kinaesthetic orientation: Görner, 1958).

To clarify this point let us briefly outline an additional set of experiments. The spider is chased away from the fly either along a rectilinear (Seyfarth and Barth, 1972) or along a curvilinear route (Seyfarth *et al.*, 1982). This is done either with spiders in which particular proprioceptors (the lyriform slit sense organs on the spider's legs) have been ablated, or in intact spiders which have received sham operations. The spiders' return paths are monitored by a closed-circuit television system and recorded on videotape.

Unlike insects, spiders do not 'march', i.e. walk in a steady, continuous pace. They tend to run in short bouts of a few steps each. In accord with this general pattern, the return paths performed by *Cupiennius* within the arena consist of several approximately straight segments, each characterized by a defined length and direction (Fig. 3.2a,b). Having reached the presumed goal the spider sharply turns around and appears to search for the lost prey by engaging in a systematic search behaviour.

A return is regarded as successful when the spider, on its way back, has reached a point 5 cm, or less, from the capture site. Given the spider's mean diagonal leg span of 10 cm this criterion will guarantee that under normal conditions the spider could have

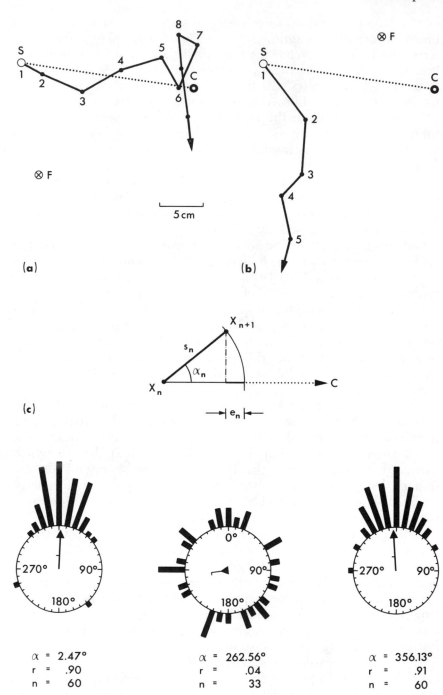

(a)

(b)

(c)

5 cm

(d)

α = 2.47°
r = .90
n = 60

(e)

α = 262.56°
r = .04
n = 33

(f)

α = 356.13°
r = .91
n = 60

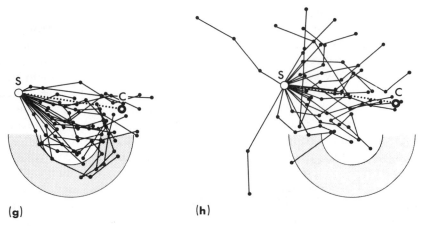

(g)   **(h)**

Figure 3.2 Homing behaviour in *Cupiennius salei*, which has been chased off a capture site. (a) Return path of an intact spider. (b) Return path of an operated animal in which all lyriform organs of all tibiae have been destroyed. C, capture site to which the spider displaced to S should finally return; F, position of relocated fly; S, start; the numbered dots mark positions at which the spider stops and turns; the dotted line indicates the direct return route. (c) Definition of turning angle α and walking error e. C, direction to capture site (goal); s_n, straight segment of walking path; X_n, X_{n+1}, positions at which the spider stops and turns. Walking errors Σe_n summed over all n segments of a return path are referred to in the text; starting angles α_1 are depicted in (d)–(f). (d–f) Starting angles α_1 of (d) intact animals, (e) operated animals, and (f) sham-operated animals. (g – h) Homing paths of (g) intact animals and (h) operated animals which have been chased curvilinearly through a semicircular corridor (shaded area). Symbols as in (a) and (b). α, mean angle of orientation; r, length of mean home vector (Batschelet, 1981); the bars on the vectors indicate significance levels of $P < 0.01$; n, number of animals tested (after Seyfarth and Barth, 1972; Seyfarth *et al.*, 1982).

touched the lost fly. While intact spiders exhibit high rates of successful runs (0.7–1.0) even when they have been chased away up to distances of 75 cm (for technical reasons larger distances could not be tested), in operated spiders the success rate drops to values below 0.25 for distances larger than 20 cm. In the experiments referred to above, 'operated' means that the spider's slit sense organs positioned near the distal joints of the femur and tibia were destroyed mechanically with a tapered tungsten needle.

It is not only the rate of success but also the starting angle (Fig. 3.2d–f) and the mean walking error (for definition see Fig. 3.2c) that differ significantly between operated and control animals. These findings demonstrate that the lyriform slit organs, groups of mechanoreceptors that respond to cuticular strain along the spider's legs (p. 85), are certainly involved in the animal's return to the place it has left a few seconds or minutes ago. More detailed conclusions can be drawn from an extension of the experiments just described. Whenever *Cupiennius* is chased off the capture site along a curvilinear rather than rectilinear path by forcing it to walk through a semicircular corridor, it takes the short-cut back to the start (Fig. 3.2g,h). While in intact spiders and sham–operated controls the success rates reach values of more than 0.85, in operated animals these rates are less than 0.50. Even in the successful cases the paths the operated spiders take to reach the previous capture site are much longer than in the controls.

The general conclusion to be drawn at this juncture is that *Cupiennius* is able to compensate for detours, that it does so by relying on idiothetic information, and that this information depends at least partly on input from the slit-organ proprioceptors.

While the tropical wandering spider, *Cupiennius*, dwells on particular types of plant and uses them as hunting grounds, the temperate-zone funnel-web spider *Agelena* (Fig. 3.3) spins horizontal webs and uses these webs to capture insect prey. In

Figure 3.3 Funnel-web spider, *Agelena labyrinthica* (female). Body length (excluding legs): 1 cm. Photograph: courtesy of P. Görner.

contrast to *Cupiennius*, which while active wanders about or waits in ambush outside its retreat, *Agelena* spends most of its time inside its retreat, a cone-shaped funnel located at the margin of the web. It is to this funnel that the spider returns immediately after each excursion, and to which it owes its name.

Agelena has been the subject of an extensive series of experiments started by Baltzer and his students (Bartels, 1929; Baltzer, 1930; Holzapfel, 1934) and continued by Görner and his colleagues (Görner, 1958; reviewed by Görner and Claas, 1985). During these 40 years of research a massive amount of data has accumulated on *Agelena*'s homing abilities and has been used to jump to conclusions about the software of the underlying neural machinery (Mittelstaedt, 1978, 1983, 1985). Nevertheless, the amount of printer's ink spent on this subject should not distract from the poor understanding we still have of the whole topic. The literature abounds with repeated new interpretations of old data. Recently, the main investigator in the field even conceded that in interpreting the factual results we might be forced to abandon the all-inclusive computational models hitherto devised (Görner, personal communication).

Certainly, a principal reason for this unsatisfactory state of affairs is the multitude of sensory stimuli *Agelena* is able to exploit while navigating to its web. Unlike the strictly nocturnal *Cupiennius*, *Agelena* is active during daytime as well, and unlike *Cupiennius*, which walks and hunts on solid ground as provided by the surface of hard leaves, *Agelena* constructs its own light hunting platform, a horizontally woven web of distinct vibrational properties. This has important consequences. In addition to proprioceptive cues, *Agelena* can use at least two additional sources of (external) information: optical cues provided by the surroundings of the web, and elasticity cues provided by the web itself.

Although various types of homing experiments have been performed in *Agelena*, one of them suffices to outline the major points. In this type of experiment a humming *Drosophila* fly cast onto the web induces the spider to leave its funnel, locate the fly, collect it, and then to return with its prey to the funnel. During its outbound (foraging) and inbound (homeward) run the spider is presented with a point light source positioned at a particular azimuthal distance from the funnel (Fig. 3.4a). Under these conditions the spider returns to its retreat on a more or less direct course; its accuracy of homing is remarkably high (Fig. 3.4b). In a subsequent test the azimuth of the light source is changed by 90° just before *Agelena* starts its homeward run. Now the spider deviates from its direct course (Fig. 3.4c). The mean deviation (ε),

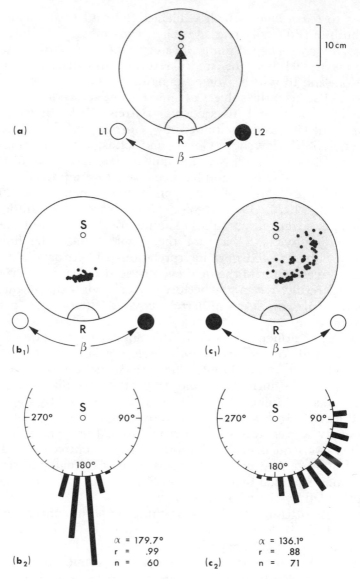

Figure 3.4 Homing behaviour in *Agelena labyrinthica*, I. (a) Experimental set-up. L1, L2, light sources separated by angular distance β (open circle: light switched on, filled circle: light switched off); R, retreat (funnel); S, start of return run. Prior to its return run each spider is lured from R to S (arrow). (b,c) Homing performances. (b) L1 switched on (as during training from R to S), (c) light switched over from L1 to L2. (b₁,c₁) Positions of the spiders after having completed 12-cm return runs; (b₂,c₂) angular distributions of starting directions. α, r, n, see Fig. 3.2 (data from Görner and Claas, 1985; Görner, unpublished).

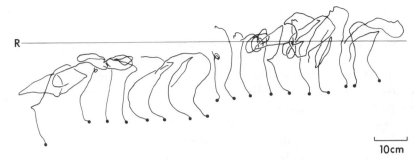

R

10cm

Figure 3.5 Homing paths of 18 *Agelena* spiders tested under the experimental paradigm outlined in Fig. 3.4c. The spiders' starting points (capture sites) are marked by black dots. Correct estimation of distance would imply that the spiders start their search behaviour after having reached the position marked by line R. Contrary to a previous claim by Dornfeldt (1972), which was repeated by Mittelstaedt (1978), such is not the case (after Görner, personal communication).

however, is considerably smaller than one would expect from the angular shift of the light source ($\beta = 90°$). Obviously, *Agelena* does not rely exclusively on visual stimuli.

There are two additional sources of information *Agelena* could use in computing its homeward course, and both would account for the observed relation $\varepsilon < \beta$: idiothetic information and information based on the structure (elasticity pattern) of the web. Either source of information would lead the spider directly back to its start ($\varepsilon = 0°$), the former by a path integration system, the latter by some kind of map, however simple, of its hunting ground. *Agelena*'s actual homing courses which are neither $\varepsilon = 0°$ nor $\varepsilon = 90°$ could then be interpreted as a compromise resulting from a continuous comparison between optical input and idiothetic input, or optical input and input from the web, or both. (Nothing has been said so far about the spider's ability to measure distances. Under the conditions used in the *Agelena* experiments distance estimation turns out to be very poor; see Fig. 3.5.)

Unfortunately, it has been exceedingly difficult to disentangle the influences of idiothetic and web-based information in *Agelena*'s overall homing performance. One way to interrupt the idiothetic path-integration system, and to empty the idiothetic information store, is to lift the spider, together with its prey, off the web and, after it has completely lost contact with the ground, put it down again at the same or at another site. If *Agelena* relied exclusively on idiothetic information, and if it had been lured to its prey in complete darkness, it should then start to walk in arbitary directions. Whereas former experiments seemed to give this impression (Görner, 1966; Moller, 1970; Dornfeldt, 1972), more recent studies indicate that in many cases the homeward direction is clearly preferred, especially when the spider has been lured to the periphery rather than the centre of the web (Görner, personal communication). Note that in general there is a higher elasticity gradient at the periphery than in the centre of the web.

The spider's use of web-based information can be inferred also from a modification of the basic type of experiment described above. In contrast to the former situation depicted in Fig. 3.4c, *Agelena* is now deprived of idiothetic information by being briefly lifted off the web (at S in Fig. 3.4c$_1$). Nevertheless, its subsequent homeward course still deviates from the 90° direction ($\varepsilon < \beta$), which *Agelena* should choose if it relied exclusively on visual stimuli. The deviation is smaller than in the former case (Fig. 3.4c), in which the spider could refer to idiothetic information as well, but it is still there (Görner and Claas, 1979, 1985). Certainly, it results from information the spider gains from the structure of the web (see also Görner, 1988). However, the question of which actual structural properties the spider detects and uses for navigation is not yet resolved. Some rather general remarks must suffice.

As already observed by Holzapfel (1934), the elasticity of the web increases with increasing distance from the retreat. Hence, on its way home, *Agelena* might either exploit this gradient of elasticity or acquire and use more elaborate knowledge about local differences in the elasticity characteristics of the web. In the light of the present data, it is not yet possible to decide between these two alternatives. Only a few experiments have been performed in which the elasticity pattern of the web has been changed. Two of them will be described. First, when a web mounted within a square-shaped frame is transformed into a diamond-like configuration, *Agelena* follows the lines of highest tension rather than returns to its retreat (Baltzer, 1930). Second, when spiders are tested on composite webs constructed from parts of natural webs, their behaviour differs from that shown on the original webs (compare Fig. 3.6b and c). Furthermore, when the composite webs are placed on meshworks of orthogonally arranged nylon threads, the spiders often follow the direction of the threads (Fig. 3.6d,e).

The bowl spider, *Frontinella pyramitela*, a sheet-web spinner, is another species in which the use of web tension as a navigational aid has been demonstrated (Suter, 1984). *Frontinella* builds bowl-shaped webs with the concave side pointing upward. It typically

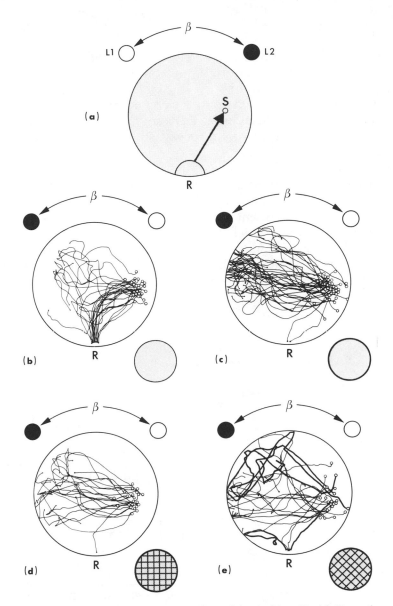

Figure 3.6 Homing behaviour in *Agelena labyrinthica*, II. (a) Experimental set-up. For symbols and conventions see Fig. 3.4a. (b–e) Return paths of individually tested spiders after the light has been switched over from L1 to L2. Elasticity pattern of web as indicated by the inset figures in the lower right: (b) natural web; (c) composite web: web composed of pieces of natural webs; (d,e) composite web placed on nylon threads strung either (d) parallel and perpendicular to the diameter of the web passing through the retreat, or (e) at an angle of 45° and 135° to that diameter (after Görner and Claas, 1985).

hangs at the lowest point on the underside of the web and returns to this preferred area, its retreat, not only after prey capture, but also after courtship activities or aggressive encounters with conspecific intruders. Tension in the web, as measured in terms of the spring constant (dyn/cm), increases systematically from the periphery to the centre of the bowl, and it is this parameter to which *Frontinella* responds.

Besides the elasticity pattern of the web and the visual landmarks in the spider's surroundings, *Agelena* can refer to yet another source of external information, namely gravity. In the experiments described above any orientation by means of gravity was ruled out by using only webs which were oriented in a strictly horizontal way, but under natural conditions most webs slope toward the retreat. Correspondingly, while homing the spider walks **down** to its funnel. Experimentally one can tilt a web in such a way that a particular corner of the square-shaped frame rather than the funnel

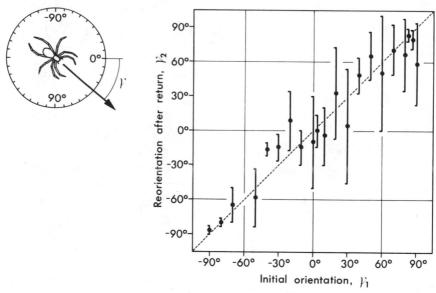

Figure 3.7 Gravity orientation in the salticid spider, *Phidippus pulcherrimus*. The spider waiting in ambush on a vertical screen fixates a fly (direction of fixation marked by arrow). After the jump is elicited by the spider, the fly is immediately removed by the experimenter. Upon return to its start the spider takes the same angular position (ordinate) which it has taken before the jump (abscissa), even thought the fly is no longer present. Mean values and standard deviations calculated from data provided by Hill (1979).

itself is at the lowest point of the whole device. Then the spider
returning with prey heads towards this lowest point (Bartels, 1929).

Jumping spiders may use gravity as a reference to re-establish
their former homing position (Fig. 3.7; Hill, 1979), but the most
likely candidates for exploiting gravitational cues in homing are the
orb-weaver spiders. This is for the simple reason that within their
radially symmetric webs their movements are confined to the two
dimensions of a vertical plane. The cross spider, *Araneus diadematus*,
normally waits at the hub for prey to enter the web, determines the
direction of the prey (by vibrational cues), runs to the prey, wraps
it in silk, cuts it out of the web, and carries it back to the hub.
Similarly, the argiopid orb spider, *Zygiella x-notata*, lives in a
funnel-like retreat connected to the capture web by a special radial
thread, which is not overlaid by the sticky capture spiral. When it
has caught its prey, *Zygiella* returns to its retreat along this special
thread. When the web is experimentally rotated 90° or 180° while
Zygiella is still wrapping its prey, the spider will later run along the
radial thread that points towards the former site of the retreat rather
than to the retreat itself (Fig. 3.8). This is all the more astounding as

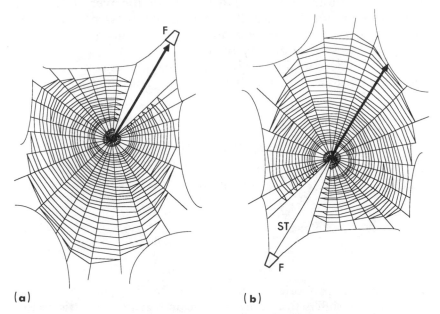

(a)   (b)

Figure 3.8 Orb web of *Zygiella x-notata* (a) in normal position, (b)
rotated by 180°. F, funnel; ST, signal thread leading from the hub to the
funnel. The arrow indicates the direction in which the spider returns to its
retreat at the periphery of the web. Based on experiments by Le Guelte
(1969) and a photograph published by Vollrath (1989).

the spider must now cross the sticky spiral hindering its path (Le Guelte, 1969). *Araneus* is misdirected in a similar way when its web is rotated during prey capture (Peters, 1932; Crawford, 1984).

Surprisingly, the three accounts cited above are the only ones discussing gravity as a possible cue used by orb-weaving spiders in returning to their retreats. Even though all three investigations convincingly show that orb weavers use cues that are independent of the web for finding their way back to the hub or the funnel, none of them provides unequivocal proof that the spiders really rely on gravity as the vertical reference (as they do in constructing their webs; Vollrath, 1988). Nevertheless, there is much circumstantial evidence that gravity actually is the decisive cue in homeward orientation. For example, *Araneus* and *Zygiella* correctly return to their starting position even in the dark.

During daytime, visual cues are used as well. In an elegant series of experiments, Crawford (1984) has shown that *Araneus* uses such cues to discriminate between the two sides of the vertical web. In fact, *Araneus* can travel on both sides of the web, and when wrapping prey it often switches from one side to the other. However, without knowing which side of the web it is on, the spider would not be able to discriminate between the correct direction to or from the hub and its mirror image as measured relative to the vertical gravity reference (Fig. 3.9).

In all cases described so far homing behaviour is restricted to two spatial dimensions, either to a horizontal plane (in *Cupiennius* and

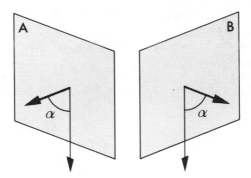

Figure 3.9 Compass orientation within the vertical plane of a spider's two-sided web suffers from mirror-image ambiguity when it is based on gravitational cues alone. The spider must know, for example, by referring to external visual stimuli, which side of the web it is on, in order to maintain a given compass course (thick arrow). α, angle between the spider's compass course and the direction of gravity (thin arrow). A and B depict the two sides of a vertical web.

Agelena) or to a vertical plane (in *Araneus* and *Zygiella*), yet many spiders can negotiate their way through three-dimensional space and nevertheless return to constant homing sites. Jumping spiders like many *Phidippus* species occupy waiting positions on herbaceous plants from which they embark in pursuit of sighted prey and to which they regularly return. During pursuit they often utilize indirect routes (detours) to attain a position from which the prey can finally be captured. While approaching their prey along a detour, they frequently loose sight of it, but in spite of these repeated open-loop intervals in their pursuit they are continually informed about their position relative to both their goal and their start. This remarkable navigational phenomenon has been well documented (Hill, 1979), but the way in which idiothetic information and information about gravitational and visual cues are assembled and used is far from being understood.

3.2.2 Crustacea

To the human investigator, studying homing behaviour in crustaceans generally means exchange of the aerial environment for the aquatic one. It is certainly for this reason that we know so much more about orientation and homing in semi-terrestrial species, for example ocypodid crabs, talitrid amphipods or tylid isopods (see reviews by Herrnkind, 1983; Rebach, 1983), than in subsea travellers such as lobsters or mantis shrimps. In fact, there are very few data on territoriality, central-place foraging and homing in marine benthic organisms, primarily due to the difficulties in obtaining the necessary long-term observations underwater. Furthermore, the human investigator is usually not familiar with the diversity of physical realms existing within the marine environment or with the uniqueness of the life-styles and behavioural capabilities of animals populating this environment.

In the previous paragraph the term 'marine' is used rather than the more general term 'aquatic'. This is to recall that the marine environment is the major habitat of decapod crustaceans, and that only a few species inhabit or spend part of their lives in freshwater or in air.

Let us first turn to the underwater environment. One of the most dominant types of horizontal to-and-fro movements in natant decapods are the offshore—inshore movements of penaeid shrimps (Allen, 1966; Hughes, 1972). However, in these 'migrations' the animals do not return to a fixed location, but to any location along a particular horizontal level running parallel to the shore (*x*-axis). They travel to and from this level by following *y*-axis courses

(y-axis orientation: Ferguson, 1967; zonal orientation: Jander, 1975; return migration in a stratified environment: Baker, 1978). Furthermore, in many species it is not the same individual that returns to the inshore habitat. In *Penaeus duorarum*, for example, the adults move offshore to spawn and the larvae of the next generation move, or are displaced, back to the intertidal zone. The same seems to hold true for the impressive offshore mass migrations of spiny lobsters, *Panulirus* and *Jasus* (Street, 1971; Berry, 1973; Herrnkind, 1980, 1985). By walking in characteristic single-file columns, so-called queues, both day and night, thousands of lobsters may traverse 30–50 km during several days of continuous marching. The functional significance of these large-scale movements as well as the guidance cues involved are not well understood (for the latter see Kanciruk and Herrnkind, 1978).

Even though most researchers have concentrated on the type of mass migration mentioned above, a few reports on long-distance homing are available. Over periods of months tagged rock lobsters, *Homarus americanus*, were recaptured more than 100 km from the tagging site and were subsequently captured again at the tagging site (Saila and Flowers, 1968; Pezzack and Duggan, 1986). About 20% of the tagged spiny lobsters, *Panulirus argus*, released 3 km from their home site were recaptured there within 4–9 days (Creaser and Travis, 1950). However, when the vanishing bearings of spiny lobsters released 2–5 km from the capture site were measured, they did not exhibit any homeward component (Herrnkind and McLean, 1971; for *Panulirus longipes* see Chittleborough, 1974).

For most of their life lobsters restrict their movements to foraging areas much less than 1 km wide. There they spend the day in rocky shelters, which they leave for nocturnal foraging trips (Fig. 3.10a,b). These trips lead them up to 300 m away from their home site to their feeding area where they move about in a meandering way. Finally, several hours before dawn, they return to their shelters, after following a short-cut route, and sometimes using the same pathway each night (*Panulirus*: Herrnkind and McLean, 1971; Herrnkind, 1980; *Homarus*: Karnofsky *et al.*, 1989). Many individuals occupy 3–4 shelters located within distances of some hundred metres (Fig. 3.10a). When chased by the investigator, they would tail-flip backwards directly towards the nearest shelter. When this shelter is experimentally blocked, they move directly to an alternative one, thus giving the impression of using some map-like information of the spatial distribution of their shelters. The use of such information is also suggested by the behaviour of male spider crabs, *Inachus phalangium* (Diesel, 1986). In these symbiotic crabs, which live in the protection of anemones, most females are

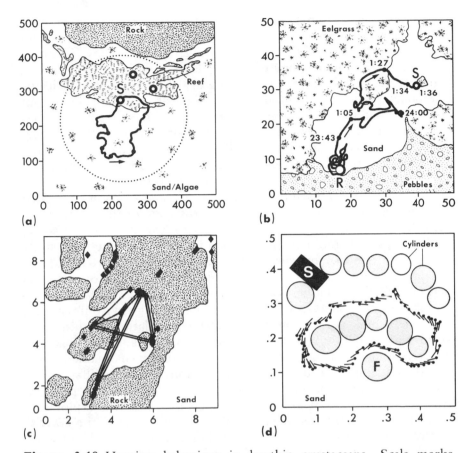

Figure 3.10 Homing behaviour in benthic crustaceans. Scale marks (numbers) in metres. (a) Spiny lobster, *Panulirus argus*. The figure depicts a foraging path of one individual that occupied three specific shelter sites, S. The lobster regularly foraged within the area indicated by the dotted line (after Herrnkind *et al.*, 1975). (b) Rock lobster, *Homarus americanus*. Homing track of an individual lobster that has been displaced from the shelter, S, and released at site R. It returned home in 2 h, 12 min. The numbers indicate local standard time (after Karnofsky *et al.*, 1989). (c) Spider crab, *Inachus phalangium*. Movements of an individually labelled male to and from particular females during a period of 39 days. The females indicated by ◆ occupied the same sites for weeks or months (after Diesel, 1986). (d) Crayfish, *Cherax destructor*. Exploratory excursion of a crayfish released within a new habitat, a tank containing a particular arrangement of plastic cylinders. Position of the animal and orientation of its body axis as taken from video recordings are depicted every second. F, feeding site (see text); S, shelter (after Varju and Sandeman, 1989).

rather site-constant, moving only short distances of less than 1 m. The males, in contrast, travel frequently between anemones harbouring females due to spawn (Fig. 3.10c). They regularly patrol an area wherein they obviously learn the positions of several females. Even when females disappear, the males still visit the anemones which harboured them. This site-constant behaviour indicates that the males are not oriented by female pheromones. Furthermore, at the beginning of their reproductive activity males move randomly to anemones with no females or with females not yet ready to spawn, but some weeks later they know the positions of the females in their area. These well-directed movements reported here for *Homarus*, *Panulirus* and *Inachus* demonstrate that benthic decapods may possess some knowledge of the home-range area they inhabit.

The previous statement is supported by the most detailed study on spatial memories in crustaceans. The Australian crayfish, *Cherax destructor*, seems to acquire a detailed tactile map of the surroundings of its shelter (Varju and Sandeman, 1989). When a reversibly blinded crayfish is released in a novel environment, for example a tank containing a particular constellation of differently sized and shaped objects, it immediately starts to explore this environment in a characteristic and stereotyped manner (Fig. 3.10d). During these exploratory excursions it continually uses its antennae to come into contact with the objects along its path. As time proceeds, the walking speed increases and the exploratory stops become more frequent and pronounced. Finally, the crayfish settles down in its shelter. When the blinded animal is then trained to a feeding site along a particular route, it will later reach this site not only along this route, but also along various other routes. It will walk in a steady pace, but stop and explore any site at which objects have been rearranged, removed or changed in size. It is hard to escape the conclusion that during its exploratory excursions in the tank the crayfish acquires some precise information about the position, size and shape of objects within its immediate home-base surroundings.

In the crayfish, such information is tactile. In other decapods it might be visual. In this regard, it is a pity that homing studies have never been performed in mantis shrimps (stomatopods). These aggressive predators are highly visual and highly mobile (Caldwell and Dingle, 1976; Cronin *et al.*, 1988; Marshall, 1988). Living in shallow waters, they forage from a permanent home (burrow) to which they routinely return. They can be observed wandering about at some distance from their burrow, but once disturbed dart back suddenly into this retreat. Thus, stomatopods 'must be able to home, but nobody has any data on how this is done' (Cronin, personal communication).

With this regrettable remark let us leave the underwater environment and turn to the intertidal zone. In evolutionary terms, the crustacean species inhabiting the interface between the terrestrial and the marine environments – the intertidal zone – are no longer marine, yet not quite terrestrial. In their space-use patterns, however, they resemble truly terrestrial animals in a number of ways, not least in the way their spatial activities can be observed and recorded by the investigator. Many intertidal decapods move up and down the beach with the flood and ebb of the tide (e.g. the shore crab *Carcinus*; Crothers, 1968), but others, such as the fiddler crabs *Uca* and *Ocypode*, stay in more restricted areas where they live in self-dug burrows. When the burrow area is covered by water,

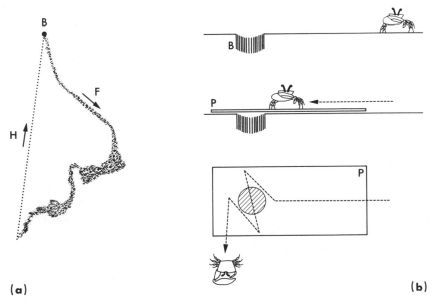

(a)   **(b)**

Figure 3.11 Nocturnal foraging and homing in fiddler crabs, *Uca*. (a) Foraging (F) and homing path (H) of *U. maracoani* as marked by the pattern of feeding pellets left by the crab on the intertidal sandflat. B, burrow. (b) Short-distance idiothetic return of *U. rapax* to its burrow, B. When the entrance to the burrow is blocked by a Perspex plate (P), the crab exhibits a characteristic zigzag search centred at the appropriate location. The experimental set-up is depicted in cross section (middle figure) and top view (lower figure) (after Hagen, 1967; (a) taken from a photograph).

the crabs remain buried, but come out to forage on the intertidal sandflats between high tides.

Even at night, and even during times of new moon, *Uca rapax* moves out from its burrow, forages over distances of about 1 m (Altevogt and Hagen, 1964) and then returns in the correct direction for the correct distance. The ability to integrate detours into the return route occurs in *U. rapax* (Hagen, 1967), *U. crenulata* (Hueftle, 1977) and probably all fiddler crabs (Fig. 3.11a). As visual, gravitational and tactile cues can be ruled out (the latter by covering the area with a perspex sheet; Fig. 3.11b), the crabs must rely on idiothetic information to monitor and integrate their foraging routes. This is most apparent in those cases in which the crabs, especially *U. tangeri*, forage along straight rather than winding routes. Upon return they do not rotate by 180° about their vertical body axis, but run home with that side of the body leading that was trailing on their way out (recall that most brachyuran decapods walk sideways; Evoy and Ayers, 1982). The possibility arises that the information acquired during the crab's outbound run is played back, simply reversed in sign, when the crab runs home.

In those species or populations that are active by day, fiddler crabs recognize their burrow areas by the surrounding visual landmarks. The extremely speedy racing crabs, *Ocypode cerato-phthalmus*, which reach running velocities of 4 m/s (Nalbach, 1987), return to their burrows after foraging at distances of up to 30 m (Hughes, 1966) or, according to a somewhat unreliable remark (Balss, 1956), 100–200 m. As these crabs occupy the same burrow for periods of 5–20 days (Hughes, 1966; Linsenmair, 1967) or even longer (Crane, 1958), landmark memories could be used for pin-pointing the entrance of the burrow. In fact, if landmarks are rearranged, the crabs search in the areas to which the landmarks have been transposed. If an artificial landmark is positioned near a burrow occupied by its owner, the crab reappearing at the surface will wait some time at the entrance, most probably to update its visual memory, before it sets out for another foraging trip. The larger an area the landmark fills in the crab's visual space, the longer the waiting time (Hagen, 1962; Langdon, 1971).

3.2.3 Insecta

(a) Migrating beyond the foraging range

Migration is the hallmark homing behaviour in birds but this is certainly not the case in insects. Of course, as has been known at least since Johnson's (1969) *magnum opus* on insect migration, at

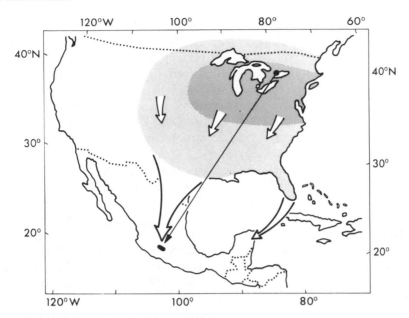

Figure 3.12 Autumn migration of the monarch butterfly, *Danaus plexippus* (eastern population), from the breeding area (shaded) to the overwintering sites in central Mexico (black). The thin arrow–headed line indicates the mark-and-recapture record (combined from Urquhart, 1960, 1976; Urquhart and Urquhart, 1979; Schmidt-Koenig, personal communication).

particular times of the year large-scale movements occur in many insect species, especially in Odonetes, orthopterans and lepidopterans (Williams, 1958; Schaefer, 1976; Cloudsley-Thompson, 1978). One of the more spectacular of these phenomena is the autumn migration of the monarch butterfly, *Danaus plexippus*, from the eastern USA and Canada to localized overwintering sites in the mountains of central Mexico (Urquhart, 1960; Brower, 1977; Fig. 3.12). Some alar-tagged specimens have been recorded up to 3000 km from their points of release, indicating that they must have travelled continually at rates of 130 km/day.

However impressive such migratory movements might appear, they hardly qualify as examples of large-scale homing performances. Individual insects do not return repeatedly to the same sites in the summer or winter territories as many birds do. Even though some monarchs flying south in the autumn may return to their summer breeding grounds (Urquhart and Urquhart, 1979), but never survive longer than a year, the majority breed new generations, and die, on the way north. Breeding has even been observed near the overwintering areas (Brower, 1961). Thus, spring migration consists of several overlapping generations. Similar migration patterns have been observed in Australian populations of *Danaus plexippus* (James, 1986) and in some other nymphalid species such as the widely distributed Painted Lady butterfly, *Vanessa cardui* (Shapiro, 1980; Myres, 1985). Again, during northward migration in spring and summer one to three intermediate generations are produced *en route*.

As far as the navigational performances are concerned, a population of migrating (airborne) insects must be regarded as a biometeorelogical system (Rainey, 1974). Indeed, aeroplane tracking and radar studies have shown that in many long-distance movements of butterflies, moths and locusts the migratory directions are largely determined by synoptic-scale wind systems (Kisimoto, 1976; Rainey, 1976; Schaefer, 1976; Kanz, 1977; Riley and Reynolds, 1979; Reid *et al.*, 1979; Drake *et al.*, 1981). Sometime in their evolutionary history these insects must have 'discovered' that properly exploiting local and seasonal wind patterns would bring them to favourable breeding grounds and thus increase their overall fitness. For example, joining the trade-wind system at the proper time carries a population of desert locusts more or less automatically into zones of wind convergence and hence of rainfall, i.e. into areas suited for breeding (Rainey, 1951). In monarch butterflies, the prevailing North American wind patterns facilitate southwestward movement in the autumn and northward or northeastward movement in the spring (Kanz, 1977).

These large-scale air-flow patterns are documented conspicuously by the passive wind-borne transport and subsequent fall-out of aerial plankton, e.g. spores of the wheat rust-fungus *Puccinia graminis*. For example, in spring masses of spores of this fungus are being deposited in Wisconsin, thousands of kilometres from their source in Texas (Gregory, 1961). The adaptive significance of the north–south 'migration' of this cereal rust-fungus is that *Puccinia* cannot survive the winter in the north or the summer in the south.

Drawing attention to aerial plankton is not to imply that migrating insects drift passively on the wind. They must actively embark on air currents as their transporting vehicle, and thus travel on such air currents adaptively rather than inadvertently. Moreover, having become airborne in the wind at the right time, a number of additional activities such as continuous wing-flapping, exploiting lift by soaring in thermals, and particular take-off and alighting patterns contribute to migratory success. Furthermore, the animal must employ some navigational means to maintain a straight course even in periods of cross-current or up-current flight. Nevertheless, I should like to argue that the difference between strong flyers such as nymphalid butterflies and passive drifters such as aphids and midges, or even fungus spores, is one of degree rather than kind. This statement is justified all the more by the fact that the routes even of some small migratory birds have been found to depend on synoptic-scale wind systems (Able, 1980; Williams and Williams, 1990).

Finally, let us recall that even the strongest flyers among the insect migrants, such as the *Danaus* and *Vanessa* species mentioned above, do not 'home' in the sense that they relocate a familiar site after long-distance displacements. In insects homing is confined to returns from short-term foraging excursions, and this is the topic to which we turn next.

(b) Returning to the central place

The best studied and, in fact, most eminent insect navigators are the social species. This is for the simple reason that social insects, especially the most highly advanced forms such as the eusocial hymenopterans (ants, bees and wasps), are central place foragers (Orians and Pearson, 1979; Stephens and Krebs, 1986). At any one time, certain individuals act as foraging machines, continually moving to and from their central place, the site of the colony, to retrieve widely scattered food items from the colony's environs. For example, a honey-bee colony confined to a 50-litre nestbox may spread its forager force consisting of about 10 000 individual scouts and recruits over an area of more than $100 \, km^2$ (Seeley, 1985). It is during these foraging endeavours that individual colony members get somewhat disconnected from the spatial coherence of the social unit and hence must employ some means of navigation to re-establish this unity, i.e. to find their way back to the centre.

Space-use patterns and homing capabilities can be studied best in the most dominant social insects, the ants. Unlike bees and wasps, their societies employ walking rather than flying foragers and thus allow for detailed cartographic surveys of the spatial activities of their forager force. Take, for example, the tropical ponerine ant, *Leptogenys ocellifera*. Several permanent routes, or trunk trails, spread out from the underground nest. Cleaned by the ants from twigs, leaves and debris they form conspicuous walkways used continuously day and night over periods of several months. At some distance from the nest individual scout ants will leave the trunk trails and search solitarily for arthropod prey. On its way

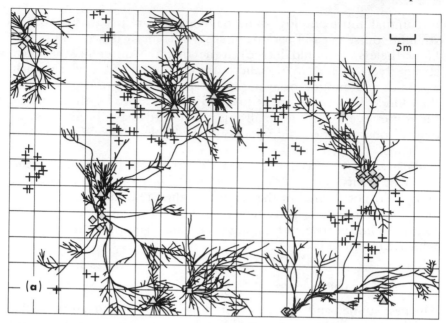

Figure 3.13 Space-use patterns in central place foragers (harvester ants, *Messor*). (a) Trail systems of *M. aegyptiacus* (□), *M. minor* (△) and *M. medioruber* (◊); Maharès, North African lowland-steppe desert. The record includes all trails that occurred during a 5-week test period (12 July–18 August). At any one time, only some of the trails were used by the colonies, and most of the more peripheral trails were used only

back to the nest, a successful scout lays a pheromone trail consisting of poison-gland and pygidial-gland secretions. While the former provide orientational cues (and are deposited by the ant already on its way out from the nest or the trunk trail), the latter stimulate recruitment. Along this recruitment trail up to 200 nest mates will then move in single file, one behind the other, out from the nest to the site of the prey, subdue it collectively and transport it to the nest (Maschwitz and Mühlenberg, 1975; Attygalle *et al.*, 1988).

Trunk-trail 'highways' are not limited to group-raiding ants like the Malayan *Leptogenys* species described above, but occur in leafcutter ants (e.g. *Atta*), harvester ants (e.g. *Pogonomyrmex* and *Messor*; Fig. 3.13a) and wood ants (*Formica*) as well. Marked by endurable chemical signposts (orientation pheromones; Hölldobler and Wilson, 1970) they may last for periods of months (Rosengren, 1971; Hölldobler, 1974, 1976). They are established and used by species in which large numbers of foragers must be recruited

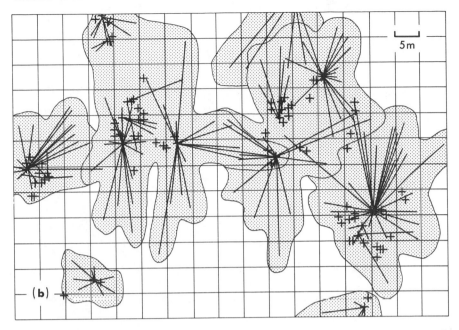

once. The symbols □, △, ◊ mark nest openings, +, nest entrance of
M. arenarius. (b) Foraging areas of different colonies of a solitarily
foraging species, *M. arenarius.* Same test area and test period as in (a). +,
Nest entrace; shaded areas, foraging ranges of individual colonies; lines
emanating from individual nest openings point towards food-finding sites
of some arbitrarily chosen foragers (Neumeyer, Funk and Wehner,
unpublished).

rapidly to patch-like distributions of food deposits, be they rich
seedfalls (in harvester ants) or aphid colonies (in wood ants).

Rather than being confined to two-dimensional space, pheromone
trails often extend into the third dimension (e.g. *Atta*: Cherrett,
1968; *Oecophylla*: Hölldobler, 1979, 1983; *Formica*: Rosengren, 1977;
Camponotus: Tilles and Wood, 1986). The giant tropical ant, *Paraponera clavata*, which nests at the base of trees in lowland neotropical forests, forages primarily in the canopy of the nest tree and
adjacent trees that are reached via canopy connections or trails on
the ground (Janzen and Carroll, 1983; Breed and Bennett, 1985).
These trails often lead to groups of blooming flowers, from where
successful foragers return carrying large drops of nectar between
their mandibles (Young and Hermann, 1980; for an instructive
illustration see Figs 3.10–2 in Hölldobler and Wilson, 1990). In most

Figure 3.14 Three-dimensional structure of a trunk-trail route in the giant tropical ant, *Paraponera clavata*. The route leads up the trunk of the nest tree (1), through the crowns of two adjacent trees (2, 3), to the ground, around the trunk of another tree (4), and then up into the canopy of the tree in which the nectar source is located (5) (modified from Breed and Bennett, 1985).

cases, as portrayed in Fig. 3.14, the course of such a trail does not reflect the most economical route between nest and food source, but exhibits a rather complex three-dimensional structure. This complex structure is the result of the colony's opportunistic foraging history: every now and then, old foraging sites are abandoned and new ones discovered. As a consequence, side branches of the trail system become incorporated into the main trunk-trail route, while previously used segments of the main route are given up. For example, when a *Paraponera* scout is baited to a location, say, 50 cm from the main foraging route and fed a concentrated 2-M sucrose solution, it returns to the main foraging route by moving slowly, dragging its gaster over the substrate and thus establishing a new pheromone trail. Subsequently, new foragers leave the main route and follow the newly established side-branch trail used by the original animal when returning from the bait (Breed and Bennett, 1985). Obviously, the ants are able to distinguish the new trail from the colony's main foraging route. Depending on the frequency with which the new and old trails are used, with time the trunk-trail flow of ants might alter its course.

While moving along pheromone trails ants do not rely exclusively on chemical cues. First, the chemical signals do not contain information about the inbound–outbound polarity of the trail (Hartwick *et al.*, 1977; Klotz, 1987; Oliveira and Hölldobler, 1989). The question of trail polarity was hotly debated at the beginning of this century (Wehner, 1990a), but it became clear in the early days of trail-pheromone studies that even if the individual chemical signposts were asymmetrical in shape (Macgregor, 1948), this asymmetry would not enable the ants to read information about the directional polarity of the trail from the individual pheromone streaks (Carthy, 1950, 1951; Wilson, 1962b). Further information is needed to resolve this ambiguity. Ants might determine the homeward direction by the mass flow of food-laden nestmates returning to the nest (Moffett, 1987), but visual landmarks have been shown to be the predominant cue (David and Wood, 1980; Klotz, 1987). Second, there is growing evidence that even trail-laying ants employ systems of route-integration drawing, among other things, upon a celestial compass (*Pogonomyrmex*: Hölldobler, 1971; *Messor*: Wehner, 1981b; *Leptothorax*: Aron *et al.*, 1988). As such systems of navigation inform the animal about its position relative to home, the ambiguity problem is solved automatically.

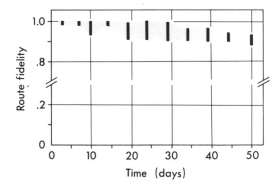

Figure 3.15 Long-lasting route fidelity in wood ants, *Formica rufa*. Ratios of labelled ants observed exclusively on their original trail (one of three trails leaving the colony's nesting site) for a period of more than 7 weeks. The ants leaving the colony along a particular trail were marked with a particular colour. Observations during the following weeks showed that nearly all workers remained faithful to 'their' trail. The bars indicate standard deviations calculated about the means from data of Rosengren and Fortelius (1986).

Finally, landmark memories seem to be used extensively to inform the ant about the particular route to take, and about the ant's current position along that route. Pronounced route fidelity (Fig. 3.15) based on visual spatial memories has been observed in a number of genera (e.g. *Formica*: Rosengren, 1971, 1977; Herbers, 1977; Rosengren and Fortelius, 1986; Möglich and Hölldobler, 1975; *Camponotus*: Traniello, 1977; Hartwick *et al.*, 1977; *Pogonomyrmex*: Hölldobler, 1976), and experiments have been designed to assess the significance of visual versus chemical cues (*Formica*: Rosengren and Pamilo, 1978; Cosens and Toussaint, 1985; Kaul, 1985; Rosengren and Fortelius, 1986; Klotz, 1987; Beugnon and Fourcassie, 1988; Fourcassie and Beugnon, 1988; *Camponotus*: Klotz, 1987; *Leptothorax*: Aron *et al.*, 1988; *Paraponera*: Breed *et al.*, 1987; Harrison *et al.*, 1989). Some interesting results have emerged from these experiments; for example that lateral displacement of artificial landmarks leads to an equivalent displacement of the foraging trail (*Formica aquilonia*: Cosens and Toussaint, 1985), or that rotation of an 'artificial forest' results in a reversal of the route-fidelity pattern (*Formica rufa*: Rosengren and Fortelius, 1986), or that the primarily nocturnal *Camponotus pennsylvanicus* exhibits a stronger tendency to follow odour marks, as compared to visual cues, than the mainly diurnal *Formica subsericea* does (Klotz, 1987). Nevertheless, due to the different species studied, as well as the different conceptual and experimental approaches taken by the investigators, it is difficult to derive any general conclusion from these results. In the most thorough study I could find on the relative importance of visual and pheromonal cues in trail-laying ants, Harrison *et al.* (1989) used a Y-shaped set-up to train foragers of *Paraponera clavata*, the giant tropical ant alluded to above, to an artificial feeder. After feeding, the ants returned to the nest depositing a pheromone trail on that branch of the fork at which the feeder was placed (left arm in Fig. 3.16a). When the pheromone trail was shifted experimentally to the other branch of the fork, naive (newly recruited) ants tended to follow the pheromone trail, while the experimental (labelled) ants travelled in the original direction as defined by the natural landmarks of the surrounding tropical rainforest (Fig. 3.16c). Only when the landmarks were screened off, did the experimental ants follow the odour trail (Fig. 3.16d).

In conclusion, while the newly recruited foragers use pheromonal cues for orientation, experienced ants preferentially use memorized visual-landmark information. Certainly, a major advantage of the preferential use of visual rather than chemical cues is increased foraging speed. While naive *Paraponera* ants relying exclusively

(a)

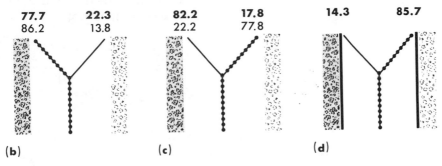

(b)   (c)   (d)

Figure 3.16 Use of chemical versus visual (landmark) cues in the neotropical ponerine ant, *Paraponera clavata*. (a) Training situation: Food, F (1-M sucrose solution), is delivered at the end of the left branch of a Y-shaped platform. N, nest; L1 and L2 symbolize the natural landmarks surrounding the experimental site. The dotted line marks the pheromone trail deposited by the returning ants. (b–d) Tests (feeder removed): (b) control, (c) pheromone trail switched to the other side of the Y-shaped platform, (d) pheromone trail switched as in (c), but in addition landmarks screened off. The numbers at the top of the figures represent the percentages of experienced ants (upper row, bold numbers) and naive ants (lower row) choosing either branch of the fork (designed on the basis of information provided by Harrison *et al.*, 1989).

on pheromonal cues move slowly back and forth across the trail (2.2 m/min), continually antennating the trail surface, the experienced ants walk faster and more directly (4.6 m/min) (Harrison *et al.*, 1989). Walking speed increases also in *Pogonomyrmex* (Hölldobler, 1971) and *Camponotus* (David and Wood, 1980) whenever the ants are able to use visual cues in addition to chemical ones. The further finding that route fidelity may be maintained at crepuscular times and even at night (*Formica*: Kaul and Korteva, 1982; Cosens and Toussaint, 1985; Rosengren and Fortelius, 1986) stimulates the question of visual contrast sensitivity in ants. At least in one study (Kaul and Korteva, 1982) it was shown that ants rely on 'canopy orientation' (Hölldobler, 1980; Kaul, 1985) to select a particular route at night.

Trunk-trails and their dendritic branching system (for cartographic representations see Hölldobler and Möglich (1980) and Fig. 3.13a) allow large numbers of foragers to be dispersed rapidly within the colony's foraging area and to effectively recruit nest mates to newly discovered food patches. As outlined above, in all such cases of trunk-trail foraging studied so far individual workers display considerable fidelity to a particular route which they use for both their foraging and return trips. However, it is not only the

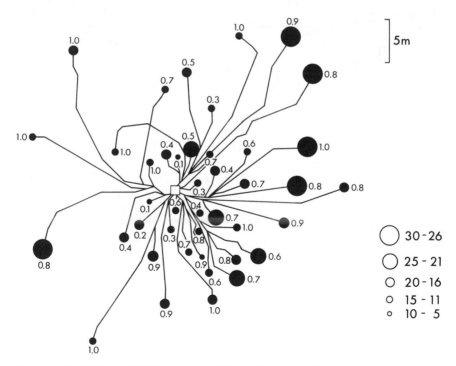

Figure 3.17 Site fidelity in individually searching ponerine ants, *Neoponera apicalis*. Map of foraging area around nest entrance (□). Foraging sites of individual ants are marked by black circles. The total number of individual excursions observed is indicated by the diameters of the circles (see inset). The numbers 0.1–1.0 represent the relative number of successful excursions, i.e. returns with prey (modified from Fresneau, 1985).

route but also the final foraging destination that may be kept constant by individual ants over considerable periods of time (e.g. Fig. 3.13b). Several individually marked carpenter ants, *Camponotus modoc*, were consistently found at the same 3-cm² area within a particular aphid colony for at least 12 days (Tilles and Wood, 1986; for similar observations in *Camponotus noveboracensis* see Ebbers and Barrows, 1980).

Path and site fidelity is not restricted to trunk-trail foragers, but occurs in solitary foragers as well. In the non-recruiting tropical species *Neoponera apicalis* individual ants develop a high degree of regional specialization (Fresneau, 1985). They may return to the same site, along the same path, over periods of at least 6 weeks (Fig. 3.17). In this strictly diurnal *Neoponera* species path fidelity persists even when any possible pheromonal cue is wiped off, indicating that the ants rely most probably on visual landmark memories in moving to and from their foraging sites. This suggestion is corroborated by the fact that *Neoponera* behaves as though lost, and moves around in searching loops, when it is displaced sideways from its familiar path or foraging site.

Such is not the case in those species of ants which are able to

Figure 3.18 Saharan desert ant, *Cataglyphis bicolor*. Body length (excluding legs): 1.2 cm. The specimen shown here is labelled individually (colour dots on head and thorax). Photograph by the author.

integrate their path and are thus not bound to retrace their outward route. The Saharan desert ants belonging to the genus *Cataglyphis* (Fig. 3.18) follow amazingly straight homeward courses which lead them over novel territory not covered during their circuitous outward journeys (Fig. 3.19). This kind of short-cut navigation immediately implies that the ant returning home cannot rely on local cues like pheromonal signposts or nearby visual landmarks; rather it must perform some kind of dead reckoning depending on internal or global external (e.g. celestial) systems of reference

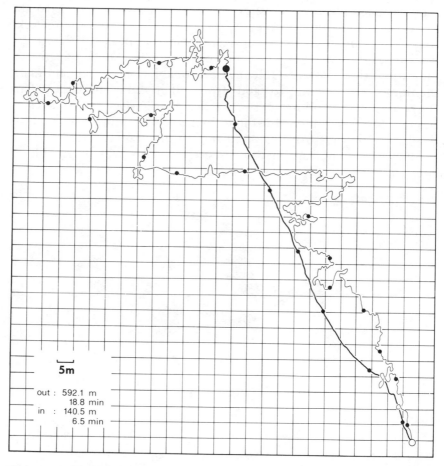

5m

out : 592.1 m
        18.8 min
in : 140.5 m
        6.5 min

Figure 3.19 Foraging and return path of a desert ant, *Cataglyphis fortis*. The thin and thick lines mark the outward and homeward path, respectively. Open circle, start of foraging trip; large filled circle, site of prey capture; small filled circles, time marks given every 60 s (from Wehner and Wehner, 1990).

(p. 87). Such dead reckoning provides the animal with a continuously updated homeward-bound vector always connecting the
animal with the starting point of its foray (p. 99). In operational
terms, the use of path- (route-) integration (dead-reckoning) systems
can be shown by displacing the ant at the start of its homeward run.
Unlike *Neoponera* (see above), *Cataglyphis* will then follow its
homeward vector by running straight in the predisplacement
direction of the nest for a distance equal to the predisplacement
distance to the nest (Wehner, 1982; Wehner and Wehner, 1986).
Having reset its route-integration system to zero, i.e. having
arrived at the fictive position of the nest, *Cataglyphis* switches on a
systematic-search programme by which it will sooner or later arrive
at the nest entrance (Wehner and Srinivasan, 1981; Wehner and
Wehner, 1986; p. 101).

Route-specific and route-integration systems are not mutually
exclusive. In trunk-trail foragers such as harvester ants individual
scouts may often leave the nest solitarily rather than follow
recruitment trails. In fact, depending on season and food availability several species rely heavily on individual foraging, e.g.
Pogonomyrmex maricopa (Hölldobler, 1974), *Messor aciculatus*
(Onoyama, 1982) and *Messor arenarius* (Fig. 3.13b). When such
individual foragers, after having found a piece of food, are displaced
to an arena devoid of any pheromonal cues and shielded from all
landmarks (Wehner, 1981b), they nevertheless head for home
(Fig. 3.20a). This behaviour clearly shows that *Messor*, similar
to *Cataglyphis*, is equipped with a route-integration system using
celestial cues for determining directions (Fig. 3.20b). When scouts
that have first followed a trunk-trail route, but have then departed
from that route for an individual search, are finally tested in the
arena, they select the direction of the point where they have left the
trail – the direction of their private exit of the colony's highway, so
to speak – rather than the direction of the nest (Fig. 3.20c). This
implies that a route-integration system is switched on at least
when the ants leave the common trail. There are some indications
(*Pogonomyrmex*: Hölldobler, 1971) that such systems are already in
operation when the ants are still following the trail.

In flying insects, foraging and homing paths cannot be traced out
in sufficient detail. This is a major drawback in using, for instance,
bees and wasps for studies in animal homing. However, there
might also be advantages. Additional conceptual approaches might
arise from the much larger foraging areas characteristic of flying
hymenopterans. The maximum foraging distances of ants lie in
the range of 200–250 m. This holds true for individual foragers
(*Cataglyphis fortis*: Wehner, 1987a) as well as group raiders and mass

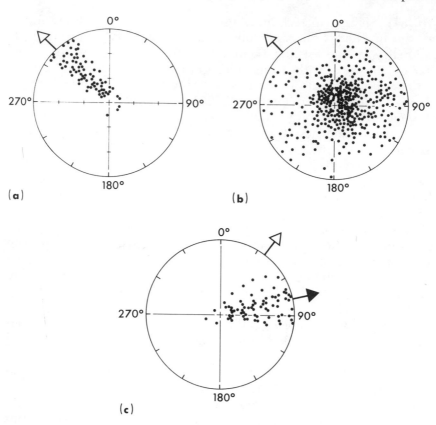

Figure 3.20 Homing behaviour based on route integration in the harvester ant, *Messor semirufus*. Having successfully completed individual foraging excursions the ants are released in the centre of a circular arena (diameter 2 m, sun and landmarks obscured). In each plot the positions of 12 ants tested individually are given every 5 cm for the first 30 cm of their paths. Open arrow, direction of nest; filled arrow, direction of the point at which the ants have left the trunk trail. (a,b) Scouts having started their individual searches directly at the nest entrance: (a) full blue sky, (b) total overcast. (c) Scouts having departed from the colony's trunk trail; full blue sky (from Wehner, 1981b).

foragers (*Eciton burchelli*: Schneirla, 1971; *E. rapax*: Burton and Franks, 1985; *Atta cephalotes*: Lewis *et al.*, 1974). In contrast, honeybees search for food up to 10 km from the hive (*Apis mellifera*: Knaffl, 1953), and orchid bees visit the same plant repeatedly along feeding routes extending up to 23 km from the nest (*Euplusia surinamensis*: Janzen, 1971). This far-ranging spatial activity stimulates the question of whether flying hymenopterans assemble and use more detailed geographical knowledge of the terrain over which they move than the walking ants are ever able to acquire.

First, as shown by detour experiments (Frisch, 1967), honey-bees use celestial-compass and route-integration systems much in the same way as described above for ants. The most thought-provoking experiments in that context have recently been performed in vespid wasps, *Polistes dominulus* (= *gallicus*) (Ugolini, 1985, 1986a,b, 1987). When individual wasps enclosed in a transparent Perspex tube were displaced along a straight course for distances of 50–2000 m and then released, they preferentially headed for home if two conditions were met. During displacement the wasps had to be able to see the sky as well as the landmark panorama floating by (Fig. 3.21a). When displaced in the dark their vanishing bearings were oriented at random (Ugolini, 1986b, 1987). Obviously, the wasps took compass readings from the sky and determined the compass direction in which they were displaced from the apparent movement of terrestrial cues within their visual field. The latter movement could be simulated by placing the test-tube containing the experimental animal between two linear black-and-white gratings that moved consistently in one direction. Then, upon release, the wasps selected the counter-direction, provided that during their simulated displacement they were presented, in addition, with celestial cues (Ugolini, 1987). Finally, out in the field again, the wasps were displaced along a multi-leg rather than a straight route (Fig. 3.21a$_2$). In this case, they tended to integrate the route they had taken involuntarily and, once released, headed for the correct home direction (Fig. 3.21b$_2$).

Second, under certain conditions bees and wasps as well as other central place foragers are able to return home even when cut off from their route-integration safety line. This can be achieved experimentally by displacing the animal in the dark from its central place (home) to arbitrary locations within its home-base area. Since Fabre (1879) and Romanes (1885) performed the first experiments of this kind in megachilid and apid bees, respectively, a heterogeneous body of field observations and experiments has accumulated (see Table 8 in Wehner, 1981a), providing some circumstantial evidence for the use of visual-landmark information in insect homing. The evidence, however, is rather indirect, and mainly based on homing time and rate of success. The enormous range of variation within these data, and the observer's ignorance of the insects' actual flight paths, render any conclusion about navigational mechanisms difficult if not impossible. For example, the return times of 400 carpenter and mining bees (*Xylocopa virginica*

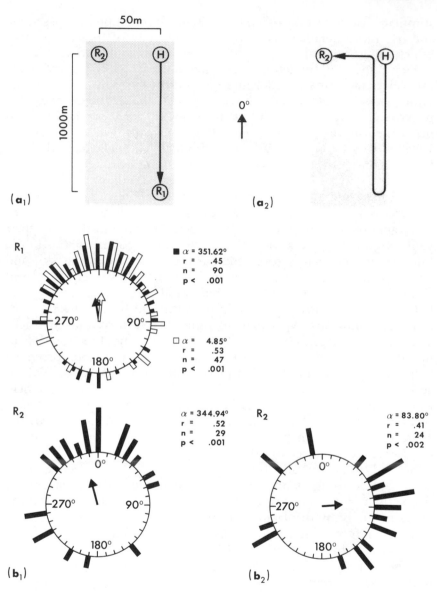

Figure 3.21 Acquisition of celestial compass information during passive displacement in the wasp, *Polistes gallicus*. (a) Experimental design. H, nest (home); R₁,₂, points of release; line with arrowhead, route of displacement. During displacement the wasps had access to terrestrial as well as celestial cues. In (a₁) the wasps released at R₂ were displaced from R₁ to R₂ in the dark. (b) Results: vanishing bearings as recorded at release sites R₁ and R₂. Diagrams b₁ and b₂ depict the results obtained under the experimental paradigms a₁ and a₂, respectively. Filled bars refer to

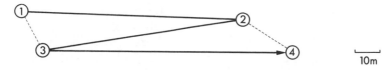

Figure 3.22 Flight path of a female euglossine bee (*Eulaema cingulata*) between four food plants. By following the route 1, 2, 3, 4 along which it had originally discovered the flowering food plants, the bee flew 240 m rather than the more economical 105 m of route 1, 3, 2, 4 (modified from Janzen, 1971).

and *Anthophora abrupta*, respectively) released 3.2 km apart from the nest varied from 20 min to 50 h (Rau, 1929, 1931). Nevertheless, there seems to be at least one consistent finding: homing success is positively correlated with foraging age (Rau, 1929; Lewtschenko, 1959; Ugolini, 1986a; but see Tengö *et al.*, 1990), and it is a likely hypothesis that individual experience as revealed by this correlation is based on visual spatial memories.

This hypothesis is strongly supported by the observation that bees and wasps, and many other insects as well (e.g. odonates: Kaiser, 1974; Ubukata, 1975), often follow distinct routes and may thus be familiar with sequences of landmark panoramas (Ferton, 1923; Baerends, 1941; Haas, 1967; Janzen, 1974; Heinrich, 1976). Along these routes individual points are visited in the order in which they were originally incorporated into the route and not in the most economical least-distance way (Ferton, 1923; Janzen, 1971; Fig. 3.22).

(c) Pin-pointing the position of the goal

Having returned from their foraging grounds, many insects exhibit extraordinary powers of pin-point navigation for finally locating

reproductive females (foundresses and auxiliaries), open bars to workers. The wasps' directional choices were recorded either as vanishing bearings in the field or within a Perspex arena. R_1, field and arena releases combined; R_2, arena releases. α, r, n, see Fig. 3.2; P, probability of random distribution (combined and modified from Ugolini, 1985, 1986b, 1987).

the micro-site of their home. This has been amply documented for
sphecid wasps – see the classical studies of Tinbergen (1932),
Baerends (1941), Iersel (1975) and others; reviewed by Wehner
(1981a) – and recently for the wasps' chrysidid brood parasites as
well (Rosenheim, 1987). Solitary bees, *Anthophora plagiata*, which
construct tube-like nests in loamy soil, descend from their return
flights precisely at the position of their nests even when their nest
entrances are covered with sheets of glass (Steinmann, 1985).
Another solitary bee, *Osmia bicornis*, does not return to its nesting
tube even after periods of many minutes, when the latter has been
shifted by as small a distance as 6 cm to the side (Steinmann, 1973;
see also Wehner, 1979). When the entrance hole of a honey-bee
colony is displaced by only 30 cm, the returning foragers will first
fly to the old site of the hole, and hover there for some time in a
dense cloud about an imaginary point in space, until they finally
locate the new site. Nevertheless, in returning from their
subsequent foraging trips they will continue to do so, i.e. to
approach first the old site and then fly from there to the new site,
for hours and sometimes even days (Butler *et al.*, 1970). Guard bees
of the meliponine species *Trigona angustula* hover stably in mid-air
close to the entrance of the nest for up to 70 min. They persistently
return to their individual hovering positions after having briefly left
them, e.g. for intercepting and chasing off robber-bee intruders.
Moreover, they maintain these positions relative to the nest
entrance even if the nest is slowly moved sideways, upward and
downward, or back and forth (Zeil and Wittmann, 1989), but fly
towards or away from the nest entrance when a striped pattern
mounted on the front face of the nest box is contracting or
expanding, respectively (Kelber and Zeil, 1990). Experiments of the
latter kind clearly indicate that in pin-point navigation insects rely
heavily on visual spatial memories (for further discussion see p.
106). It is only after they have finally arrived at the nest entrance
that nest-specific olfactory cues, e.g. 'nest-exit pheromones', might
become important in identifying, rather than locating, the
individual nesting tube (Steinmann, 1976; Anzenberger, 1986) or
the concolonial entrance hole (Hölldobler and Wilson, 1986; Jessen
and Maschwitz, 1986).

3.3 MECHANISMS

In homing, animals can resort to two principally different ways of
orientation by using either **egocentric** (self-centred) or **geocentric**
(terrain-centred) systems of reference. In the former case positional
information is obtained by some kind of route integration (dead

reckoning), which provides the animal with a continuous representation of its spatial position relative to its starting point (home). In the latter case the animal gains positional information from the relative location of its start within the environmental framework of its home-range area. Rather than relying on information collected continually *en route* about the animal's own movements (egocentred information), an animal using geocentred information could episodically take a positional fix and thus rely on information collected **on site**. While any system of navigation that depends on egocentred information is inherently subject to cumulative errors, such errors could be minimized and finally reset to zero by the additional use of geocentred information.

3.3.1 Egocentric systems: route integration

Many an arthropod returns to its starting point by choosing the direct (short-cut) route rather than retracing its circuitous outbound path. This capacity of route integration is amply documented in Figs 3.2g, 3.11a and 3.19 for spiders, crabs and insects, respectively. It implies that the animal is able to continuously compute its present location from its past trajectory, so that it is continually informed about its home vector, i.e. the vector pointing from its present location towards home.

(a) Idiothetic course control

As described above (pp. 12, 48), spiders can integrate their routes by purely idiothetic means, drawing upon information provided by proprioceptors. These proprioceptors are cuticular strain receptors, the so-called lyriform slit sense organs, located near the joints of the spider's legs (Fig. 3.23). If these organs are destroyed experimentally, the homing abilities of nocturnal spiders, *Cupiennius*, are gone (Seyfarth and Barth, 1972; Seyfarth *et al.*, 1982). There are several indications that in *Agelena* idiothetic information is used as well, but due to the expcrimental strategies applied so far hard proof is difficult to come by in this spider (Görner, 1988).

With the development of techniques to measure forces, pressures, and strains *in vivo*, the sensory function of the cuticular strain detectors has been studied in substantial detail. Measurements of the strains occurring in different parts of the exoskeleton during free locomotion have shown that the lyriform organs are located exactly in those parts of the leg which during locomotion exhibit the largest amount of deformation (Barth, 1986). According to the rather stereotyped stepping pattern of walking spiders, in which

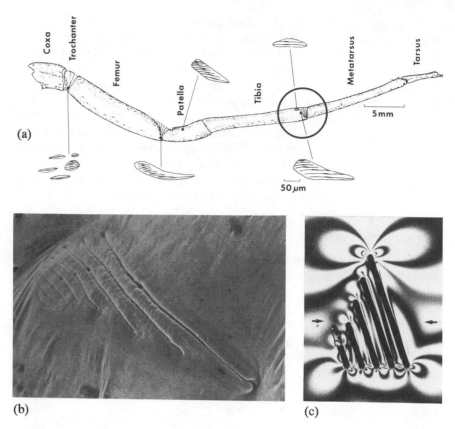

(a)

(b)   (c)

Figure 3.23 Cuticular strain detectors used in idiothetic homing behaviour of the wandering spider, *Cupiennius salei*. (a) Frontal view of leg indicating positions of lyriform organs. (b) Lyriform organ of the tibia, encircled in (a). (c) Deformation of a model of closely arranged slits studied by tension optical methods. The model of the lyriform organ is loaded as indicated by the arrows. The deformation pattern shown by the isochromatics around the organ model is identical with the pattern measured directly with displacement transducers ((a) from Barth and Libera, 1970; (b) SEM picture by K. Hammer and E.A. Seyfarth; (c) from Blickhan *et al.*, 1982).

step length is the most constant feature of leg movement (Land, 1972), there might be a direct 'neural line' from the cuticular proprioceptors to those parts of the nervous system in which idiothetic information is handled and used.

(b) Celestial compass: reading light patterns in the sky

Information used by an animal's route-integration system can be provided by external stimuli as well. The use of geomagnetic cues (Papi *et al.*, 1978; Wiltschko *et al.*, 1978) or visual flow fields (Saint-Paul, 1982) has been discussed in birds. The prime example, however, of animals that home by route integration and employ their visual systems to accomplish decisive aspects of this task, are the insects. They use a celestial compass for measuring angular displacements, and it is the analysis of this compass that represents one of the most striking recent examples of combined research in the fields of neurobiology and behaviour. (For reviews on which the following account is mainly based see Wehner and Rossel, 1985; Rossel and Wehner, 1986; Rossel, 1989; Wehner, 1989a,b.)

The bee's and ant's celestial compass exploits two sources of information: the direct light from the sun (Santschi, 1911) and the scattered light from the sky (Santschi, 1923; Frisch, 1949). Both are used by the insect to determine a reference direction to which all rotatory movements of the animal are related. This reference direction is the solar meridian, the perpendicular from the sun down to the horizon. (Equally, it could be the anti-solar meridian, i.e. the meridian opposite to the solar meridian.)

Light patterns in the sky include spectral and polarization gradients extending across the entire celestial hemisphere. Both are detected and used by the insect, the latter by means of a specialized part of the retina located at the uppermost dorsal margin of the eye (POL area), the former by the remainder of the dorsal retina. As the polarization gradients in the sky are steeper than the spectral gradients, it does not come as a surprise that the homing accuracy of *Cataglyphis* is significantly higher when the ant can use the polarized light in the sky rather than the spectral gradients, or the sun, alone. Hence let us now turn to the insect's most effective means of keeping record of its angular turns: its polarized-light compass.

The pattern of linearly polarized light in the sky is depicted in Fig. 3.24a,b (for details see Wehner and Rossel, 1985). This figure shows the spatial distribution of the angles of polarization for two elevations of the sun (24° and 60° above the horizon). Two features are immediately apparent: first, there is a marked symmetry plane formed by the solar and the anti-solar meridian; second, the pattern of polarization changes its intrinsic spatial properties as the sun changes its elevation above the horizon. Consequently, animals active at different times of the day experience different celestial

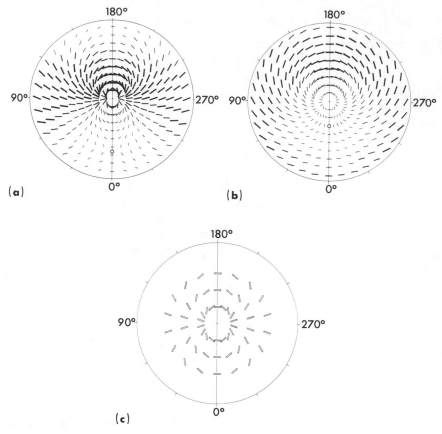

Figure 3.24 (a,b) Two-dimensional representation of the pattern of polarized light in the sky. Elevation of sun (white disc) is (a) 24° and (b) 60°; the zenith is in the centres of the figures. The direction and size of the filled bars mark the angle and degree (percentage) of polarization, respectively. (c) Generalized representation of the insect's celestial map as derived from behavioural experiments in bees and ants. The open bars indicate where in the sky the insect assumes any particular angle of polarization to occur (from Wehner and Wehner, 1990).

patterns, but mirror-image symmetry is an invariable feature of all these patterns.

What does the insect know about these patterns of polarized light in the sky? The answer that emerged from elaborate series of behavioural experiments in both bees and ants (see references cited on p. 87) is amazingly simple: the insect is programmed with a stereotyped celestial 'map', or template, in which the particular

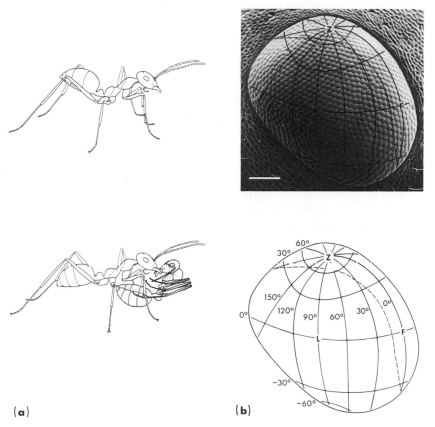

Figure 3.25 (a) Constant head position (pitch angle) of an ant (*Cataglyphis bicolor*) walking either without load (above) or carrying heavy load, a live nestmate (below). (b) Projection of the coordinates of visual space on to the compound eye of *Cataglyphis bicolor*. F, forward direction; L, lateral direction; Z, zenith. The 0° line passing through F and L marks the horizon. Bar: 100 μm (modified from Wehner, 1982).

angles of polarization occur at fixed azimuthal positions relative to the solar meridian (Fig. 3.24c). As neuroanatomical and electro-physiological studies suggest, this map resides in the retina of the POL area, i.e. that part of the insect's eye that is both necessary and sufficient for the detection of polarized light patterns. In this area those photoreceptors that act as polarization analysers form a 'matched sensory filter' (Wehner, 1987b) matched in its spectral and spatial properties to an average pattern of polarized light. First, as far as the spectral match is concerned, the polarization analysers are ultraviolet receptors. In functional terms, this makes a lot of sense because due to the physics of light scattering within the earth's atmosphere those parts of the sky that are most highly polarized also exhibit the highest relative amount of short-wavelength radiation. Second, the polarization analysers are spatially arranged in a way that is similar to the geometry of the celestial map as derived from behavioural experiments. By and large, it mimics the distribution of the celestial angles of polarization.

A necessary prerequisite for this matched filter to work as a compass is its proper alignment with the celestial hemisphere (Fig. 3.25). In fact, irrespective of the load it carries, *Cataglyphis* maintains a constant angular (pitch) position of its head (Fig. 3.25a). When trained to walk on tilted surfaces, *Cataglyphis* compensates by counter-tilting its head. Only when the experimental tilt exceeds the limit up to which the ant is able to properly adjust its head position, do navigational errors occur that are in accord with the misalignment between the POL area and the celestial hemisphere (unpublished work).

A convenient way of using the matched sensory filter is to scan the sky by sweeping the retinal array of analysers across the celestial array of angles of polarization. By scanning the sky, i.e. rotating about its vertical body axis, the animal will transfer the compass information provided by the skylight pattern from the space domain to the time domain (Wehner, 1989a,b). In walking ants, such rotatory scanning movements can actually be observed (Wehner and Wehner, 1990; Wehner et al., 1992).

As the internal array of analysers is hard-wired and stays in place (Fig. 3.24c), but the external pattern of polarization is dynamic and changes with the height of the sun (Fig. 3.24a,b), the match between both patterns cannot be complete. However, this is not too serious a drawback. If the task the insect's celestial compass had to solve amounted to the problem of determining the symmetry plane of the sky by applying some kind of scanning mechanism, a partially-matched filter exhibiting mirror-image properties would suffice. Even if the reference direction as determined by the matching process should deviate from the actual symmetry plane of the sky by a certain constant amount (e.g. if the insect had access only to parts of the whole pattern of light from the sky), the matched-filter system would work. This is for the simple reason that it does not matter at all to what internal reference the animal is finally relating its angular turns, as long as it always uses the same reference (Wehner, 1992).

(c) Celestial compass: compensating for the rotation of the sky

Whatever compass mechanism it applies, the insect must solve yet
another and much more fundamental problem. Owing to the daily
westward movement of the sun across the sky, the symmetry
plane, and with it the whole pattern of polarization, rotates about
the zenith. How animals cope with this problem became the subject
of heated debates soon after the sun compass had been discovered
(Brun, 1914; Santschi, 1923), but positive evidence that insects
(Frisch, 1950), crustaceans (Papi, 1955), spiders (Papi et al., 1957),
and birds as well (Kramer, 1950), are actually able to time-
compensate the movement of the sun, was provided only much
later.

The sun moves along its arc with uniform speed (15°/h), but this
is not the case for the horizontal component of the sun's position,
the sun's azimuth (the azimuthal position of the solar meridian).
Consequently, the reference point of the celestial compass moves
along the horizon with non-uniform speed. Its rate of movement
depends upon three variables: time of day, time of year, and geographical latitude. These dependencies are expressed quantita-
tively in the so-called sun-azimuth/time functions (ephemeris
functions) depicted, for a given latitude, in Fig. 3.26b. As indicated
by the early experiments of Frisch (1950), Jander (1957), Renner
(1957) and Meder (1958), bees and ants are informed, more or less
accurately, about the spatial and temporal pattern of the sun's
azimuthal movement. To acquire this information they must read
time from an internal clock and correlate the time-linked positions
of the sun with an earthbound system of reference. That an internal
clock is at work can be demonstrated by testing foragers in which
the internal clock has been time-shifted (Beier and Lindauer, 1970),
or which have been trained at one geographical longitude and then
tested at another (Renner, 1959). In both cases, bees do not orient in
their correct home direction, thus indicating that they do not
determine true north by referring to an earthbound compass system
based, for example, on the earth's magnetic field (which bees are
able to perceive: Martin and Lindauer, 1977; de Jong, 1982).

As already shown by these early investigations, bees rely on cues
provided by the rotating celestial hemisphere. But what they know
about the systematics of this rotation, and how this knowledge is
acquired by evolutionary or individual experience, remains elusive.
A recent investigator even concluded, from an admittedly limited
set of experiments, that bees did not use any long-term knowledge
about the sun's rate of movement but merely extrapolated from the
most recently observed rate of movement (Gould, 1980) computed

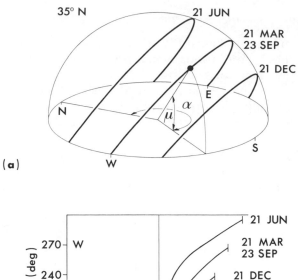

(a)

(b)

Figure 3.26 (a) The daily movement of the sun across the celestial hemisphere as it occurs at a geographical latitude of 35°N. The sun's arc is given for the summer solstice (21 June), the vernal and autumnal equinox (21 March, 23 September) and the winter solstice (21 December). α and μ depict azimuth and elevation of sun, respectively. N, E, S, W, north, east, south, west. (b) Ephemeris functions correlating the sun's azimuth with the time of day. SR, sunrise; SS, sunset.

by a running-average processing system (Gould, 1984). This extrapolation hypothesis could easily be disproven for both ants (Wehner and Lanfranconi, 1981) and bees (Dyer, 1987), and was already at variance with data presented previously by Renner (1959) and New

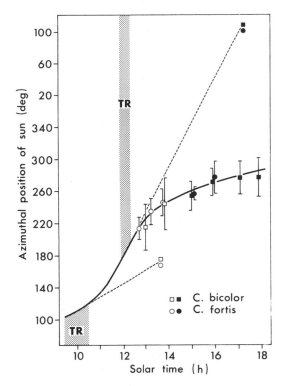

Figure 3.27 Time compensation in *Cataglyphis* ants (*C. bicolor, C. fortis*). Ants were trained either in the morning or at noon (at times TR, see shaded areas) and tested either in the early afternoon (open symbols) or in the late afternoon (filled symbols), respectively. In the meantime they were kept in the dark. The symbols (mean values and standard deviations) indicate at which azimuthal position the ants expect the sun to occur. Solid curve, actual position of sun (ephemeris function) for 35°N, 23–29 July; dashed lines, hypothetical position of sun if the ants had linearly extrapolated the sun's rate of movement from the time when they had last seen the sky (modified from Wehner and Lanfranconi, 1981).

and New (1962). However, the mere formulation of this somewhat curious hypothesis is indicative of the poor knowledge we yet have on this important aspect of animal homing. Even at the present time of writing, there is only one species, the desert ant *Cataglyphis fortis*, for which a fine-grain reconstruction of an animal's internal ephemeris function is available (Wehner and Lanfranconi, 1981; Figs 3.27 and 3.28).

How is it possible to reconstruct an animal's 'knowledge' about the diurnal course of the sun? Let us assume that the ants are trained

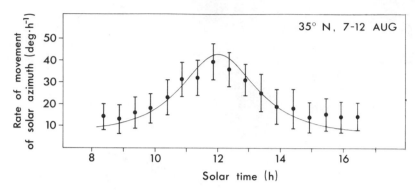

Figure 3.28 Derivative of ephemeris function (solid line: 35°N, 7–12 August) and the ants' estimate of the sun's rate of movement (mean values and standard deviations; *Cataglyphis fortis*) (from Müller and Wehner, unpublished).

to a given direction in the morning (when the sun's azimuth moves slowly), are then kept in the dark for some hours, and are finally released at noon (when the sun's azimuth moves fast). According to the linear-extrapolation hypothesis cited above, the ants treated this way should underestimate the movement of the sun and thus deviate to the right from their home direction. Similarly, ants trained at noon and tested in the late afternoon should overestimate the sun's rate of movement and thus deviate to the left of their homeward courses. In neither case, however, do the ants significantly deviate from their home direction. They compensate quite exactly for the movement of the sun (Fig. 3.27).

But how exactly do they really compensate? Once the extrapolation hypothesis is refuted, this question can be answered by narrowing down the time window for which the ants are kept in the dark to, say, 1 h. For example, separate groups of ants trained at 8:00, 9:00, 10:00 h, etc. are tested at 9:00, 10:00, 11:00 h, etc., respectively. From the angle by which they deviate from their true homeward course one can compute the azimuthal position in which they assume the sun to occur at any particular time of day. The results of one series of experiments performed at a given geographical latitude and a given time of year are presented in Fig. 3.28. The data points depict the derivative of the ant's internal ephemeris function $\omega = \alpha \cdot t^{-1}$ vs time of day) rather than this function itself (α vs time of day). As the data points show, the ants are quite well informed about the actual local variation of the sun's rate of movement. For example, they do not assume a constant rate of movement (15°/h). In detail, however, they tend to under-estimate the sun's rate of movement when it is high (at noon), and overestimate this rate when it is low (at dawn and dusk). Work is in progress in which ants are tested at other times of the year, e.g. at the summer solstice when the ephemeris function (and its derivative) is the steepest, and at other geographical latitudes. The results should tell us whether ants foraging at different times of the year and in different geographical areas use internal ephemeris functions that vary correspondingly.

How does the insect acquire its knowledge about the rotation of the sky? The most likely hypothesis is that bees and ants form detailed memories of the diurnal course of the sun's azimuthal movement by using local landmarks as a system of reference (New and New, 1962; Wehner and Lanfranconi, 1981). Direct evidence for this hypothesis can be taken from an ingenious type of experiment exploiting the honey-bee's dance communication system (Dyer and Gould, 1981; Dyer, 1987; see Fig. 3.29). Bees which have been trained at site 1 to fly along a prominent landmark route (line of trees) to reach a feeding station are displaced, together with the hive, to a new site (site 2), in which a similar route of landmarks is aligned in a different compass direction (varying by the angle β from the former one). At site 2 two feeding stations are established which coincide either in their compass direction (compass station) or in their landmark-based direction (landmark station) with the situation the bees have experienced at site 1. As already shown by Frisch and Lindauer (1954), in this case nearly all bees follow the prominent landmark route even when it is not arranged in the proper compass direction, and thus fly to the landmark, rather than the compass, station. Now an intriguing question arises: What direction do these bees indicate inside the hive under overcast conditions?

To understand the rationale behind the experiment, first note that in the bee's dance communication system foraging directions are indicated by the angular orientation of recruitment dances performed on the vertical comb, and that the dance angles are related to a celestial rather than terrestrial system of reference: upward-pointing dances (0°) tell the recruits to fly in the direction of the sun's azimuth (Frisch, 1967). Further imagine that the experiments at site 2 are performed at local noon when the sun is in the south and thus in the direction of the landmark station as seen from the hive. Under these conditions the bees visiting the landmark station would dance straight up (0°) on the comb (Fig. 3.29b) provided that the sun and/or polarized light patterns are available. But how do they behave on overcast days? Surprisingly,

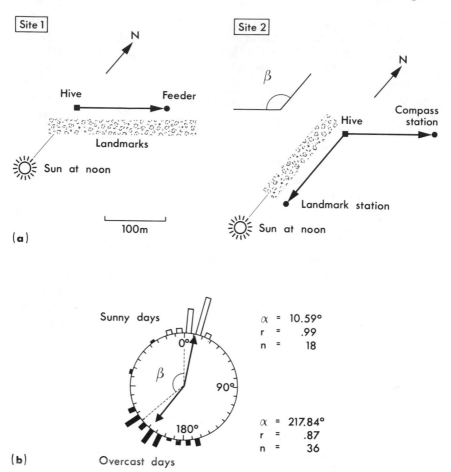

Figure 3.29 'Landmark-shift' experiments performed in honey-bees, *Apis mellifera*. (a) Topographical design of experiment. Bees trained under a sunny sky at site 1 were moved, with the hive, to site 2 where they were tested on either sunny or completely overcast days. For further explanations see text. β, angle between lines of landmarks at sites 1 and 2. (b) Dance angles of bees which first had been trained to the feeder at site 1 and then visited the landmark station at site 2 at which they were tested on both sunny and overcast days, α, r, n, see Fig. 3.2 (combined and modified from Dyer and Gould, 1981).

their dance angles are not distributed at random (as one might have expected due to the lack of celestial cues during the bees' foraging flights). Instead the bees dance uniformly in a direction that deviates by the angle β from the 0° direction (Fig. 3.29b). Obviously, they set their dance angles according to a memory of the sun's position relative to the landmarks at site 1. As they confuse the lines of landmarks at the two sites, the position of the sun inferred by the bees from the landmarks at site 2 is wrong: it deviates by exactly that angular amount (β) from the true position of the sun by which the direction of the landmarks at site 2 differs from that at site 1. Hence, the bees must have memorized the sun's position at site 1 relative to the landmarks there.

This leaves the question of how this spatio–temporal correlation occurs. At present one can only hypothesize that the animal linearly interpolates between successive memorized positions of the sun and thus 'constructs' its ephemeris function by fitting together a number of time-linked spatial data. Support for this hypothesis comes from experiments in which bees and ants are tested under conditions in which they cannot determine the sun's azimuth directly, either because the sun passes within a few degrees of the zenith (as it occurs on certain days in the tropics at noon: New and New, 1962) or because the sun is below the horizon (at night: Wehner and Lanfranconi, 1981; Wehner, 1982). In both cases the data suggest that the insect time-compensates the sun's move-ment at a linear rate between two adjacent known positions. Nevertheless, the hypothesis that the insect is programmed with some general knowledge about the shape of the ephemeris function is still to be tested.

(d) Distance estimation

The celestial compass provides information about **angular dis-placements**. In addition, the insect needs a ruler to determine the **linear** components of locomotion. As to the physiological hard-ware of this ruler, it is a long-standing hypothesis that honey-bees assess distances by measuring energy expenditure on their foraging flights (Heran and Wanke, 1952; Heran, 1956, 1977; Scholze et al., 1964). Bees that had flown upwards on a slope indicated longer distances (by exhibiting shorter wagging-run durations) than bees that had flown the same distance downwards. The difference could not be explained by either the speed or the duration of the flights. However, when the variability of the data (Esch, 1978; Brandstetter et al., 1988) and of the environmental (e.g. wind) conditions is set against the small differences to be expected in the bees' energy expenditures, it becomes compellingly clear that the 'energy hypothesis' is not yet supported by sufficient experimental evidence. A recent study did not even reveal any differences in the estimation of distances between bees whose

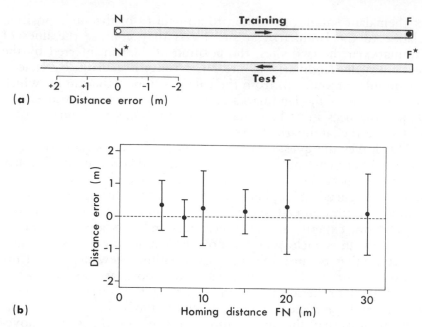

(a)

(b)

Figure 3.30 Distance estimation in desert ants, *Cataglyphis fortis*. (a) Experimental set-up: training and test channels. For cross-section see inset of Fig. 3.31a. (b) Results. $N = 214$ (modified from Müller, 1989).

energy expenditures had been increased by additional loads, and control bees (Neese, 1988). As the latter author suggests, stretch receptors within the wall of the honey-filled crop might provide the flying bee with information about 'fuel consumption' and thus distance flown.

In walking ants, energy expenditure is certainly not the means by which distances are measured. Irrespective of the load they carry, ants assess their walking distances with amazing accuracy (Fig. 3.30). This result was obtained in *Cataglyphis* foragers trained and tested within linear channels. In these channels the ants terminated their uninterrupted straight return runs by a 180° turn followed by oscillating to-and-fro movements about the fictive position of the nest. The distances between the first 180°-turning points and the fictive position of the nest did not differ significantly from zero. The slight tendency to overshoot, by 20 cm in the mean, was independent of the homing distance and might already be part of the subsequent search pattern.

If it is not energy expenditure by which ants estimate distance, what might it be? An old idea dating back to Piéron (1904) suggests that the ants might be counting the number of steps they take to cover a particular distance. Even though the length, as well as the frequency, of the ant's steps increases with increasing walking speed (Zollikofer, 1988), at any particular foraging trip an individual *Cataglyphis* ant maintains a constant rate of movement (Wehner and Srinivasan, 1981). Information about the number of steps could be provided by afferent or efference-copy signals.

For an ant, another way of estimating distances would be to visually record the movement of the image of the ground over which it walks. As each particular ant moves at a constant height above ground, the rate of retinal image motion provides a reliable cue of the distance travelled.

(e) Integration

Any mechanism of route integration requires that the animal processes all its consecutive movements – all its rotatory and translatory components of locomotion – by, for example, integrating velocity signals or double integrating acceleration signals over time. The latter strategy is realized technically in the Inertial Navigation System used in aeronautics (Mayne, 1974). It was discussed (Darwin, 1873; Barlow, 1964) but later dismissed (Keeton, 1974) for birds, but it might be applied by some mammals, especially rodents (gerbils: Mittelstaedt and Mittelstaedt, 1980; hamsters: Etienne *et al.*, 1986; rats: Potegal, 1987), in which inertial information is provided by the vestibular system. Some insects (dipterans) possess special mechanoreceptor organs (halteres) by which they can perceive angular acceleration (Pringle, 1948), but they use this information for stabilizing their body position during flight (Sandeman and Markl, 1980; Hengstenberg, 1988) rather than for integrating their flight courses.

Stimulated by the work on homing in ants and spiders (pp. 48, 78), Mittelstaedt has proposed (most recently in 1985) that route-integrating animals perform some kind of vector summation, e.g. use a Cartesian-based system of coordinates to integrate their components of motion along two mutually perpendicular directions. Even though advertised in a long series of papers starting in the early 1960s, this vector-sum hypothesis has never been tested experimentally. Most surprisingly, the basic question of whether animals are at all able to compute the mean home vector accurately has never been asked until very recently (Wehner and Wehner, 1986; Müller and Wehner, 1988). In these studies *Cataglyphis* ants were trained to walk along angular trajectories (Fig. 3.31a) and then released within an open test field. There

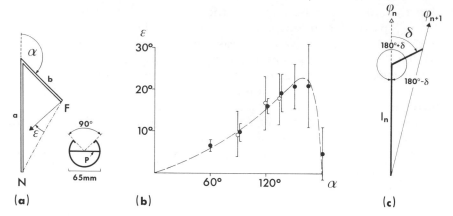

Figure 3.31 Route integration in desert ants, *Cataglyphis fortis*. (a) Experimental set-up. Ants were trained along the angular route a, b ($a = 10$ m, $b = 5$ m) from the nest (N) to a feeder (F) and then released in the open test field. There they deviated systematically by the error angle ε from the true homeward course FN. Inset: cross-section of training channel. (b) Error angle ε as a function of turning angle α. The error angles ($\pm$ standard deviation) do not depend on whether the ants use sun-compass (open circles) or polarization-compass information (filled circles). (c) Computational model; see text and dashed curve in (b) (modified from Müller and Wehner, 1988).

they immediately set out in what they 'thought' were their homeward directions. These homeward courses, however, deviated significantly and systematically from the true homeward directions. In a two-leg training system, the error angle ε is a function of the turning angle α and (not shown here) the ratio of the lengths of the two legs.

The mere fact that systematic navigational errors occur immediately rules out the possibility that the ants perform a vector-sum computation. Apparently they do not solve the integration problem in its complete form. A computational model that perfectly describes the ants' behaviour (see dashed line in Fig. 3.31b) assumes that the ants compute some kind of distance-weighted arithmetic mean of all angles steered. In particular, the ant is supposed to add some measure of the angle δ between its nth step and the direction of the mean vector pertaining to its $(n - 1)$th step (Fig. 3.31c) to this previous vector, and in so doing scales down all successive angular contributions in proportion to the distance it has moved away from the nest (Müller and Wehner, 1988). In the ant's

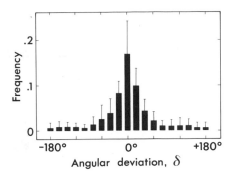

Figure 3.32 Turning angles δ (for definition see Fig. 3.31c) performed by *Cataglyphis fortis* during foraging. Data taken from digitized paths of 19 ants. The histogram represents mean values and standard deviations (from Wehner and Wehner, 1990).

rcal foraging life, this approximate rule-of-thumb solution works sufficiently well. First, while searching for food the ants avoid those angular turns δ that would induce large sampling errors (Fig. 3.32). Second, due to the symmetrical frequency distribution of the turns δ, no systematic errors build up in the ant's route-integration system (Wehner and Wehner, 1990).

The model derived here for *Cataglyphis* ants also describes the results of some detour experiments performed many years ago in bees (*Apis*) and spiders (*Cupiennius, Agelena*). In fact, in all these experimental studies (Fig. 3.33) in which the animals relied on idiothetic or visual systems of navigation, systematic homing errors occurred. To the investigators these errors were enigmatic, but when the algorithm outlined above is applied to the experimental trajectories used in these former studies, the errors found by the previous authors are in full accord with what our computational model predicts.

(f) Systematic search

When its route-integration system has been reset to zero, the animal should have arrived again at the starting point of its journey, its home. However, due to navigational errors that might have accumulated during the entire round trip, this is not necessarily the case, and the animal might well miss its start. It could then engage in a random search for true home, i.e. wander about like a particle in Brownian motion. Such a random search as discussed for migratory vertebrates, both fish (Saila and Shappy, 1963) and birds

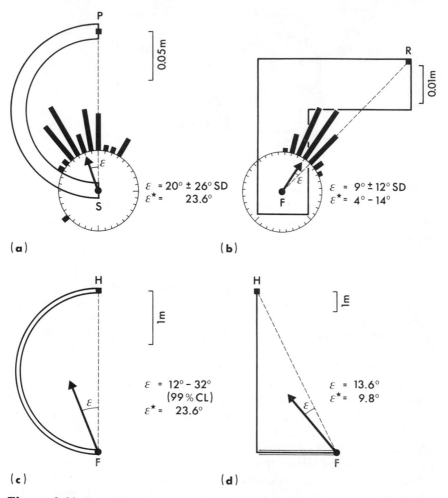

Figure 3.33 Route integration in arthropods. Training along semicircular or rectangular detours PS (a), RF (b), HF (c,d). The actual deviation of the animal's return course from the direct course is denoted by ε. The deviation expected according to the model described in the text is denoted by ε* (computed by Müller, 1989). (a) Wandering spider, *Cupiennius salei*. See also Fig. 3.2g. Data from Seyfarth *et al.* (1982). (b) Funnel-web spider, *Agelena labyrinthica*. As the experimental device allows for a rather wide range of detour trajectories RF, the possible values of ε* range as indicated. Data from Görner (1985). (c) Walking honey-bees, *Apis mellifera*. Data from Bisetzky (1957). (d) Flying honey-bees. Data from Lindauer (1963). In (c) and (d) the direction ε was determined by following the bees' communication dances at H. The homing direction at F is the 180° counter-direction. CL, confidence limits; F, food; H, hive; P, prey; R, retreat (funnel); S, start; SD, standard deviation.

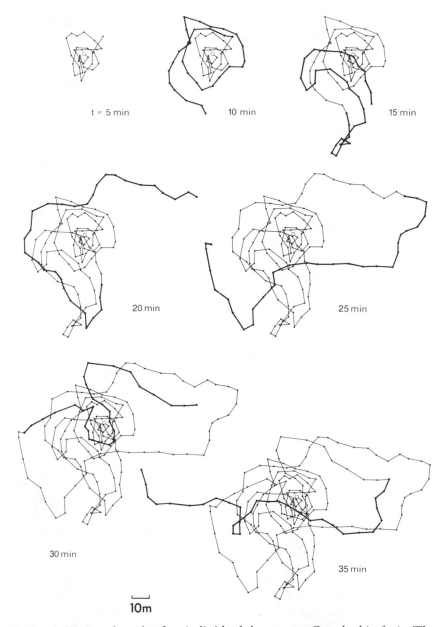

Figure 3.34 Search path of an individual desert ant, *Cataglyphis fortis*. The fictive position of the nest is indicated by the open circle. Time marks (filled circles) are given every 10 s. The sequence of diagrams shows how the search pattern develops as search time (t) proceeds. The thick lines mark the route taken by the ant in the last 5 min of each recording (from Wehner and Wehner, 1986).

(Wilkinson, 1952), will always be successful, given unlimited time. In real life, however, there are inevitable time constraints imposed on the homing animal, so that a premium will be put on any type of systematic search by which the probability of relocating the starting point can be increased beyond the random-search level.

Indeed, once they are lost, many arthropods resort to a systematic search programme. The only group of animals in which the geometry of this search pattern has been studied comprehensively, and in which the rationale underlying the spatial layout of the search programme has been unravelled, are desert ants (*Cataglyphis*: Wehner and Srinivasan, 1981; Wehner and Wehner, 1986). The ant's search pattern consists of a number of loops of ever increasing size, starting and ending at, or near to, the origin, where the data store for the return trip has been emptied (Fig. 3.34). Thus, the ants spend most of their time searching at locations where home is most likely to be. The search density rapidly decreases with increasing distance from the origin. A similar result has been obtained for desert isopods, *Hemilepistus*, searching for their family-owned burrows (Hoffmann, 1983a,b). Furthermore, the spatial layout of the ant's search pattern – the shape of the search density profile – is matched to the size of the navigational errors to be expected from the animal's dead-reckoning system. The greater the distance from which the animal has returned from foraging, the larger the errors that could have accumulated during the animal's round trip, and, indeed, the wider and less peaked is the animal's search density profile (Fig. 3.35). Hence, when *Cataglyphis* has reached the starting point of its journey, but the nest entrance is not there, it does not simply switch on a rigidly preordained search programme. Instead, it uses a flexible strategy in which the shape of the search density profile is adapted to the target density function which describes the fictive position of the nest as computed by the animal's route-integration system.

A closer inspection of the ant's search pattern reveals that a 'cryptic spiral' (Wehner and Wehner, 1986) is hidden behind the system of loops and turns. In fact, the ant's search pattern can be simulated sufficiently well when one assumes that the ant (1) searches along a spiral path, which every now and then changes its sense of rotation, (2) interrupts its spiral search after a given time and returns to the centre (that means what it thinks is the centre), and (3) after having arrived at the centre spirals out again along a somewhat wider spiral course (Müller and Wehner, in preparation). By iterating between returning to the centre and spiralling out at an ever widening pace the ant creates the type of search density profile depicted in Fig. 3.35. As far as the navigational requirements are

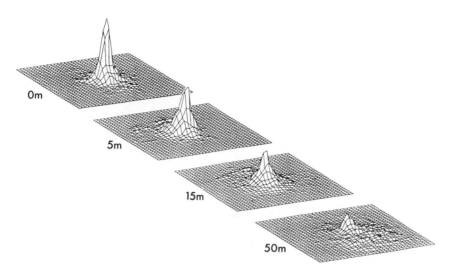

Figure 3.35 Search-density profiles of *Cataglyphis fortis* for different return distances (0, 5, 15 and 50 m). The centre of each distribution is located at the fictive position of the nest (Bernasconi, Nieuwlands and Wehner, unpublished).

concerned, performing a spiral walk means maintaining a constant angle $\delta > 90°$ (for definition of δ see Fig. 3.31c) between the actual orientation of the ant's body and the vector pointing to the start. The larger the angle δ, the wider the spiral.

But why is the spiral course cryptic rather than obvious? First, during each particular spiral phase of the search a given angle δ_i is not strictly kept constant, but varies according to a normal distribution about a mean δ_i. This can be deduced directly from an analysis of the ants' actual paths. Second, the ant's dead-reckoning system continually operating during the search is based on an approximate rather than exact solution of the route-integration problem (p. 100). As is easily borne out by a computer simulation, these additional characteristics of the search programme transform the spiral segments of the search into a system of origin-centred loops of ever increasing size. In the end, an effective behavioural programme emerges from a combination of noisy and approximate subroutines.

A strongly deterministic spiral search lacking the resets and returns to the centre described above would be the least time-demanding mode of search behaviour. However, following a strictly spiral course is appropriate only for a perfect scanning machine. In real life, however, this is an extremely dangerous solution, because once the target is missed, owing to some sudden inattention of the navigator, there is no chance to hit it again.

3.3.2 Geocentric systems

After having ventured considerable distances away from their nests, burrows, or foraging sites, many arthropods regularly return to the start. As described above, any egocentric (dead-reckoning) system of homeward navigation is prone to cumulative errors. Resorting to a systematic search programme is the animal's final emergency plan designed to cope with such errors. A more efficient way of resetting the dead-reckoning errors to zero and finally pin-pointing the start is the additional use of 'geographic' information as provided by visual landmarks or self-deposited chemical signposts. As a historical aside it might be mentioned that it was Darwin who alluded not only to egocentric systems of navigation (Darwin, 1873), but also to geocentric systems as used by patrolling bumblebee males (Darwin, 1888; see Wehner, 1981a, p. 460). For the use of egocentric and geocentric systems of navigation in man, the reader is invited to consult Finney (1976) and Kayton (1989), respectively.

(a) Landmark memories

In spite of the small absolute size of their eyes, most arthropods are endowed with large visual fields providing the animal with full panoramic vision (Wehner, 1981a; Wehner and Srinivasan, 1984). However, owing to their small body sizes arthropods lack stereopsis and focusing mechanisms which are used in vertebrates to estimate depth. It is only while moving and thus experiencing retinal image motion that they can gain information about the third dimension.

Furthermore, in many instances of homing behaviour insects treat three-dimensional sets of landmarks just as if these sets were two-dimensional panoramas. They then confound distance and size. A telling example that insects form two-dimensional memory images, or photographic 'snapshots' (Neisser, 1966; Cartwright and Collett, 1983), is provided by a simple experiment performed in *Cataglyphis* (Fig. 3.36). When these desert ants are trained to a spot midway between two cylindrical landmarks, they store an image of the visual scene as viewed from the goal, and later search midway between the markers whenever the apparent (angular) sizes of the landmarks coincide with what they have experienced during training: for example, when the markers twice the training size are

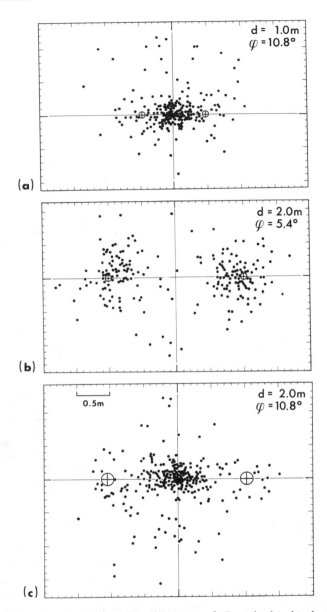

Figure 3.36 Search-density distributions of *Cataglyphis bicolor* trained to locate its nest entrance midway between two identical cylinders. The positions of the cylinders are indicated by the crosses within the open circles. d, distance of cylinders from fictive position of nest; φ, angular height of cylinders as seen from the nest entrance. The black dots show the positions of 24 ants periodically during 5-min searches (modified from Wehner and Räber, 1979).

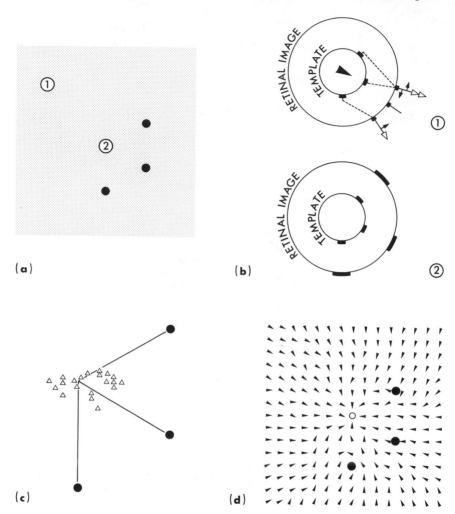

Figure 3.37 Matching-to-memory in honey-bees, *Apis mellifera*. Experiments and model. (a) Training of bees to a food source (site 2) characterized by a set of three cylindrical landmarks (filled circles). (b) Matching of retinal image to snapshot (template). A full match is achieved only at site 2. At other places, e.g. at site 1, the bee moves so as to decrease the mismatch between current image and template. In the model, any particular black sector of the template is paired (black bars on inner circle) with the closest black sector of the current retinal image (black bars on outer circle). Two unit vectors are associated with each pair, one for rotational movements (small filled arrows), the other for translational movements (small open arrows). These unit vectors decrease the differences in retinal position and size, respectively, of the corre-

separated by twice the training distance (Fig. 3.36c). The fact that the arrangement of landmarks in Fig. 3.36c yields a less peaked search density profile than that in Fig. 3.36a is due to the less pronounced motion-parallax cues experienced by the ants in the double-distance arrangement. On the other hand, the ants do not search at the expected site when the landmarks appear smaller (Fig. 3.36b) or larger by changes in either distance (Fig. 3.36b) or size.

Similar results are obtained in experiments with honey-bees trained to locate a food source by using no more than the apparent sizes and bearings of local landmarks (Fig. 3.37a,c). Similarly to an ant, a bee seems to compare its memorized snapshot (or 'template') of the landmarks with the current retinal image and to move so that this retinal image transforms to match its memory. If, for example, a dark sector of the template is positioned to the left of an appropriate sector on the retina, the bee rotates to the right; if it is smaller than the appropriate sector on the retina, it moves forward. A unit vector is associated with each of these rotational and translational components of movement, and all the unit vectors are summed (Fig. 3.37b). If a computer model is designed that works according to these computational rules, the flight directions indicated by the model bee (Fig. 3.37d) coincide sufficiently well with the directions actually taken by the experimental animals (Cartwright and Collett, 1983).

There is one important prerequisite for the model to work. Template and retinal image must be oriented with respect to an external system of coordinates most probably provided by distant landmarks. It is interesting to note in this context that celestial cues are not used as a system of reference. Even desert ants, for whom the ever blue sky arching over their foraging grounds provides

sponding marks in the template and the retinal image. Summing up all unit vectors leads to the direction in which the bee should move (large filled arrow in the centre). (c) Results of experiments in which either landmark size or landmark distance was varied. For training array see (a). Each data point ($\triangle$) represents the peak of the search density profile of an individual bee. (d) Behaviour of a model bee that searches according to the rules outlined in (b). The arrows depict the movements of the model bee at each position on the grid; see large black arrow in (b). Finally, the model bee arrives at the location indicated by the open circle, site 2 in (a) (modified and combined from Cartwright and Collett, 1982, 1983).

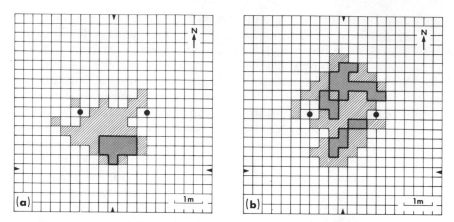

Figure 3.38 Search-density distribution of *Cataglyphis bicolor* trained to locate its nest entrance. The latter is marked by two cylindrical landmarks (black circles) positioned at an angular distance of 60° to the north of the nest. The black arrows point towards the fictive position of the nest. (a) Test area near the training area; distant landmarks coincide in training and test area. (b) Test area several hundred metres away from training area; skyline differs from that in training area. In both cases the full blue sky is available to the ants (modified from Wehner *et al.*, 1983).

conspicuous compass information (p. 87), do not use their celestial compass in order to decide whether the goal is, say, north or south of an otherwise ambiguous array of landmarks (Fig. 3.38). Apparently, celestial cues positioned at infinity are used exclusively in egocentric, rather than geocentric, navigation. Ambiguities existing in a given array of landmarks can be resolved only by referring to distant landmark panoramas (Wehner *et al.*, 1983).

The distant visual surroundings might also be used to retrieve the local landmark memories. As at some distance from the goal the retinal image of the local landmarks may match rather poorly the memorized snapshots previously taken at the goal, landmark memories must be primed by some contextual cue (Collett and Kelber, 1988). One type of cue certainly involves the zero reset of the route-integration system. On their homeward journeys *Cataglyphis* ants pay attention to local landmarks defining their final goal, the nest, only when their dead-reckoning store is nearly emptied. They ignore nest-specific arrangements of landmarks when they encounter these markers before they have 'reeled off' their home vector. The latter situation can be created experimentally by either displacing the landmarks closer to the start of the ant's homeward trajectory or displacing the ant, at the beginning of its homeward journey, part way towards its final destination. In both types of experiment the ants walking home along their vector route get 'trapped' by the landmarks the earlier/ the farther the landmarks are away from the route (Wehner *et al.*, unpublished data). This is a sensible strategy to adopt because distant landmarks cause retinal image motion that changes only slightly as the ant proceeds along its route, and thus defines the wider area of the goal. Nearby landmarks inducing high-motion parallax are ignored during the early stages of the homing process, but they are finally needed to pin-point the goal. If the ant recalled its close landmark memories earlier on its way home, it might well get trapped by the multitude of similar close landmark arrangements which it will certainly encounter along its homeward route. Hence, while returning home by gradually emptying its vector-navigation store the ant seems to retrieve a series of landmark memories that refer to increasingly larger rates of retinal image motion.

A somewhat similar hypothesis was put forward by Cartwright and Collett (1987) in a theoretical paper about the possible use of mental maps by honey-bees (p. 119). Bees relocating a food source to which they have been trained are considered to use two types of landmark memories in succession: first a distance-filtered snapshot from which close landmarks are excluded and which can thus guide the bee to the neighbourhood of the goal, and second an unfiltered snapshot including close landmarks and thus guiding the bee to the exact location of the goal.

So far we have discussed the possibility that insects return to their home by using two-dimensional snapshots of the landscape seen from their point of departure, but how do they take these snapshots in the first place? There are many casual observations that bees (*Bombus* and *Apis*) and wasps (*Vespula* and *Bembix*) perform 'orientation flights' or 'locality studies' (reviewed by Wehner, 1981a), but a detailed analysis of how landmark memories are actually acquired has been performed only recently in sphecid wasps (*Cerceris*: Zeil, in preparation). On departure from their nests, the wasps perform systematic orientation flights consisting of stereotyped behavioural sequences. While continually facing the nest entrance, they fly in ever increasing arcs around that tiny hole in the ground to which they must later return (Fig. 3.39a). During these circling flights the wasps reverse their flight directions at regular intervals by counter-rotating in such a way that they fixate the goal at azimuthal positions of $+60°$ or $-60°$ (Fig. 3.39b). Furthermore, while systematically increasing their horizontal distance from the nest and their height above ground they always

(a)

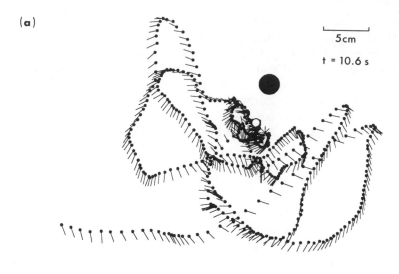

5cm

t = 10.6 s

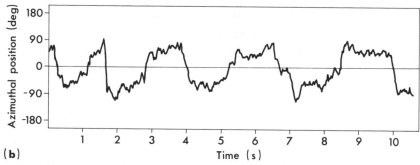

(b)

Figure 3.39 Orientation flight of the wasp, *Cerceris rybyensis.* (a) Horizontal view. Position and orientation of the wasp are given every 20 ms. Head is marked by a small black dot. Open circle, nest entrance; large filled circle, cylindrical landmark (2.2 cm wide, 6.3 cm high). (b) Retinal position of nest entrance during the sequence of the orientation flight shown in (a) (modified from Zeil, in preparation).

see the goal at an elevation of about 45° below the horizon. Finally, they increase flight velocity with distance such as to maintain a constant rate of turning (100–200°/s), i.e. to cover equal angular amounts relative to the goal in equal periods of time.

This stereotyped behavioural programme underlying the wasp's orientation flights transforms the three-dimensional pattern of landmarks around the nest entrance into a systematically struc-tured two-dimensional field of retinal image flow. Owing to the characteristics of the flight pattern just described the retinal position of the goal, and the positions of the landmarks close to the goal, will not change as the animal flies through an arc, but the retinal images of distant landmarks will move systematically: in the direction opposite to the one in which the animal flies, and the faster the farther away the animal is from the goal.

These considerations bring us to the question of what visual cues are actually available to the animal and what algorithms are employed by it. As we have just seen, in acquiring snap-shot information the animal seems to exploit a particular retinal topography of image-motion stimuli. Insects are already known to use motion parallax to distinguish the relative distances of objects (Lehrer *et al.*, 1988; Srinivasan *et al.*, 1989; Kirchner and Heusipp, 1990), but this does not yet answer the question of what information they extract from these flow-field images, and later use for re-orientation. Upon return, young honey-bee foragers tend to approach the nest site from the same direction from which they have faced it during their orientation flights (Vollbehr, 1975), so that during homing they might recall, in reversed order, the same sequence of landmark memories they have acquired upon departure. Furthermore, we know that, at least in some instances, information about relative distances is retained in the bee's memory pictures (Cheng *et al.*, 1987), but what degree of spatial filtering, either two-dimensional (Anderson, 1977) or three-dimensional (Cartwright and Collett, 1987), occurs before these pictures are stored, remains to be determined.

A number of parametric tests performed with bees and ants indicate that the snapshots taken upon departure consist of relatively unprocessed images silhouetted against the background. The homing insect seems to gauge the fit between the current and the memorized eidetic images (Wehner, 1972; Collett and Land, 1975; Wehner and Räber, 1979; Hölldobler, 1980). In this matching process even small differences in angular height and width of landmarks are detected. The height of landmarks incorporated into the snapshots can be as low as 5° (in ants and bees: Wehner and Räber, 1979; Gould, 1987), or even less (1.6° in crabs: Nalbach, 1987). There is no simple way of deducing such limits of detectability from the angular resolution of the eye. Detectability is mainly a matter of contrast sensitivity rather than of retinal resolution (Wehner and Srinivasan, 1984). This is borne out by the observation made in crabs (Kraus and Tautz, 1981) that the spatial limit of detecting an object strongly depends on the contrast between the object and its background.

Nevertheless, some general correlations between the structure of an animal's visual sphere (Fig. 3.40d) and the visual tasks the animal

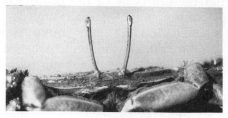

(a)

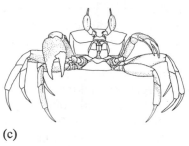

(b)

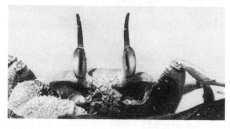

(c)

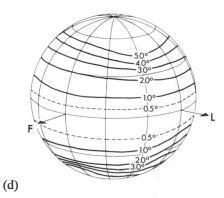

(d)

must accomplish in navigation can usually be drawn from a closer optical examination of compound eyes. One of the most thorough studies performed in this context is that of Zeil *et al.* (1986) on shore-living crabs. In the stalked eyes of these semi-terrestrial crabs (Fig. 3.40a–c) pronounced band-like zones of especially high visual acuity (resolution and contrast sensitivity) occur. While running, the crabs keep these 'visual streaks' perfectly aligned with the horizon along which they can make amazingly accurate discriminations between differently sized skyline landmarks (Nalbach, 1987).

(b) Trail pheromones

Retracing one's steps, i.e. following the same route during both the outbound and the inbound trip, is certainly the most straightforward and computationally least demanding way of homeward navigation. As described in section 3.3.2c, the route can be memorized by storing visual landmark images, but it can also be marked by chemical signposts.

Chemical (pheromone) trails are most ubiquitous and have been studied most intensively in ants. Since the classical papers of Carthy (1950, 1951), Sudd (1957, 1959), Wilson (1962a,b) and Hangartner and Bernstein (1964) research has focused mainly on trail laying as an effective means for mass recruitment. However, as is now becoming apparent, ants use long-lasting orientation pheromones in addition to the more short-lived recruitment pheromones. For example, in *Pogonomyrmex* the former are produced in the Dufour's gland, the latter in the poison gland (Hölldobler and Wilson, 1970). In one species of *Messor* that has been studied in this context both signals originate from the same source, the Dufour's gland (Hahn and Maschwitz, 1985). In *Camponotus*, hindgut material functions as an orientation cue, while poison-gland secretions provide a recruitment signal (Traniello, 1977). Ants of the genus *Leptogenys*,

Figure 3.40 Stalked eyes and visual streaks in shore-living crabs, which use skyline cues for navigation. (a) *Macrophthalmus setosus*; (b,c) *Ocypode ceratophthalmus*. (d) Visual sphere of the latter species. Interommatidial distance (receptor spacing) is indicated by the thick isolines. F and L, forward and lateral direction, respectively. ((a,b) Courtesy of J. Zeil; (c,d) from Wehner, 1987b; data in (d) from *et al.*, 1986).

whose foraging system has been described on p. 69, use pygidial-
gland secretions to stimulate recruitment, and secretions of the
poison gland to aid in route-based navigation (Attygalle *et al.*, 1988;
for information on other ponerine species see Hölldobler, 1984;
Jessen and Maschwitz, 1986; Breed *et al.*, 1987). During foraging
raids *Onychomyrmex*, a member of the primitive ponerine tribe
Amblyoponini, lays powerful recruitment trails with a trail
substance originating from a sternal gland (Hölldobler *et al.*, 1982),
while tarsal-gland secretions provide a homing signal (Hölldobler
and Palmer, 1989).

In the neotropical ant *Paraponera clavata* (pp. 71 and 74) individual
workers use pheromone trails as orientational devices even when
they do not recruit (Breed *et al.*, 1987). Furthermore, the
orientation pheromones deposited by non-recruiting workers are
discriminatable individually by concolonial foragers (Breed and
Harrison, 1987). Similar observations have been made in another
ponerine (Jessen and Maschwitz, 1985, 1986) and a myrmicine
species (Maschwitz *et al.*, 1986). They complement former work on
colony-specific (*Oecophylla longinoda*: Hölldobler and Wilson, 1977;
Lasius neoniger: Traniello, 1980; *Atta cephalotes*: Hölldobler and
Wilson, 1986) and species-specific trail pheromones (*Solenopsis* spp.:
Wilson, 1962a; *Lasius* spp.: Hangartner, 1967; *Pogonomyrmex* spp.:
Regnier *et al.*, 1973). However promising these findings are, the
study of orientation trails has just begun, and little can be said at
present about the role they play in the ant's overall homing
performance.

Trail pheromones derive from a number of exocrine glands
located mostly on the ant's metasoma (Fig. 3.41a), but also in the
tibia (*Crematogaster*: Fletcher and Brand, 1968; Leuthold, 1968)
and basitarsus (*Onychomyrmex*: Hölldobler and Palmer, 1989) of
the hindlegs (Fig. 3.41b). In many cases in which the secretions
of trail-pheromone glands have been characterized chemically
(Attygalle and Morgan, 1985) they have turned out to be blends
of compounds rather than single pheromones. For example, the
poison gland producing the orientation pheromone of *Leptogenys
diminuta* contains, among other substances, (4S)-4-methyl-3-hept-
anon (I), (3R,4S)-4-methyl-3-heptanol (II), and 8-pentadecanol
(III) (Attygalle *et al.*, 1988). 4-Methyl-3-heptanol (II) can exist as
four optical isomers comprised of two pairs of diastereomers –
(3R,4R), (3S,4S), (3S,4R), (3R,4S) – but only the latter represents
the natural poison-gland compound. It is also this racemic mixture
(II) that exhibits high activity in releasing trail-following behaviour.
The role of the other two compounds (I, III) is still unknown. The
recruitment pheromone of the pygidial gland is *cis*-isogeraniol

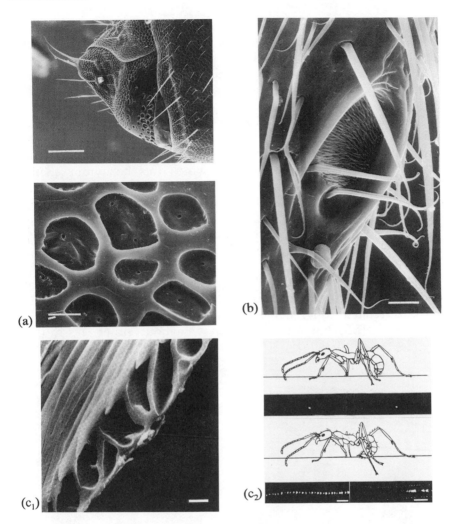

Figure 3.41 Trail pheromones in ants: cuticular specifications around the openings of the pheromone glands, trail laying behaviour, and trail marks. (a) Sternal glands of weaver ants, *Oecophylla longinoda*. Upper figure: position of the openings of the gland on the ventral surface of the terminal metasomal segment. Bar: 100 μm. Lower figure: gland openings. Bar: 10 μm. (b) Tarsal gland of a ponerine ant, *Onychomyrmex* sp. Basitarsus of hind leg. The gland is located in the centre. Bar: 20 μm. (c₁) Cuticular reservoirs for pygidial gland secretions in *Leptogenys diminuta*. Bar: 5 μm. (c₂) *L. diminuta* depositing orientation pheromone from poison gland (upper figure) and recruitment pheromone from pygidial gland (lower figure). Insets: trail marks. Bar: 0.2 mm (combined and modified from Hölldobler and Wilson, 1977; Hölldobler and Palmer, 1989; Maschwitz and Steghaus-Kovac, 1991).

(Attygalle *et al.*, 1991). In the multi-component (poison-gland) trail pheromone of *Tetramorium meridionale*, the most complex mixture yet identified as an insect trail pheromone, indole and four alkylpyrazines act synergistically: the former as the trail-recruitment component, the latter as the trail-orientation cue (Jackson *et al.*, 1990).

Very little is known about the sensory side of trail-pheromone navigation. As known already from earlier experiments, in which trail-following behaviour broke down when the ants' antennae were cut off (Forel, 1886), or in which this behaviour changed dramatically when the antennae were amputated unilaterally or crossed (Hangartner, 1967), antennal chemoreceptors are responsible for the detection of trail substances (Fig. 3.42). However, what types of chemoreceptory sensilla are involved in the analysis of either orientation or recruitment pheromones has not been studied in any species. In *Atta*, in which workers of different size classes specialize on different behavioural tasks, size-related variations in exocrine glands (Wilson, 1980) seem to be paralleled by size-related differences in certain types of olfactory sensilla (Jaisson, 1972); but nothing is known about either the response

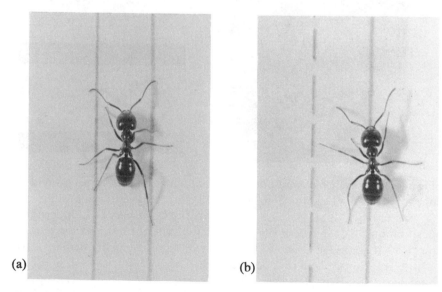

(a)   (b)

Figure 3.42 Walking pattern of ants (*Lasius fuliginosus*) along an artificial double trail of recruitment pheromone applied to a Perspex sheet. The pheromone concentration is (a) equal in both trails or (b) smaller by a factor of 2 in the left-hand trail. In the latter case the ants select the more strongly scented right-hand trail (from Hangartner, 1967).

characteristics of chemoreceptor cells specialized for the detection of particular trail pheromones or the neural processing of pheromonal information. As the techniques are at hand (Kaissling, 1987) and as the insect's olfactory pathway has recently become a focus of intensive research (Arnold *et al.*, 1985; Rospars, 1988; Homberg *et al.*, 1989; Kanzaki, 1989), combined neurophysiological and behavioural studies on olfactory processing of trail-signal information should have high priority in future work. This is recommended all the more as in this type of homing behaviour the behavioural modes in question are well defined: ants can distinguish between different individual, different intercolonial and different interspecific trails (references in Wehner and Wehner, 1990), between orientation and recruitment pheromones (Hölldobler, 1971; Maschwitz and Mühlenberg, 1975) as well as between trunk-trail and side-branch markings (Breed and Bennett, 1985).

Navigation by means of only scent trails is an intrinsically slow process, especially when the trail has been laid by a single ant. Neither does such a trail provide an unvariably continuous chemical guideline (Wilson, 1962a; Hangartner, 1969; Lane, 1977), nor does the ant following this trail walk at a steady continuous pace (Harrison *et al.*, 1989). To use memorized visual landmarks rather than self-deposited chemical signposts is a more demanding though more efficient means of route-based navigation. A scent trail tells the animal merely that, and not where, it is on the route. Unlike the mucous trails of gastropods (Stirling and Hamilton, 1986), the pheromone trails of ants do not even provide information about the inbound–outbound polarity of the trail (p. 73).

(c) Map-like behaviour

Do ants and bees integrate the information they have obtained while moving along vectors (section 3.3.1e) and routes (section 3.3.2a) into what one could call the mental analogue of a topographical map? Despite the recent claim that honey-bees use such cognitive maps in navigating to homing and feeding sites (Gould, 1986; Gould and Towne, 1987), this question cannot be answered in the affirmative. Gould's results can be explained conveniently by simpler computational abilities that ants and bees are already known to possess (Dyer and Seeley, 1989; Wehner and Menzel, 1990). More interestingly, they could not be confirmed in a number of recent investigations that were specifically designed to reproduce these findings.

However transient the ardour about the honey-bee's cognitive map might have been, it is intriguing to inquire about why it

could have arisen. First, talking about cognitive maps has recently become fashionable among experimental psychologists and
behavioural neurobiologists working mainly on mammals and birds
(reviewed in Gallistel, 1989). If one considers the bee's brain to be
merely a stripped-down version of the human brain, the jump
on the bandwagon is somewhat inevitable. Second, there are indeed some awe-inspiring homing performances in arthropods,
especially in social insects. Without sufficient experimental scrutiny
such behaviour might well look as if the animal used some kind
of map.

Let us now try to define what a cognitive map is meant to be. In
a broad sense, it can be defined as the internal representation of the
geometric relations among noticeable points in the animal's
environment. In operational terms, an animal using such a map

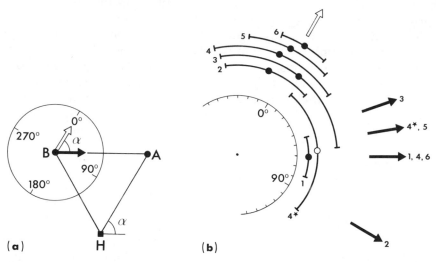

Figure 3.43 Test of map hypothesis in honey-bees, *Apis mellifera*. (a)
Experimental set-up. H, hive; A, training site; B, release site. Bees to be
released were captured at H prior to their foraging flight HA. The open
and filled arrows depict the bees' vanishing directions as predicted by the
compass and map hypothesis, respectively. In different experiments
distances HA = HB were 150–400 m and angles α 40–90°. (b) Results: the
circles and bars indicated the mean vanishing directions and standard
deviations, respectively. As in (a), the open and filled arrows are the
predictions due to the compass and map hypothesis. Data from: 1, Gould
(1987); 2, Wehner and Menzel (1990); 3, Menzel *et al.* (1990); 4, Dyer
(unpublished); for difference between 4 and 4* see text; 5, Wehner *et al.*
(1990); 6, Seeley (unpublished).

must be able to determine its position relative to home, or any other charted point, even when it has been displaced inadvertently to an arbitrary location within its environment. For example, if honey-bees previously trained to a feeding site are captured at the hive prior to their departure, carried to another site and released there, they should vanish straight towards the feeding station. This is exactly what they did in Gould's (1986) experiments (Fig. 3.43b, data set 1), but failed to do this in later experiments performed by five independent research groups at five different locations (Fig. 3.43, data sets 2–6 and, in addition, Rossel and Troje, personal communication). In the vast majority of releases the bees selected the predisplacement compass course that would have led them to the feeding station had they departed from the hive. There is only one exception (Fig. 3.43b, data set 4★): at one of Dyer's (1991) release sites the bees headed for the feeding station rather than followed the compass course. At first glance this result seems to support the map hypothesis, but on closer inspection of the arrangement of landmarks within the area in which the expcriment was performed, it does not. At the release site in question large-scale features of the landscape defining the route from the hive to the feeding site were visible to the bees. From this finding as well as from a number of control experiments Dyer (1991, p. 244) has concluded that even the exceptional result mentioned above 'provides decisive evidence against the hypothesis that the bees used mental maps'.

A number of additional experimental paradigms have been applied and used by the authors cited in the caption of Fig. 3.43 (references 2–6). Taken together the results of all these experiments are consistent with the hypothesis that bees while returning home to their central place use vectors and routes rather than maps. In no case did they show the ability to compute the short-cut between two familiar sites when displaced arbitrarily to one of them. This held true even if the bees had been trained deliberately in a way that would have facilitated the assemblage and use of landmark-based maps. For example, when they had been trained from the hive (H) to fly to, and return from, two feeding stations (A and B), they did not compose a map even of the restricted area HAB delineated by the two familiar routes HA and HB (Wehner *et al.*, 1990). Furthermore, even highly experienced bees failed to behave as if they had cognitive maps (Dyer, 1991).

The same conclusion can be drawn from experiments performed with *Cataglyphis* ants. When displaced to arbitrary sites even within their close nest environs, the ants search around at random until they discover a familar route (Fig. 3.44). Finally, it follows from

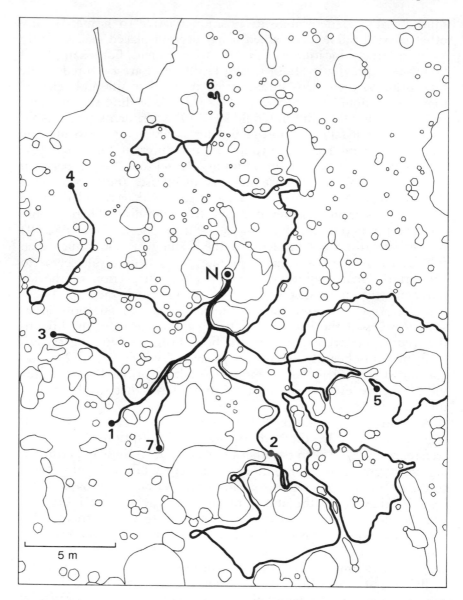

Figure 3.44 Homing paths of one individual ant (*Cataglyphis bicolor*) displaced from the nest (N) and released successively at sites 1–7. Prior to the experiment the ant had regularly foraged in the direction of site 1. The landmarks indicated in the map are small desert shrubs 0.1–1.2 m high (from Wehner, 1990b).

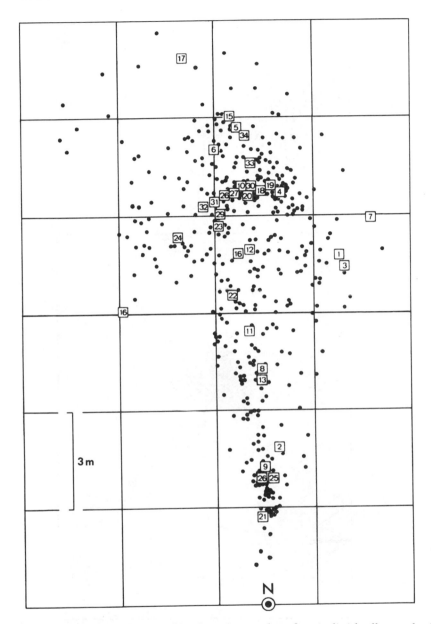

Figure 3.45 Spatial layout of 34 foraging paths of an individually marked ant (*Cataglyphis bicolor*), recorded over a period of 5 days. Fixes of the ant's positions (black dots) are taken every 30 s. N, nest. The sites where the ant found food items (arthropod carcasses) are indicated by squares provided with the numbers of the foraging trips (modified from Wehner *et al.*, 1983).

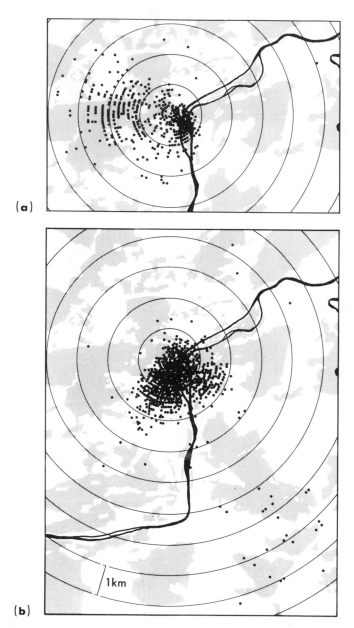

Figure 3.46 Foraging ranges of one particular colony of honey-bees (*Apis mellifera*), during two successive months: (a) June; (b) July. The foraging sites were recorded by reading the bees' recruitment dances in the hive during at least 11 days in each month. Black line, river; shaded areas, woodland (Dillier and Wehner, unpublished).

the foraging ecology of ants (Fig. 3.45) and bees (Fig. 3.46) that these insects are most unlikely to know the spatial arrangements of landmarks within their nest environs sufficiently well to work out novel routes within the colony's foraging range. At any one time, individual bees (reviewed in Wehner, 1981a) as well as the forager force of the whole colony (Visscher and Seeley, 1982; Fig. 3.46) use only relatively few food-source patches, and the colony constantly readjusts its allocation of foragers to various profitable sites (Seeley, 1986). This implies that during its entire foraging life any particular bee will encounter neither the need nor the possibility to sample the entire area around the central place of the colony, and thus to acquire a coherent map-like knowledge of the distribution of landmarks within this area. Apparently, the insect's systems of navigation have not been shaped by natural selection to assemble complete spatial representations of the arrangements of landmarks within the animal's foraging range, and to perform geometric manipulations within such cognitive maps.

3.4 SUMMARY AND CONCLUSIONS

What are the central issues, fundamental concepts, and broad implications of the research done in the field of arthropod navigation? In the context of this chapter the central point in question revolves not so much around the sensory cues involved in homeward navigation but around the way in which the navigator organizes the required sensory information to localize itself in environmental space.

What seems to emerge from the previous discourse on phenomena (p. 47) and mechanisms (p. 84) of homing in small-brain navigators, be they spiders, crabs, or insects, is a fairly coherent story. Central place foragers which leave their central place for distances of up to a million body lengths and thus frequently travel over completely unknown territory are always linked to the start by what could conveniently be dubbed a rigidly tightened thread of Ariadne. By integrating information on angles steered and distances travelled into a mean home vector, i.e. applying route-integration (dead-reckoning) strategies, the animal provides itself with a continually updated representation of the direction in which the starting point lies and how far away it is. Such vector information is acquired and used within an egocentric (self-centre) system of reference. The vector leading back to the start is attached, so to speak, to the animal rather than the terrain over which it moves. Such dead reckoning is the principal and predominant way of navigation in both chelicerate and mandibulate, walking and flying, short-distance and long-distance travellers, irrespective of whatever sensory system is involved in obtaining the necessary information about the angular and linear components of motion. Idiothetic information in wandering spiders or celestial compass information in desert ants and honey-bees, to mention just some of the key species on which recent research has focused, is handled and used **exclusively** within an egocentric frame of reference.

Any egocentric system, however, has snags. It is not only prone to cumulative errors, but also incapable of accounting for passive displacements. Furthermore, during an animal's excursion it must run incessantly rather than operate intermittently. Using a map, or geocentric system of navigation, would release the animal from any of these constraints, e.g. enable it to compute the direct homeward course even after inadvertent displacements to arbitrary points within its charted area.

Do arthropods use map-based information? First, even in long-distance travellers like monarch butterflies (p. 67) there is no need to invoke the concept of a map based on a network of geophysical gradients (grid map). Second, landmark-based familiar-area maps have recently been proposed for honey-bees, but the evidence is not conclusive (p. 121). This is not to say that arthropods do not use landmarks for navigation. In fact, they have been shown to rely on terrestrial signposts, be they tactile (p. 64), chemical (pp. 71 and 115) or visual (pp. 73, 81, and 106), but we cannot claim yet that they incorporate these site-specific features into some kind of topographic map. Bees and ants, for example, learn visual landmarks by taking 'snapshots' of the visual panoramas around the points they frequently visit (p. 106) and later use this landmark information in a matching-to-memory mode of behaviour. In this context, they can follow distinct routes (pp. 74, 111) by sequentially retrieving a number of memory traces, or stored snapshots. It is intriguing to inquire about the possibility that the assemblage and use of landmark-based route information is a first step in freeing the animal of its egocentric frame of reference.

There is still another possibility by which the animal might acquire map-based information. As in dead reckoning, arthropods are able to compute vectors pointing from their current position towards home, they might be able to 'lay down' their egocentred vectors on the ground, link them to familiar landmarks and then use these 'geostabilized' vectors to design a vector map (mosaic map *sensu* Füller *et al.*, 1983; Wallraff, 1974; Cartwright and Collett, 1987). Either hypothesis, albeit unsupported yet by experimental evidence, is attractive enough to stimulate future research.

Finally, the research potential offered by arthropods extends beyond the approaches discussed so far. It allows one to study not only the navigational strategies but also the underlying neurobiological mechanisms, i.e. the software as well as the hardware of the behaviour observed. In arthropods, more emphasis has been placed on investigations of the sensory and neural basis of homing than has been performed in any other group of animals. In vertebrates, for example, research on homing behaviour has not yet produced the mutually reinforcing evidence from behavioural and physiological approaches and thus has not yet reached the state of the art set, for example, by the analysis of electrolocation in fish or echolocation in bats (see reviews in Laughlin, 1989). In arthropods, on the other hand, the use of cuticular strain receptors (pp. 49 and 85), pheromone receptors (p. 118), or dichroic photoreceptors (p. 89) in different kinds of homeward navigation offers some promising fields of integrative neuroethological research.

There is one more general message to be derived from studies on arthropod homing: computation and implementation cannot be separated and considered independently. For too long and too strictly research on animal navigation has focused on a top-down approach in mainly emphasizing abstract problem analysis rather than neurobiological implementation. As we now learn from a number of studies, e.g. from the analysis of the hymenopteran skylight compass, the particular problems to be solved by the insect navigator are defined largely by the potentialities and constraints of the neural machinery accomplishing a given task. In this context, the task itself might be defined quite differently in animal and man. If certain boundary conditions were met, for example if the animal were able to rely on certain assumptions about a constant sensory world, or if it performed a certain type of locomotor activity within this world, the computational task could be simplified, often substantially, and it is this simplified task, rather than the all-inclusive problem envisaged by the human investigator, to which the animal had to find its proper evolutionary responses. In unravelling these responses it is essential to inquire about the subtleties of both the animal's behavioural repertoire and the underlying neural computations. As judged from past experience outlined in this chapter, arthropods seem to be advantageous in offering a good chance for proceeding successfully along both lines.

REFERENCES

Able, K.P. (1980) Mechanisms of orientation, navigation, and homing, in *Animal Migration, Orientation and Navigation* (ed. S.A. Gauthreaux), Academic Press, New York, pp. 283–373.

Accademia d'Italia (1934) *Le Opere di Lazzaro Spallanzani*, Vol. 3, U. Hoepli, Milano.

Anderson, A.M. (1977) A model for landmark learning in the honey-bee. *J. Comp. Physiol.*, **114**, 335–55.

Allen, J.A. (1966) The rhythms and population dynamics of decapod Crustacea. *Oceanogr. Mar. Biol. A. Rev.*, **4**, 247–65.

Altevogt, R. and Hagen, H.v. (1964) Über die Orientierung von *Uca tangeri* im Freiland. *Z. Morph. Oekol. Tiere*, **53**, 636–56.

Anzenberger, G. (1986) How do carpenter bees recognize the entrance of their nests? An experimental investigation in a natural habitat. *Ethology*, **71**, 54–62.

Arnold, G., Masson, C. and Budharugsa, S. (1985) Comparative study of the antennal lobes and their afferent pathway in the worker bee and the drone (*Apis mellifera*). *Cell Tiss. Res.*, **242**, 593–605.

Aron, S., Deneubourg, J.L. and Pasteels, J.M. (1988) Visual cues and trail-following idiosyncrasy in *Leptothorax unifasciatus*: an orientation process during foraging. *Insectes Sociaux*, **35**, 355–66.

Attygalle, A.B. and Morgan, E.D. (1985) Ant trail pheromones. *Adv. Insect Physiol.*, **18**, 1–30.

Attygalle, A.B., Steghaus-Kovac, S., Ahmed, V.U. *et al.* (1991) cis-Isogeraniol, a recruitment pheromone of the ant *Leptogenys diminuta*. *Naturwissenschaften*, **78**, 90–2.

Attygalle, A.B., Vostrowsky, O., Bestmann, H.J., Steghaus-Kovac, S. and Maschwitz, U. (1988) (3R,4S)-4-Methyl-3-heptanol, the trail pheromone of the ant *Leptogenys diminuta*. *Naturwissenschaften*, **75**, 315–17.

Baerends, G.P. (1941) Fortpflanzungsverhalten und Orientierung der Grabwespe *Ammophila campestris*. *Tijdschr. Entomol. Deel*, **84**, 68–275.

Baker, R.R. (1978) *The Evolutionary Ecology of Animal Migration*, Hodder and Stoughton, London.

Balss, H. (1956) Decapoda. Ökologie, in *Bronn's Klassen und Ordnungen des Tierreichs*, Akademische Verlagsgesellschaft, Leipzig, pp. 1285–476.

Baltzer, F. (1930) Über die Orientierung der Trichterspinne *Agelena labyrinthica* nach der Spannung des Netzes. *Rév. Suisse Zool.*, **37**, 363–9.

Barlow, J.S. (1964) Inertial navigation as a basis for animal navigation. *J. Theor. Biol.*, **6**, 76–117.

Bartels, M. (1929) Sinnesphysiologische und psychologische Untersuchungen an der Trichterspinne *Agelena labyrinthica*. *Z. Vergl. Physiol.*, **10**, 527–93.

Barth, F.G. (1986) Vibrationssinn und vibratorische Umwelt von Spinnen. *Naturwissenschaften*, **73**, 519–30.

Barth, F.G. and Libera, W. (1970) Ein Atlas der Spaltsinnesorgane von *Cupiennius salei*. Chelicerata (Araneae). *Z. Morph. Tiere*, **68**, 343–69.

Barth, F.G. and Seyfarth, E.-A. (1979) *Cupiennius salei*. Araneae in the highland of central Guatemala. *J. Arachnol.*, **7**, 255–63.

Barth, F.G., Seyfarth, E.-A., Bleckmann, H. and Schüch, W. (1988) Spiders of the genus *Cupiennius* (Araneae, Ctenidae). I. Range distribution, dwelling plants, and climatic characteristics of the habitats. *Oecologia*, **77**, 187–93.

Batschelet, E. (1981) *Circular Statistics in Biology*, Academic Press, London.

Beier, W. and Lindauer, M. (1970) Der Sonnenstand als Zeitgeber für die Biene. *Apidologie*, **1**, 5–28.

Berry, P.F. (1973) The biology of the spiny lobster *Palinurus delagoae*, off the coast of Natal, South Africa. *Oceanogr. Res. Inst. (Durban) Invest. Rep.*, **31**, 1–27.

Beugnon, G. and Fourcassie, V. (1988) How do red wood ants orient during diurnal and nocturnal foraging in a three dimensional system? II. Field experiments. *Insectes Sociaux*, **35**, 106–24.

Bisetzky, A.R. (1957) Die Tänze der Bienen nach einem Fussweg zum Futterplatz. *Z. Vergl. Physiol.*, **40**, 264–88.

Blickhan, R., Barth, F.G. and Ficker, E. (1982) Biomechanics in a sensory system (strain detection in the exoskeleton of arthropods). *Proc. Int. Conf. Exp. Stress Analysis*, **7**, 223–34.

Brandstetter, M., Crailsheim, K. and Heran, H. (1988) Provisioning of food in the honey bee before foraging. *Biona Report*, **6**, 129–48.

Breed, M.D. and Bennett, B. (1985) Mass recruitment to nectar sources in *Paraponera clavata*: a field study. *Insectes Sociaux*, **32**, 198–208.

Breed, M.D., Fewell, J.H., Moore, A.J. and Williams, K.R. (1987) Graded recruitment in a ponerine ant. *Behav. Ecol. Sociobiol.*, **20**, 407–11.

Breed, M.D. and Harrison, J.M. (1987) Individually discriminable recruitment trails in a ponerine ant. *Insectes Sociaux*, **34**, 222–6.

Brines, M.L. and Gould, J.L. (1979) Bees have rules. *Science*, **206**, 571–3.

Brower, L.P. (1961) Studies on the migration of the monarch butterfly. I. Breeding population of *Danaus plexippus* and *Danaus gilippus berenice* in South Central Florida. *Ecology*, **42**, 76–83.

Brower, L.P. (1977) Monarch migration. *Nat. Hist.*, **86**, 40–52.

Brun, R. (1914) *Die Raumorientierung der Ameisen*, G. Fischer, Jena.

Burton, J.L. and Franks, N.R. (1985) The foraging ecology of the army ant *Eciton rapax*: an ergonomic enigma? *Ecol. Entomol.*, **10**, 131–41.

Butler, C.G., Fletcher, D.J.C. and Watler, D. (1970) Hive entrance finding by honeybee (*Apis mellifera*) foragers. *Anim. Behav.*, **18**, 78–91.

Caldwell, R.L. and Dingle, H. (1976) Stomatopods. *Scient. Am.*, **234**(1), 80–9.

Carthy, J.D. (1950) Odour trails of *Acanthomyops fuliginosus*. *Nature*, **166**, 154.

Carthy, J.D. (1951) The orientation of two allied species of British ant. *Behaviour*, **3**, 275–318.

Cartwright, B.A. and Collett, T.S. (1982) How honey bees use landmarks to guide their return to a food source. *Nature*, **295**, 560–4.

Cartwright, B.A. and Collett, T.S. (1983) Landmark learning in bees. Experiments and models. *J. Comp. Physiol.*, **151**, 521–43.

Cartwright, B.A. and Collett, T.S. (1987) Landmark maps for honeybees. *Biol. Cybern.*, **57**, 85–93.

Cheng, K., Collett, T.S., Pickhard, A. and Wehner, R. (1987) The use of visual landmarks by honeybees: bees weight landmarks according to their distance from the goal. *J. Comp. Physiol. A*, **161**, 469–75.

Cherrett, J.M. (1968) The foraging behavior of *Atta cephalotes* (Hymenoptera, Formicidae). I. Foraging pattern and plant species attacked in tropical rain forest. *J. Anim. Ecol.*, **37**, 387–403.

Chittleborough, R.G. (1974) Home range, homing and dominance in juvenile western rock lobsters. *Aust. J. Mar. Freshwat. Res.*, **25**, 227–34.

Cloudsley-Thompson, J.L. (1978) *Animal Migration*, G.P. Putnam, New York.

Collett, T.S. and Kelber, A. (1988) The retrieval of visuospatial memories by honeybees. *J. Comp. Physiol. A.*, **163**, 145–50.

Collett, T.S. and Land, M.F. (1975) Visual spatial memory in a hoverfly. *J. Comp. Physiol. A.*, **100**, 59–84.

Cosens, D. and Toussaint, N. (1985) An experimental study of the foraging strategy of the wood ant *Formica aquilonia*. *Anim. Behav.*, **33**, 541–52.

Crane, J. (1958) Aspects of social behavior in fiddler crabs, with special reference to *Uca maracoani*. *Zoologia*, **43**, 113–30.

Crawford, J.D. (1984) Orientation in a vertical plane: the use of light cues by an orb-weaving spider, *Araneus diadematus*. *Anim. Behav.*, **32**, 162–71.

Creaser. E.P. and Travis, D. (1950) Evidence of a homing instinct in the Bermuda spiny lobster. *Science*, **112**, 169–70.

Cronin, T.W., Nair, J.N. and Doyle, R.D. (1988) Ocular tracking of rapidly moving visual targets by stomatopod crustaceans. *J. Exp. Biol.*, **138**, 155–79.

Crothers, J.H. (1968) The biology of the shore crab *Carcinus maenas*. II. The life of the adult crab. *Field Stud.*, **2**, 579–614.

Darwin, C. (1873) Origin of certain instincts. *Nature*, **7**, 417–18.

Darwin, F. (1888) *The Life and Letters of Charles Darwin Including an Autobiographical Chapter*, J. Murray, London.

David, C.T. and Wood, D.L. (1980) Orientation to trails by a carpenter ant, *Camponotus modoc* (Hymenoptera: Formicidae), in a giant sequoia forest. *Can. Ent.*, **112**, 993–1000.

Diesel, R. (1986) Optimal mate searching strategy in the symbiotic spider crab, *Inachus phalangium* (Decapoda). *Ethology*, **72**, 311–28.

Dornfeldt, K. (1972) Die Bedeutung der Haupt- und Nebenaugen für die photomenotaktische Orientierung der Trichterspinne *Agelena labyrinthica* mit Hilfe einer Lichtquelle. *PhD Thesis*, Freie Universität Berlin.

Drake, V.A., Helm, K.F., Readshaw, J.L. and Reid, D.G. (1981) Insect migration across Bass Strait during spring: a radar study. *Bull. Ent. Res.*, **71**, 449–66.

Dyer, F.C. (1987) Memory and sun compensation by honey bees. *J. Comp. Physiol. A*, **160**, 621–33.

Dyer, F.C. (1991) Bees acquire route-based memories but not cognitive maps in a familiar landscape, *Anim. Behav.*, **41**, 239–46.

Dyer, F.C. and Gould, J.L. (1981) Honey bee orientation: a backup system for cloudy days. *Science*, **214**, 1041–2.

Dyer, F.C. and Seeley, T.D. (1989) On the evolution of the dance language. *Am. Nat.*, **133**, 580–90.

Ebbers, B.C. and Barrows, E.M. (1980) Individual ants specialize on particular aphid herds (Hymenoptera: Formicidae; Homoptera: Aphididae). *Proc. Ent. Soc. Wash.*, **82**, 405–7.

Esch, H. (1978) On the accuracy of the distance message in the dance of honey bees. *J. Comp. Physiol. A*, **123**, 339–47.

Etienne, A.S., Maurer, R., Saucy, F. and Teroni, E. (1986) Short-distance homing in the golden hamster after a passive outward journey. *Anim. Behav.*, **34**, 696–715.

Evoy, W.H. and Ayers, J. (1982) Locomotion and control of limb movements, in *The Biology of Crustacea*, Vol. 4 (eds D.C. Sandeman and H.L. Atwood), Academic Press, New York, pp. 62–105.

Fabre, J.H. (1879) *Souvenirs Entomologiques* (1 Série), C. Delagrave, Paris.

Fabre, J.H. (1882) *Souvenirs Entomologiques* (2 Série), C. Delagrave, Paris.

Ferguson, D.E. (1967) Sun-compass orientation in anurans, in *Animal Orientation and Navigation* (ed. R.M. Storm), Oregon State University Press, Corvallis, pp. 21–34.

Ferton, C. (1923) *La Vie des Abeilles et des Guêpes*, E. Chiron, Paris.

Finney, B.R. (1976) Pacific navigation and voyaging, in *Polynesian Society Memoir No. 39* (ed. B.R. Finney), Polynesian Society Inc., Wellington.

Fletcher, D.J.C. and Brand, J.M. (1968) Source of the trail pheromone and method of trail laying in the ant *Crematogaster peringueyi*. *J. Insect Physiol.*, **14**, 783–8.

Forel, A. (1886) Etudes myrmécologiques en 1886. *Ann. Soc. Entomol. Belg.*, **30**, 131–215.

Fourcassie, V. and Beugnon, G. (1988) How do red wood ants orient when foraging in a three-dimensional system? I. Laboratory experiments. *Insectes Sociaux*, **35**, 92–105.

Frantsevich, L.I., Govardovski, V., Gribakin, F. *et al.* (1977) Astro-orientation in *Lethrus* (Coleoptera, Scarabaeidae). *J. Comp. Physiol.*, **121**, 253–71.

Fresneau, D. (1985) Individual foraging and path fidelity in a ponerine ant. *Insectes Sociaux*, **32**, 109–16.

Frisch, K. v. (1949) Die Polarisation des Himmelslichts als orientierender Faktor bei den Tänzen der Bienen. *Experientia*, **5**, 142–8.

Frisch, K. v. (1950) Die Sonne als Kompass im Leben der Bienen. *Experientia*, **6**, 210–22.

Frisch, K. v. (1967) *The Dance Language and Orientation of Bees*, Harvard University Press, Cambridge, Mass.

Frisch, K. v. and Lindauer, M. (1954) Himmel und Erde in Konkurrenz bei der Orientierung der Bienen. *Naturwissenschaften*, **41**, 245–53.

Füller, E., Kowalski, U. and Wiltschko, R. (1983) Orientation of homing in pigeons: compass orientation vs piloting by familiar landmarks. *J. Comp. Physiol.*, **153**, 55–8.

Gallistel, C.R. (1989) Animal cognition: the representation of space, time and number. *A. Rev. Psychol.*, **40**, 155–89.

Görner, P. (1958) Die optische und kinästhetische Orientierung der Trichterspinne *Agelena labyrinthica*. *Z. Vergl. Physiol.*, **41**, 111–53.

Görner, P. (1966) Über die Koppelung der optischen und kinästhetischen Orientierung bei den Trichterspinnen *Agelena labyrinthica* und *Agelena gracilens*. *Z. Vergl. Physiol.*, **53**, 253–76.

Görner, P. (1988) Homing behavior of funnel web spiders (Agelenidae) by means of web-related cues. *Naturwissenschaften*, **75**, 209–11.

Görner, P. and Claas, B. (1979) The influence of the web on the directional orientation in the funnel-web spider *Agelena labyrinthica*. *Verh. Dtsch. Zool. Ges.*, **72**, 316.

Görner, P. and Claas, B. (1985) Homing behavior and orientation in the funnel-web spider, *Agelena labyrinthica*, in *Neurobiology of Arachnids* (ed. F.G. Barth), Springer, Berlin, pp. 275–97.

Gould, J.L. (1980) Sun compensation by bees. *Science*, **207**, 545–7.

Gould, J.L. (1984) Processing of sun-azimuth information by honey bees. *Anim. Behav.*, **32**, 149–52.

Gould, J.L. (1986) The locale map of honey bees: do insects have cognitive maps? *Science*, **232**, 861–3.

Gould, J.L. (1987) Landmark learning by honey bees. *Anim. Behav.*, **35**, 26–34.

Gould, J.L. and Towne, W.F. (1987) Evolution of the dance language. *Am. Nat.*, **130**, 317–38.

Gregory, P.H. (1961) *The Microbiology of the Atmosphere*, Plant Science Monographs, L. Hill, London.

Haas, A. (1967) Vergleichende Verhaltensstudien zum Paarungsschwarm der Hummeln (*Bombus*) und Schmarotzerhummeln (*Psithyrus*). *Z. Tierpsychol.*, **24**, 257–77.

Hagen, H.O. v. (1962) Freilandstudien zur Sexual- und Fortpflanzungs-biologie von *Uca tangeri* in Andalusien. *Z. Morph. Ökol. Tiere*, **51**, 611–725.

Hagen, H.O. v. (1967) Nachweis einer kinästhetischen Orientierung bei *Uca rapax*. *Z. Morph. Ökol. Tiere*, **58**, 301–20.

Hahn, M. and Maschwitz, U. (1985) Foraging strategies and recruitment behaviour in the European harvester ant *Messor rufitarsis*. *Oecologia*, **68**, 45–51.

Hangartner, W. (1967) Spezifität und Inaktivierung des Spurpheromons von *Lasius fuliginosus* und Orientierung der Arbeiterinnen im Duftfeld. *Z. Vergl. Physiol.*, **57**, 103–36.

Hangartner, W. (1969) Trail laying in the subterranean ant, *Acanthomyops interjectus*. *J. Insect Physiol.*, **15**, 1–4.

Hangartner, W. and Bernstein, S. (1964) Über die Geruchsspur von *Lasius fuliginosus* zwischen Nest und Futterquelle. *Experientia*, **20**, 392–3.

Harrison, J.F., Fewell, J.H., Stiller, T.M. and Breed, M.D. (1989) Effects of experience on use of orientation cues in the giant tropical ant. *Anim. Behav.*, **37**, 869–71.

Hartwick, E., Friend, W. and Atwood, C. (1977) Trail-laying behaviour of the carpenter ant, *Camponotus pennsylvanicus* (Hymenoptera: Formicidae). *Can. Ent.*, **109**, 129–36.

Heinrich, B. (1976) The foraging specializations of individual bumblebees. *Ecol. Monogr.*, **46**, 105–28.

Hengstenberg, R. (1988) Mechanosensory control of compensatory head roll during flight in the blowfly *Calliphora erythrocephala*. *J. Comp. Physiol. A*, **163**, 151–65.

Heran, H. (1956) Ein Beitrag zur Frage nach der Wahrnehmungsgrundlage bei der Entfernungsweisung der Bienen. *Z. Vergl. Physiol.*, **38**, 168–218.

Heran, H. (1977) Entfernungsweisung der Bienen unter extremen Flugbedingungen (Einfluss der Luftdichte und der Belastung auf das Tanztempo). *Insectes Sociaux*, **24**, 272.

Heran, H. and Wanke, L. (1952) Beobachtungen über die Entfernungsmeldung der Sammelbienen. *Z. Vergl. Physiol.*, **34**, 383–93.

Herbers, J.M. (1977) Behavioral constancy in *Formica obscuripes* (Hymenoptera: Formicidae). *Ann. Ent. Soc. America*, **70**, 485–6.

Herrnkind, W.F. (1965) Investigations concerning homing, directional orientation, and insight in the sand fiddler crab, *Uca pugilator*. MD Thesis, University of Miami.

Herrnkind, W.F. (1980) Spiny lobsters: patterns of movement, in *Biology and Management of Lobsters*, Vol. 1 (eds B. Phillips and J. Cobb), Academic Press, New York, pp. 349–407.

Herrnkind, W.F. (1983) Movement patterns and orientation, in *The Biology of Crustacea*, Vol. 7 (eds F.J. Vernberg and W.B. Vernberg), Academic Press, New York, pp. 41–105.

Herrnkind, W.F. (1985) Evolution and mechanisms of mass single-file migration in spiny lobster: synopsis. *Contr. Mar. Sci.*, Suppl., **27**, 197–211.

Herrnkind, W.F. and McLean, R. (1971) Field studies of homing, mass emigration, and orientation in the spiny lobster, *Panulirus argus*. *Ann. N.Y. Acad. Sci.*, **188**, 359–77.

Herrnkind, W.F., van der Walker, J. and Barr, L. (1975) Population dynamics, ecology, and behavior of the spiny lobster, *Panulirus argus*, of St. John, U.S. Virgin Islands: habitation and pattern of movements. Results of the Tektite Program, Vol. 2. *Bull. Nat. Hist. Mus. L.A. County*, **20**, 31–45.

Hill, D.E. (1979) Orientation by jumping spiders of the genus *Phidippus* (Araneae: Salticidae) during the pursuit of prey. *Behav. Ecol. Sociobiol.*, **5**, 301–22.

Hoffmann, G. (1983a) The random elements in the systematic search behavior of the desert isopod *Hemilepistus reaumuri*. *Behav. Ecol. Sociobiol.*, **13**, 81–92.

Hoffmann, G. (1983b) The search behavior of the desert isopod

Hemilepistus reaumuri as compared with a systematic search. *Behav. Ecol. Sociobiol.*, **13**, 93–106.

Hölldobler, B. (1971) Homing in the harvester ant *Pogonomyrmex badius*. *Science*, **171**, 1149–51.

Hölldobler, B. (1974) Home range orientation and territoriality in harvesting ants. *Proc. Natl Acad. Sci., U.S.A.*, **71**, 3274–7.

Hölldobler, B. (1976) Recruitment behavior, home range orientation and territoriality in harvester ants, *Pogonomyrmex*. *Behav. Ecol. Sociobiol.*, **1**, 3–44.

Hölldobler, B. (1979) Territoriality in ants. *Proc. Am. Phil. Soc. Philadelphia*, **123**, 211–18.

Hölldobler, B. (1980) Canopy orientation: a new kind of orientation in ants. *Science*, **210**, 86–8.

Hölldobler, B. (1981) Foraging and spatiotemporal territories in the honey ant *Myrmecocystus mimicus* (Hymenoptera: Formicidae). *Behav. Ecol. Sociobiol.*, **9**, 301–14.

Hölldobler, B. (1983) Territorial behavior in the green tree ant (*Oecophylla smaragdina*). *Biotropica*, **15**, 241–50.

Hölldobler, B. (1984) Evolution in insect communication, in *Insect Communication* (ed. T. Lewis), Academic Press, London, pp. 349–77.

Hölldobler, B., Engel, H. and Taylor, R.W. (1982) A new sternal gland in ants and its function in chemical communication. *Naturwissenschaften*, **69**, 90–1.

Hölldobler, B. and Möglich, M. (1980) The foraging system of *Pheidole militicida* (Hymenoptera: Formicidae). *Insectes Sociaux*, **27**, 237–64.

Hölldobler, B. and Palmer, J.M. (1989) A new tarsal gland in ants and the possible role in chemical communication. *Naturwissenschaften*, **76**, 385–6.

Hölldobler, B. and Wilson, E.O. (1970) Recruitment trails in the harvester ant, *Pogonomyrmex badius*. *Psyche*, **77**, 365–99.

Hölldobler, B. and Wilson, E.O. (1977) Colony-specific territorial pheromone in the African weaver ant *Oecophylla longinoda*. *Proc. Natl Acad. Sci., U.S.A.*, **74**, 2072–5.

Hölldobler, B. and Wilson, E.O. (1986) Nest area exploration and recognition in leafcutter ants (*Atta cephalotes*). *J. Insect Physiol.*, **32**, 143–50.

Hölldobler, B. and Wilson, E.O. (1990) *The Ants*, Belknap Press of Harvard University Press, Cambridge, Mass.

Holzapfel, M. (1934) Die nicht-optische Orientierung der Trichterspinne *Agelena labyrinthica*. *Z. Vergl. Physiol.*, **20**, 55–116.

Homberg, U., Christensen, T.A. and Hildebrand, J.G. (1989) Structure and functions of the deuterocerebrum in insects. *A. Rev. Entomol.*, **34**, 477–501.

Hueftle, K. (1977) Near-orientation in the homing of the fiddler crab, *Uca crenulata*. MS Thesis, San Diego State University, San Diego, California.

Hughes, D. (1966) Behavioural and ecological investigations of the crab *Ocypode ceratophthalmus* (Crustacea: Ocypodidae). *J. Zool.*, **150**, 129–43.

Hughes, D.A. (1972) On the endogenous control of tide-associated displacements of pink shrimp, *Penaeus duorarum. Biol. Bull.*, **142**, 271–80.

Iersel, J.J.A. v. (1975) The extension of the orientation system of *Bembix rostrata* as used in the vicinity of its nest, in *Function and Evolution in Behaviour* (eds G. Baerends, C. Beer and A. Manning), Clarendon, Oxford, pp. 142–68.

Jackson, B.D., Keegans, S.J., Morgan, E.D., Cammaerts, M.C. and Cammaerts, R. (1990) Trail pheromone of the ant *Tetramorium meridionale. Naturwissenschaften*, **77**, 294–6.

Jaisson, P. (1972) Solare et determinismo del comportamiento en las hormigas del género *Atta. Fol. Entomol. Mexicana*, **23–24**, 108–10.

James, D.G. (1986) Studies on the migration of *Danaus plexippus* (Lepidoptera: Nymphalidae) in the Sydney area. *Aust. Ent. Mag.*, **13**, 27–31.

Jander, R. (1957) Die optische Richtungsorientierung der Roten Waldameise (*Formica rufa*). *Z. Vergl. Physiol.*, **40**, 162–238.

Jander, R. (1975) Ecological aspects of spatial orientation. *A. Rev. Ecol. Syst.*, **6**, 171–88.

Janzen, D.H. (1971) Euglossine bees as long-distance pollinators of tropical plants. *Science*, **171**, 203–5.

Janzen, D.H. (1974) The deflowering of Central America. *Nat. Hist.*, **83**, 48–53.

Janzen, D.H. and Carroll, C.R. (1983) *Paraponera clavata* (bala, giant tropical ant), in *Costa Rican Natural History* (ed. D.H. Janzen), University of Chicago Press, Chicago, pp. 752–3.

Jessen, K. and Maschwitz, U. (1985) Individual specific trails in the ant *Pachycondyla tesserinoda* (Formicidae, Ponerinae). *Naturwissenschaften*, **72**, 549–50.

Jessen, K. and Maschwitz, U. (1986) Orientation and recruitment behavior in the ponerine ant *Pachycondyla tesserinoda*: laying of individual-specific trails during tandem-running. *Behav. Ecol. Sociobiol.*, **19**, 151–5.

Johnson, C.G. (1969) *Migration and Dispersal of Insects by Flight*, Methuen, London.

Jong, D. de (1982) Orientation of comb building by honeybees. *J. Comp. Physiol.*, **147**, 495–501.

Kaiser, H. (1974) Verhaltensgefüge und Temporalverhalten der Libelle *Aeschna cyanea* (Odonata). *Z. Tierpsychol.*, **34**, 398–429.

Kaissling, K.E. (1987) *R.H. Wright Lectures on Insect Olfaction*, Simon Fraser University, Burnaby, B.C., Canada.

Kanciruk, P. and Herrnkind, W.F. (1978) Mass migration of spiny lobster, *Panulirus argus* (Crustacea: Palinuridae): behavior and environmental correlates. *Bull. Mar. Sci.*, **28**, 601–23.

Kanz, J.E. (1977) The orientation of migrant and non-migrant monarch butterflies. *Psyche*, **84**, 120–41.

Kanzaki, R. (1989) Physiology and morphology of higher–order neurons in olfactory pathways of the moth brain: pheromone-processing neurons in the protocerebrum, in *Neural Mechanisms of Behavior* (eds J.

Erber, R. Menzel, H.J. Pflüger and D. Todt), G. Thieme, Stuttgart, pp. 253–4.

Karnofsky, E.B., Atema, J. and Elgin, R.H. (1989) Field observations of social behavior, shelter use, and foraging in the lobster, *Homarus americanus. Biol. Bull.*, **176**, 239–46.

Kaul, R.M. (1985) Some optical orientation characters in *Formica rufa* (Hymenoptera, Formicidae). *Zool. J., Moscow*, **64**, 1823–8 (in Russian).

Kaul, R.M. and Korteva, G.A. (1982) Night orientation of ants *Formica rufa* (Hymenoptera: Formicidae) upon movement on routes. *Zool. Zh.*, **61**, 1351–8 (in Russian).

Kayton, M. (1989) *Navigation — Land, Sea, Air, and Space*, IEEE Press, New York.

Keeton, W.T. (1974) The orientational and navigational basis of homing in birds. *Adv. Study Behav.*, **5**, 47–132.

Kelber, A. and Zeil, J. (1990) A robust procedure for visual stabilization of hovering flight position in guard bees of *Trigona* (*Tetragonisca*) *angustula* (Apidae, Meliponinae). *J. Comp. Physiol. A*, **167**, 569–77.

Kirchner, W.H. and Heusipp, M. (1990) Freely flying honeybees use retinal image motion and motion parallax in visual course control. *Proc. Neurobiol. Conf. Göttingen*, **18**, 84.

Kisimoto, R. (1976) Synoptic weather conditions inducing long-distance immigration of planthoppers, *Sogatella furcifera* and *Nilaparvata lugens. Ecol. Entomol.*, **1**, 95–109.

Klotz, J.H. (1987) Topographic orientation in two species of ants (Hymenoptera: Formicidae). *Insectes Sociaux*, **34**, 236–51.

Knaffl, H. (1953) Über die Flugweite und Entfernungsmeldung der Bienen. *Z. Bienenforsch.*, **2**, 131–40.

Kramer, G. (1953) Wird die Sonnenhöhe bei der Heimfindeorientierung verwertet? *J. Orn.*, **94**, 201–19.

Kraus, H.J. and Tautz, J. (1981) Visual distance-keeping in the soldier crab, *Mictyris platycheles* (Grapsoidea: Mictyridae). A field study. *Mar. Behav. Physiol.*, **6**, 123–33.

Land, M.F. (1972) Stepping movements made by jumping spiders during turns mediated by the lateral eyes. *J. Exp. Biol.*, **57**, 15–40.

Lane, A. (1977) Recrutement et orientation chez la fourmi *Leptothorax unifasciatus*: Rôle de la piste et des tandems. PhD Thesis, University of Dijon.

Langdon, J. (1971) Shape discrimination and learning in the fiddler crab *Uca pugilator*. PhD Thesis, Florida State University, Tallahassee, Florida.

Laughlin, S.B. (ed.) (1989) Principles of sensory coding and processing. *J. Exp. Biol.*, **146**, 1–322.

Le Guelte, L. (1969) Learning in spiders. *Am. Zool.*, **9**, 145–52.

Lehrer, M., Srinivasan, M.V., Zhang, S.W. and Horridge, G.A. (1988) Motion cues provide the bee's visual world with a third dimension. *Nature*, **332**, 356–7.

Leuthold, R.H. (1968) Recruitment to food in the ant *Crematogaster ashmeadi. Psyche*, **75**, 233–48.

Lewis, T., Pollard, G.V. and Dibley, G.C. (1974) Rhythmic foraging in the leaf-cutting ant *Atta cephalotes* (Formicidae: Attini). *J. Anim. Ecol.*, **43**, 129–41.

Lewtschenko, I.A. (1959) The return of bees to the hive. *Russ. Pchelov*, **36**, 38–40 (in Russian).

Lindauer, M. (1963) Kompassorientierung. *Ergeb. Biol.*, **26**, 158–81.

Linsenmair, K.E. (1967) Konstruktion und Signalfunktion der Sandpyramide der Reiterkrabbe *Ocypode saratan* (Decapoda Brachyura Ocypodidae). *Z. Tierpsychol.*, **24**, 403–56.

Lubbock, J. (1889) *On the Senses, Instincts, and Intelligence of Animals with Special Reference to Insects*, K. Paul Trench, London.

Macgregor, E.G. (1948) Odour as a basis for orientated movement in ants. *Behaviour*, **1**, 267–96.

Marshall, N.J. (1988) A unique colour and polarization vision system in mantis shrimps. *Nature*, **333**, 557–60.

Martin, H. and Lindauer, M. (1977) Der Einfluss des Erdmagnetfeldes auf die Schwereorientierung der Honigbiene. *J. Comp. Physiol.*, **122**, 145–87.

Maschwitz, U., Lenz, S. and Buschinger, A. (1986) Individual specific trails in the ant *Leptothorax affinis* (Formicidae: Myrmicinae). *Experientia*, **42**, 1173–4.

Maschwitz, U. and Mühlenberg, M. (1975) Zur Jagdstrategie einiger orientalischer *Leptogenys*-Arten (Formicidae: Ponerinae). *Oecologia*, **20**, 65–83.

Maschwitz, U., Steghaus-Kovac, S. (1991) Individualismus versus Kooperation (Hymenoptera: Formicidae). *Naturwissenschaften*, **78**, 103–13.

Mayne, R. (1974) A systems concept of the vestibular organs, in *Handbook of Sensory Physiology*, Vol. VI/2 (ed. H.H. Kornhuber), Springer, Berlin, pp. 493–580.

Meder, E. (1958) Über die Einberechnung der Sonnenwanderung bei der Orientierung der Honigbiene. *Z. Vergl. Physiol.*, **40**, 610–41.

Melchers, M. (1967) Der Beutefang von *Cupiennius salei* (Ctenidae). *Morph. Ökol. Tiere*, **58**, 321–46.

Menzel, R., Chittka, L., Eichmüller, S. *et al.* (1990) Dominance of celestial cues over landmarks disproves map-like orientation in honey bees. *Z. Naturforsch.*, **45C**, 723–6.

Mittelstaedt, H. (1978) Kybernetische Analyse von Orientierungsleistungen, in *Kybernetik 1977* (eds G. Hauske and E. Butenandt), R. Oldenbourg, München, pp. 144–95.

Mittelstaedt, H. (1983) The role of multimodal convergence in homing by path integration. *Fortschr. Zool.*, **28**, 197–212.

Mittelstaedt, H. (1985) Analytical cybernetics of spider navigation, in *Neurobiology of Arachnids* (ed. F.G. Barth), Springer, Berlin, pp. 298–316.

Mittelstaedt, M.L. and Mittelstaedt, H. (1980) Homing by path integration in a mammal. *Naturwissenschaften*, **67**, 566.

Mittelstaedt, H. and Mittelstaedt, M.L. (1973) Mechanismen der Orientierung ohne richtende Aussenreize. *Fortschr. Zool.*, **21**, 46–58.

Moffett, M.W. (1987) Ants that go with the flow: a new method of orientation by mass communication. *Naturwissenschaften*, **74**, 551–3.

Möglich, M. and Hölldobler, B. (1975) Communication and orientation during foraging and emigration in the ant *Formica rufa*. *J. Comp. Physiol. A*, **101**, 275–88.

Moller, P. (1970) Die systematischen Abweichungen bei der optischen Richtungsorientierung der Trichterspinne *Agelena labyrinthica*. *Z. Vergl. Physiol.*, **66**, 78–106.

Müller, M. (1989) Mechanismus der Wegintegration bei *Cataglyphis fortis* (Hymenoptera, Insecta). PhD Thesis, University of Zürich.

Müller, M. and Wehner, R. (1988) Path integration in desert ants, *Cataglyphis fortis*. *Proc. Natl. Acad. Sci., U.S.A*, **85**, 5287–90.

Myres, M.T. (1985) A southward return migration of painted lady butterflies, *Vanessa cardui*, over southern Alberta in the fall of 1983, and biometeorological aspects of their outbreaks into North America and Europe. *Can. Field Nat.*, **99**, 147–55.

Nalbach, H.O. (1987) Neuroethologie der Flucht von Krabben (Decapoda: Brachyura). PhD Thesis, University of Tübingen.

Neese, V. (1988) Die Entfernungsweisung der Sammelbiene: ein energetisches und zugleich sensorisches Problem. *Biona Report*, **6**, 1–15.

Neisser, U. (1966) *Cognitive Psychology*, Appleton-Century-Crofts, New York.

New, D.A.T. and New, J.K. (1962) The dances of honeybees at small zenith distances of the sun. *J. Exp. Biol.*, **39**, 271–91.

Oliveira, P.S. and Hölldobler, B. (1989) Orientation and communication in the neotropical ant *Odontomachus bauri*. *Ethology*, **83**, 154–66.

Onoyama, K. (1982) Foraging behavior of the harvester ant *Messor aciculatus*, with special reference to foraging sites and dial activity of individual ants. *Japan J. Ecol.*, **32**, 453–61.

Orians, G.H. and Pearson, N.E. (1979) On the theory of central place foraging, in *Analysis of Ecological Systems* (eds D.J. Horn, G.R. Stairs and R.D. Mitchell), Ohio State University Press, Columbus, pp. 155–77.

Papi, F. (1955) Experiments on the sense of time in *Talitrus saltator* (Crustacea: Amphipoda). *Experientia*, **11**, 201–2.

Papi, F. (1959) Sull'orientamento astronomico in specie del gen. *Arctosa* (Araneae, Lycosidae). *Z. Vergl. Physiol.*, **41**, 481–9.

Papi, F., Ioale, P., Fiaschi, V., Benvenuti, S. and Baldaccini, N.E. (1978) Pigeon homing: cues detected during outward journey influence initial orientation, in *Animal Migration, Navigation, and Homing* (eds K. Schmidt-Koenig and W.T. Keeton), Springer, Berlin, pp. 65–77.

Papi, F. and Pardi, L. (1953) Ricerche sull'orientamento di *Talitrus saltator* (Crustacea: Amphipoda). II. Sui fattori che regolano la variazione dell'angolo di orientamento nel corso del giorno. L'orientamento di nottel. L'orientamento diurno di altre popolazioni. *Z. Vergl. Physiol.*, **35**, 490–518.

Papi, F., Pardi, L., Serretti, L. and Parrini, S. (1957) Nuove ricerche sull'orientamento e il senso del tempo di *Arctosa perita* (Araneae: Lycosidae). *Z. Vergl. Physiol.*, **39**, 531–61.

Pardi, L. (1960) Innate components in the solar orientation of littoral amphipods. *Cold Spring Harbor Symp. Quant. Biol.*, **25**, 395–401.

Pardi, L. and Papi, F. (1953) Ricerche sull'orientamento di *Talitrus saltator* (Crustacea, Amphipoda). I. L'orientamento durante il giorno in una popolazione del litorale tirrenico. *Z. Vergl. Physiol.*, **35**, 459–89.

Peters, H.M. (1932) Experimente über die Orientierung der Kreuzspinne *Epeira diademata* im Netz. *Zool. Jb. Allg. Zool. Physiol.*, **51**, 239–88.

Pezzack, D. and Duggan, D.R. (1986) Evidence of migration and homing of lobsters (*Homarus americanus*) on the Scotian shelf. *Can. J. Fish. Aquat. Sci.*, **43**, 2206–11.

Pieron, H. (1904) Du rôle du sens musculaire dans l'orientation de quelques espèces de fourmis. *Bull. Inst. Gen. Psychol.*, **4**, 168–86.

Potegal, M. (1987) The vestibular navigation hypothesis: a progress report, in *Cognitive Processes and Spatial Orientation in Animal and Man* (eds P. Ellen and C. Thinus-Blanc), Martinus Nijhoff, Dordrecht, pp. 28–34.

Pricer, J.L. (1908) The life history of the carpenter ant. *Biol. Bull. Mar. Biol. Lab. Woods Hole*, **14**, 177–218.

Pringle, J.W.S. (1948) The gyroscopic mechanism of the halteres of Diptera. *Phil. Trans R. Soc. Lond. B*, **233**, 347–84.

Rainey, R.C. (1951) Weather and the movement of locust swarms: a new hypothesis. *Nature*, **168**, 1057–60.

Rainey, R.C. (1974) Biometeorology and insect flight: some aspects of energy exchange. *A. Rev. Entomol.*, **19**, 407–39.

Rainey, R.C. (1976) Flight behaviour and features of the atmospheric environment, in *Insect Flight* (ed. R.C. Rainey), *Symp. R. Entomol. Soc. Lond.*, **7**, 75–112.

Rau, P. (1929) Experimental studies in the homing of carpenter and mining bees. *J. Comp. Physiol.*, **9**, 35–70.

Rau, P. (1931) Additional experiments on the homing of carpenter and mining bees. *Comp. Psychol.*, **12**, 257–61.

Rebach, S. (1983) Orientation and migration in Crustacea, in *Studies in Adaptation. The Behavior of Higher Crustacea* (eds S. Rebach and D.W. Dunham), John Wiley, New York, pp. 217–64.

Regnier, F., Nieh, M. and Hölldobler, B. (1973) The volatile Dufour's gland component of the harvester ants *Pogonomyrmex rugosus* and *P. barbatus*. *J. Insect Physiol.*, **19**, 981–92.

Reid, D.G., Wardhaugh, K.G. and Roffey, J. (1979) Radar studies of insect flight at Benalla, Victoria, in February 1974. *Tech. Pap. Div. Ent. CSIRO.*, **16**, 1–21.

Renner, M. (1957) Neue Versuche über den Zeitsinn der Honigbiene. *Z. Vergl. Physiol.*, **40**, 85–118.

Renner, M. (1959) Über ein weiteres Versetzungsexperiment zur Analyse des Zeitsinns und der Sonnenorientierung der Honigbiene. *Z. Vergl. Physiol.*, **42**, 449–83.

Riley, J.R. and Reynolds, D.R. (1979) Radar-based studies of the migratory flight of grasshoppers in the middle Niger area of Mali. *Proc. R. Soc. Lond.*, **204**, 67–82.

Romanes, G.J. (1885) Homing faculty of Hymenoptera. *Nature*, **32**, 630.

Rosengren, R. (1971) Route fidelity, visual memory and recruitment behaviour in foraging wood ants of the genus *Formica* (Hymenoptera, Formicidae). *Acta Zool. Fennica*, **133**, 1–106.

Rosengren, R. (1977) Foraging strategy of wood ants (*Formica rufa* group). I. Age polyethism and topographic traditions. *Acta Zool. Fennica*, **149**, 1–30.

Rosengren, R. and Fortelius, W. (1986) Ortstreue in foraging ants of the *Formica rufa* group – hierarchy of orienting cues and long-term memory, *Insectes Sociaux*, **33**, 306–37.

Rosengren, R. and Pamilo, P. (1978) Effect of winter timber felling on behaviour of foraging wood ants (*Formica rufa* group) in early spring. *Memorabilia Zool.*, **29**, 143–55.

Rosenheim, J.A. (1987) Host location and exploitation by the clepto-parasitic wasp *Argochrysis armilla*: the role of learning (Hymenoptera: Chrysididae). *Behav. Ecol. Sociobiol.*, **21**, 401–6.

Rospars, J.P. (1988) Structure and development of the insect antenno-deutocerebral system. *Int. J. Insect Morphol. Embryol.*, **17**, 243–94.

Rossel, S. (1989) Polarization sensitivity in compound eyes, in *Facets of Vision* (eds D.G. Stavenga and R.C. Hardie), Springer, Berlin, pp. 298–316.

Rossel, S. and Wehner, R. (1986) Polarization vision in bees. *Nature*, **323**, 128–31.

Rovner, J.S. and Knost, S.J. (1974) Post-immobilization wrapping of prey by lycosid spiders of the herbaceous stratum. *Psyche*, **81**, 398–415.

Saila, S.B. and Flowers, J.M. (1968) Movements and behaviour of berried female lobsters displaced from offshore areas to Narragansett Bay, Rhode Island. *J. Cons. Perm. Int. Explor. Mer.*, **31**, 342–51.

Saila, S.B. and Shappy, R.A. (1963) Random movement and orientation in salmon migration. *J. Cons. Int. Explor. Mer.*, **28**, 153–66.

Saint-Paul, U. v. (1982) Do geese use path integration for walking home?, in *Avian Navigation* (eds F. Papi and H.G. Walraff), Springer, Berlin, pp. 298–307.

Sandeman, D.C. and Markl, H. (1980) Head movements in flies (*Calliphora*) produced by deflexion of the halteres. *J. Exp. Biol.*, **85**, 43–60.

Santschi, F. (1911) Observations et remarques critiques sur le mécanisme de l'orientation chez les fourmis. *Rév. Suisse Zool.*, **19**, 305–38.

Santschi, F. (1913) Comment s'orientent les fourmis. *Rév. Suisse Zool.*, **21**, 347–426.

Santschi, F. (1923) L'orientation sidérale des fourmis, et quelques considérations sur leurs différentes possibilités d'orientation. *Mém. Soc. Vaudoise Sci. Nat.*, **4**, 137–75.

Schaefer, G.W. (1976) Radar observations of insect flight, in *Insect Flight* (ed. R.C. Rainey), *Symp. R. Entomol. Soc. Lond.*, **7**, 157–97.

Schneirla, T.C. (1971) *Army Ants: a study in social organization* (ed. H.R. Topoff), Freeman, San Francisco.

Scholze, E., Pichler, H. and Heran, H. (1964) Zur Entfernungsschätzung der Bienen nach dem Kraftaufwand. *Naturwissenschaften*, **51**, 69–70.

Seeley, T.D. (1985) *Honeybee Ecology. A Study of Adaptation in Social Life*, Princeton University Press, Princeton.

Seeley, T.D. (1986) Social foraging by honeybees: how colonies allocate foragers among patches of flowers. *Behav. Ecol. Sociobiol.*, **12**, 343–54.

Seyfarth, E.A. (1980) Daily patterns of locomotor activity in a wandering spider. *Physiol. Entomol.*, **5**, 199–206.

Seyfarth, E.A. and Barth, F.G. (1972) Compound slit sense organs on the spider leg: mechanoreceptors involved in kinesthetic orientation. *J. Comp. Physiol.*, **78**, 176–91.

Seyfarth, E.A., Hergenröder, R., Ebbes, H. and Barth, F.G. (1982) Idiothetic orientation of a wandering spider: compensation of detours and estimates of goal distance. *Behav. Ecol. Sociobiol.*, **11**, 139–48.

Shapiro, A.M. (1980) Evidence for a return migration of *Vanessa cardui* in northern California (Lepidoptera: Nymphalidae). *Pan Pacif. Entomol.*, **56**, 319–22.

Srinivasan, M.V., Lehrer, M., Zhang, S.W. and Horridge, G.A. (1989) How honeybees measure their distance from objects of unknown size. *J. Comp. Physiol. A*, **165**, 605–13.

Steinmann, E. (1973) Über die Nahorientierung der Einsiedlerbienen *Osmia bicornis* und *O. cornuta* (Hymenoptera, Apoidea). *Mitt. Schweiz. Entomol. Ges.*, **46**, 119–22.

Steinmann, E. (1976) Über die Nahorientierung solitärer Hymenopteren: Individuelle Markierung der Nesteingänge. *Mitt. Schweiz. Ent. Ges.*, **49**, 253–8.

Steinmann, E. (1985) Die Wand-Pelzbiene *Anthophora plagiata*. *Jber. Natf. Ges. Graubünden*, **102**, 137–42.

Stephens, D.W. and Krebs, J.R. (1986) *Foraging Theory*, Princeton University Press, Princeton, New Jersey.

Stirling, D. and Hamilton, P.V. (1986) Observations on the mechanism of detecting mucous trail polarity in the snail *Littorina irrorata*. *Veliger*, **29**, 31–7.

Street, R.J. (1971) Rock lobster migration off Otago. *N.Z. Comm. Fish*, **71**, 16–17.

Sudd, J.H. (1957) Communication and recruitment in Pharaoh's ant, *Monomorium pharaonis*. *Anim. Behav.*, **5**, 104–9.

Sudd, J.H. (1959) Interaction between ants on a scent trail. *Nature*, **183**, 1588.

Suter, R.B. (1984) Web tension and gravity as cues in spider orientation. *Behav. Ecol. Sociobiol.*, **16**, 31–6.

Tengö, J., Schöne, H. and Chmurzynski, J. (1990) Homing in the digger wasp *Bembix rostrata* (Hymenoptera, Sphecidae) in relation to sex and stage. *Ethology*, **86**, 47–56.

Tilles, D.A. and Wood, D.L. (1986) Foraging behavior of the carpenter ant, *Camponotus modoc* (Hymenoptera: Formicidae), in a giant sequoia forest. *Can. Ent.*, **118**, 861–7.

Tinbergen, N. (1932) Über die Orientierung des Bienenwolfes (*Philanthus triangulum*). *Z. Vergl. Physiol.*, **16**, 305–34.

Traniello, J.F.A. (1977) Recruitment behavior, orientation, and the organization of foraging in the carpenter ant *Camponotus pennsylvanicus* (Hymenoptera: Formicidae). *Behav. Ecol. Sociobiol.*, **2**, 61–79.

Traniello, J.F.A. (1980) Colony specification in the trail pheromone of an ant. *Naturwissenschaften*, **67**, 361–2.

Ubukata, H. (1975) Life history and behavior of a corduliid dragonfly, *Cordulia aenea amurensis*. II. Reproductive period with special reference to territoriality. *J. Fac. Sci. Hokkaido Univ., Ser. VI. Zool.*, **19**, 812–33.

Ugolini, A. (1985) Initial orientation and homing in workers of *Polistes gallicus* (Hymenoptera, Vespidae). *Z. Tierpsychol.*, **69**, 133–40.

Ugolini, A. (1986a) Homing ability in *Polistes gallicus* (Hymenoptera, Vespidae). *Monitore Zool. Ital.*, **20**, 1–15.

Ugolini, A. (1986b) Homing in female *Polistes gallicus* (Hymenoptera, Vespidae), in *Orientation in Space* (ed. G. Beugnon), Privat, IEC, Toulouse, pp. 57–62.

Ugolini, A. (1987) Visual information acquired during displacement and initial orientation in *Polistes gallicus*. *Anim. Behav.*, **35**, 590–5.

Urquhart, F.A. (1960) *The Monarch Butterfly*, University of Toronto Press, Toronto.

Urquhart, F.A. (1976) Found at last: the monarch's winter home. *Nat. Geogr.*, **150**, 160–73.

Urquhart, F.A. and Urquhart, N.R. (1979) Breeding areas and overnight roosting locations in the northern range of the monarch butterfly (*Danaus plexippus plexippus*) with a summary of associated migratory routes. *Can. Field-Nat.*, **93**, 41–7.

Varju, D. and Sandeman, D. (1989) Tactile learning in a new habitat and spatial memory in the crayfish *Cherax destructor*. *Proc. Neurobiol. Conf. Göttingen*, **17**, 21.

Visscher, P.K. and Seeley, T.D. (1982) Foraging strategy of honeybee colonies in a temperate deciduous forest. *Ecology*, **63**, 1790–801.

Vollbehr, J. (1975) Zur Orientierung junger Honigbienen bei ihrem ersten Orientierungsflug. *Zool. Jb. allg. Zool. Physiol.*, **79**, 33–69.

Vollrath, F. (1988) Spiral orientation of *Araneus diadematus* orb webs built during vertical rotation. *J. Comp. Physiol. A*, **162**, 413–19.

Vollrath, F. (1989) Spider orientation, in *Orientation and Navigation – Birds, Humans and other Animals* (ed. M. Bygorne), Royal Institute of Navigation, London, pp. 1–5.

Wallraff, H.G. (1974) *Das Navigationssystem der Vögel. Ein theoretischer Beitrag zur Analyse ungeklärter Orientierungsleistungen*. R. Oldenbourg, München.

Wehner, R. (1972) Dorsoventral asymmetry in the visual field of the bee, *Apis mellifica*. *J. Comp. Physiol.*, **77**, 256–77.

Wehner, R. (1979) Mustererkennung bei Insekten: Lokalisation und Identifikation visueller Objekte. *Verh. Dtsch. Zool. Ges.*, **72**, 19–41.

Wehner, R. (1981a) Spatial vision in arthropods, in *Handbook of Sensory Physiology*, Vol. VII/6c (ed. H. Autrum), Springer, Berlin, pp. 287–616.

Wehner, R. (1981b) Astronomischer Kompass bei duftspurlegenden Ameisen (*Messor*). *Jb. Akad. Wiss. Lit. Mainz*, **81**, 112–16.

Wehner, R. (1982) Himmelsnavigation bei Insekten. Neurophysiologie und Verhalten. *Neujahrsbl. Naturforsch. Ges. Zürich*, **184**, 1–132.

Wehner, R. (1987a) Spatial organization of foraging behaviour in individually searching desert ants, *Cataglyphis* (Sahara Desert) and *Ocymyrmex* (Namib Desert), in *From Individual to Collective Behavior in Social Insects* (eds J.M. Pasteels and J.-L. Deneubourg), Birkhäuser, Basel, pp. 15–42.

Wehner, R. (1987b) 'Matched filters' – neural models of the external world. *J. Comp. Physiol. A*, **161**, 511–31.

Wehner, R. (1989a) Neurobiology of polarization vision. *TINS*, **12**, 353–9.

Wehner, R. (1989b) The hymenopteran skylight compass: matched filtering and parallel coding. *J. Exp. Biol.*, **146**, 63–85.

Wehner, R. (1990a) On the brink of introducing sensory ecology: Felix Santschi (1872–1940) – Tabib-en-Neml. *Behav. Ecol. Sociobiol.*, **27**, 295–306.

Wehner, R. (1990b) Do small-brain navigators use cognitive maps? *Proc. Neurobiol. Conf. Göttingen*, **18**, 30.

Wehner, R. (1992) Visuelle Navigation: Kleinstgehirtrategien. *Verh. Dtsch. Zool. Ges.*, **84**, 89–104.

Wehner, R., Bleuler, S., Nievergelt, C. and Shah, D. (1990) Bees navigate by using vectors and routes rather than maps. *Naturwissenschaften*, **77**, 479–82.

Wehner, R. and Flatt, I. (1972) The visual orientation of desert ants, *Cataglyphis bicolor*, by means of terrestrial cues, in *Information Processing in the Visual System of Arthropods* (ed. R. Wehner), Springer, Berlin, pp. 295–302.

Wehner, R., Fukushi, T. and Wehner, S. (1992) Rotatory components of movement in high speed dessert ants, *Cataglyphis bombycina*. *Proc. Neurobiol. Conf. Göttingen*, **20**, 303.

Wehner, R., Harkness, R.D. and Schmid-Hempel, P. (1983) Foraging strategies in individually searching ants, *Cataglyphis bicolor* (Hymenoptera: Formicidae). *Akad. Wiss. Lit. Mainz, Math.-Naturwiss. Kl.*, Fischer, Stuttgart.

Wehner, R. and Lanfranconi, B. (1981) What do the ants know about the rotation of the sky? *Nature*, **293**, 731–3.

Wehner, R. and Menzel, R. (1990) Do insects have cognitive maps? *A. Rev. Neurosci.*, **13**, 403–14.

Wehner, R. and Räber, F. (1979) Visual spatial memory in desert ants, *Cataglyphis bicolor* (Hymenoptera, Formicidae). *Experientia*, **35**, 1569–71.

Wehner, R. and Rossel, S. (1985) The bee's celestial compass – a case study in behavioural neurobiology. *Fortschr. Zool.*, **31**, 11–53.

Wehner, R. and Srinivasan, M.V. (1981) Searching behaviour of desert ants, genus *Cataglyphis* (Formicidae, Hymenoptera). *J. Comp. Physiol. A*, **142**, 315–38.

Wehner, R. and Srinivasan, M.V. (1984) The world as the insect sees it, in *Insect Communication* (ed. T. Lewis), Academic Press, London, pp. 29–47.

Wehner, R. and Wehner, S. (1986) Path integration in desert ants. Approaching a long-standing puzzle in insect navigation. *Monitore Zool. Ital.*, **20**, 309–31.

Wehner, R. and Wehner, S. (1990) Insect navigation: use of maps or Ariadne's thread? *Ethol. Ecol. Evol.*, **2**, 27–48.

Wigglesworth, V.B. (1987) Is the honey-bee conscious? *Antenna, Lond.*, **11**, 130.

Wilkinson, D.H. (1952) The random element in bird 'navigation'. *J. Exp. Biol.*, **29**, 532–60.

Williams, C.B. (1958) *Insect Migration*, Collins, London.

Williams, T.C. and Williams, J.M. (1990) The orientation of transoceanic migrants, in *Bird Migration. Physiology and Ecophysiology* (ed. E. Gwinner), Springer, Berlin, pp. 7–21.

Wilson, E.O. (1962a) Chemical communication among workers of the fire ant, *Solenopsis saevissima*. I. The organization of mass-foraging. *Anim. Behav.*, **10**, 134–47.

Wilson, E.O. (1962b) Chemical communication among workers of the fire ant, *Solenopsis saevissima*. II. An information analysis of the odour trail. *Anim. Behav.*, **10**, 148–58.

Wilson, E.O. (1980) Caste and division of labor in leafcutter ants (Hymenoptera: Formicidae: *Atta*). I. The overall pattern in *A. sexdens*. *Behav. Ecol. Sociobiol.*, **7**, 143–56.

Wiltschko, R., Wiltschko, W. and Keeton, W.T. (1978) Effect of outward journey in an altered magnetic field on the orientation of young homing pigeons, in *Animal Migration, Navigation, and Homing* (eds K. Schmidt-Koenig and W.T. Keeton), Springer, Berlin, pp. 152–62.

Young, A.M. and Hermann, H.R. (1980) Notes on foraging of the giant tropical ant, *Paraponera clavata* (Hymenoptera: Formicidae: Ponerinae). *J. Kansas Entomol. Soc.*, **53**, 35–55.

Zeil, J., Nalbach, G. and Nalbach, H.O. (1986) Eyes, eye stalks and the visual world of semi-terrestrial crabs. *J. Comp. Physiol. A*, **159**, 801–11.

Zeil, J. and Wittmann, D. (1989) Visually controlled station-keeping by hovering guard bees of *Trigona (Tetragonisca) angustula* (Apidae, Meliponinae). *J. Comp. Physiol. A*, **165**, 711–18.

Zollikofer, C. (1988) Vergleichende Untersuchungen zum Laufverhalten von Ameisen (Hymenoptera: Formicidae). PhD Thesis, University of Zurich.

Chapter 4
Fishes

Thomas P. Quinn and Andrew H. Dittman

4.1 INTRODUCTION

Fishes are the most speciose group of vertebrates and display a extraordinary variety of life history patterns. Many fish species are migratory, particularly those larger than about 60 cm (Roff, 1988). Fishes may migrate from marine feeding areas to spawn in freshwater (anadromy; e.g. shad, salmon, striped bass, lamprey). Others, such as eels, rear in freshwater but return to the sea to spawn and are termed catadromous. There are also species, referred to as amphidromous, which migrate between fresh- and saltwater for purposes other than reproduction. Based on present information, McDowall (1988) listed 110 anadromous species, 56 catadromous species and 61 amphidromous species.

While diadromous species (those migrating between freshwater and marine environments) such as salmon and eels are famous, the majority of migratory fishes confine their movements to fresh- or salt-water. The migrations are generally seasonal, from feeding grounds to spawning grounds and back (Harden Jones, 1968; Quinn and Brodeur, 1991). Drift by larval forms is a common component of migration patterns (Zijlstra, 1988). Seasonal migrations may cover considerable distances at sea (Harden Jones, 1968) and in freshwater (Lowe-McConnell, 1975). In addition to seasonal migrants, some species undertake daily migrations from feeding areas to shelter sites (Ogden and Ehrlich, 1977; Bray, 1980) or to breeding areas (Myrberg *et al.*, 1988). In coastal waters, many species display daily onshore–offshore movement patterns (Sciarrotta and Nelson, 1977) and many mesopelagic fishes (those inhabiting the open oceans' midwater zone) make extensive daily vertical migrations (Clarke and Wagner, 1976).

Animal Homing. Edited by Floriano Papi. Published in 1992 by Chapman & Hall, 2–6 Boundary Row, London SEI 8HN. ISBN 0 412 36390 9.

Fisheries scientists have developed a wide variety of tools for studying fish migrations, including many forms of tagging, telemetry, analysis of catch per unit of effort, and population identification by means of scale pattern analysis, parasite fauna, meristic and morphometric characteristics, and biochemical genetic markers. The movement patterns of many fishes have been described in considerable detail. However, in spite of their abundance and the importance of migration in their capture and conservation, most such fishes have been subjected to little experimental study. Consequently, the mechanisms used by fishes to accomplish their diverse migrations are not well understood.

Laboratory studies have documented the thresholds of various 'conventional' sensory systems such as vision and olfaction and have demonstrated the capacity of at least certain fishes to detect more unusual stimuli such as electric (Kalmijn, 1978) and magnetic fields (Quinn, 1980; Walker, 1984), polarized light patterns (Hawryshyn et al., 1990), and ultraviolet light (Hawryshyn and Beauchamp, 1985). In spite of these studies, our knowledge of the mechanisms of fish migration is limited by two factors. First, there are too few studies which have actually teased apart the roles of the myriad stimuli which are present in field situations. Second, there is a disproportionate amount of information on certain fishes, particularly the family Salmonidae, and our inferences about fish migration rely too heavily on experiments with these notable but perhaps not representative species.

This chapter reviews the mechanisms of homing in fishes. Homing is a component of migration in many fishes but apparently non-migratory species will often return if displaced from their home range. In this chapter, we accept McCleave's (1967) classification of homing into three categories, with some modification. He distinguished homing after artificial displacement, homing to a site of previous spawning, and homing to a natal site. To these we add homing to a familiar foraging or refuge site, as there is no reason to define homing only in the context of reproduction. This classification scheme, including both natural and experimental homing, follows that outlined by Papi (Chapter 1) but excludes such migration patterns as larval drift and retention (Miller et al., 1985).

Our approach will be to (1) discuss homing in the context of fish life history patterns, (2) examine the mechanisms underlying homing in selected, well-studied species representing the different types of homing, and (3) provide a more detailed review of migration and homing in Pacific salmon (Oncorhynchus spp.). This genus is selected for special consideration owing to the quality and quantity of information on their homing.

4.2 THE ROLE OF HOMING IN THE LIFE HISTORIES OF FISHES

Little more than descriptive information is available about the patterns of migration and homing in most species and even that information is sometimes rather sketchy. Direct observation, tagging or telemetry may enable us to determine the locations of individuals quite precisely in small streams or coral reefs. However, many marine fishes are difficult to capture, tag and release or inhabit areas not suitable for observations. The larval forms of pelagic fishes are too small and numerous to be tagged in any case. Nevertheless, it is important to bear in mind the variety of migration patterns and habitats of fishes before focusing on the relatively small number of species which have been studied in detail. This section attempts to synthesize information on diverse groups of fishes into three general categories: feeding migrations, reproductive migrations and movements within the home range.

4.2.1 Feeding migrations: daily

A number of fishes undergo daily migrations between feeding and refuge areas. In some cases there is no information on the extent of individual movement, only on the changes in distribution of the group. For example, many benthic fishes move shoreward and towards the surface at night and offshore and deeper in the day. These movements are evidenced by changes in catch per unit of effort of fishing gear.

One such case of diel movements is the ratfish (*Hydrolagus colliei*) in Puget Sound, Washington (Quinn, 1980). There is an increase in catch of ratfish in shallow water at night. This migration is comprised principally of small ratfish, as the larger individuals seem to stay in shallow water by day to a greater extent than small ones (Fig. 4.1). Many other nearshore species, freshwater as well as marine, also display daily onshore–offshore migrations which are apparently not related to reproduction (Helfman, 1981). Most of these temperate water species reproduce on a seasonal basis and the diel movements are probably between shallow feeding areas and deep refuges from predation.

It might stretch the term to regard such movements as homing. There is no evidence that individuals have specific feeding or refuge areas. The onshore–offshore movements are probably timed by a combination of endogenous circadian rhythm and changing light levels. The movements need no orientation mechanism besides a vertical (pressure) response and a tendency to stay near the bottom.

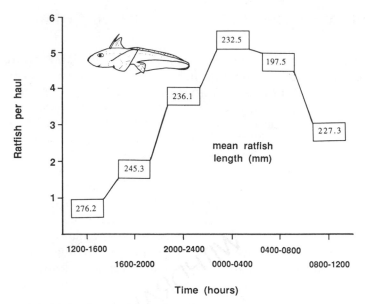

Figure 4.1 Diel changes in abundance and average size of ratfish (*Hydrolagus colliei*) in shallow water (5–55 m), as indicated by catch per 5 min haul with a small otter trawl. Data averaged over six 24-h surveys at West Point, Puget Sound, Washington (from Quinn *et al.*, 1980).

A number of fish species make precise movements between feeding and refuge areas. Two such cases, the tropical grunts (Haemulidae) and scalloped hammerhead shark (*Sphyrna lewini*), will be discussed in detail in section 4.3. However, there are many less well-studied species with apparently similar patterns. Many coral reef fishes take refuge in the structure of coral at specific sites and then move to broadly defined feeding areas each day or night (Hobson, 1973). For example, individual parrotfish (*Scarus* spp.) migrate along fixed routes from their daily feeding territories to night-time shelter sites (Ogden and Buckman, 1973; Dubin and Baker, 1982). Other fishes such as the atherinid (*Pranesus pinguis*) and copper sweeper (*Pempheris schomburgki*) form relatively inactive schools in the shelter of reefs during the day. At very specific light levels, these fishes migrate to feeding areas over 1 km away, to return the next morning (Hobson and Chess, 1973; Gladfelter, 1979). These movements allow the fish to take advantage of the shelter from predation provided by the structurally complex coral and forage in more productive areas. The ecological significance of these migrations is underlined by the findings of Robblee and

Zieman (1984) that seagrass (*Thalassia*) beds have essentially separate nocturnal and diurnal fish communities.

In the case of these fishes, their return to specific sites constitutes homing but little is known about the mechanisms. Reese (1989) has observed foraging movements by butterfly fish (Chaetodontidae) within their home ranges along predictable paths. Such movements seem to be guided by learned landmarks, because 'when a coral head is removed the fish look for it in its former location.' This supports Hobson's (1973) view that visual landmarks play an important role in the daily migrations of reef fishes. However, observation and experimental work with parrotfish (*Scarus guacamaia* and *S. coelestinus*) indicated that their migration is guided by a sun compass (Winn *et al.*, 1964).

While most studies of diel feeding movements have been conducted on small fishes moving in structured environments, evidence indicates that some large, pelagic species may also home from open water to shallow banks. Ultrasonic telemetry of skipjack tuna (*Katsuwonus pelamis*) revealed fish staying on Kaula Bank, Hawaii, during the day and making nightly journeys of 25–106 km each night into open water (Yuen, 1970). More recently, Holland *et al.* (1990a) reported that yellowfin (*Thunnus albacares*) and bigeye (*T. obesus*) tuna left floating artificial structures as well as coastal areas in the evening and returned to them from open water in the morning (Fig. 4.2). The tuna generally swam closer to the surface at night but routinely made dives and ascents of over 100 m. The horizontal movements were rapid, direct, and took place in open water where the fish would be unable to see the shore or bottom. The fish showed evidence of both the ability to move in a homeward direction and to swim in straight paths for hours at a time. While no conclusions can be drawn about the mechanisms guiding these nocturnal movements, conditioning experiments have demonstrated that yellowfin tuna can detect magnetic fields (Walker, 1984). Apparently oriented pelagic movements are not restricted to tunas, as swordfish (*Xiphias gladias*), blue sharks (*Prionace glauca*) and Pacific blue marlin (*Makaira nigricans*) also are known to swim for hours or days in straight lines in open water (Carey and Robison, 1981; Carey and Scharold, 1990; Holland *et al.*, 1990b).

4.2.2 Feeding migrations: seasonal

While well-known fishes such as salmon migrate from broadly defined, pelagic feeding areas to specific spawning sites, there are marine fishes which apparently home to specific feeding areas from

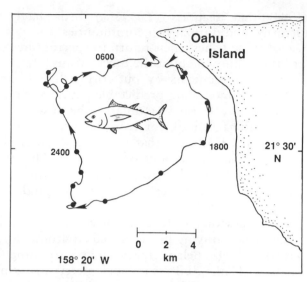

Figure 4.2 Movements of a yellowfin tuna (*Thunnus albacares*) tracked by ultrasonic telemetry off Oahu, Hawaii. Dots represent successive hours of the track (fish no. 8303 from Holland *et al.*, 1990a).

their breeding grounds. Extensive tagging studies have been conducted on Atlantic cod (*Gadus morhua*) in the waters off Newfoundland, Canada (Templeman, 1974). In 1954–55, 18,822 cod were tagged, mostly at nearshore feeding sites in summer (e.g. Bonavista; Fig. 4.3). There were 4151 cod recaptured at least 1 year after tagging. Although nine individuals moved very long distances, 50% were recaptured within 74 km of the tagging site and 21% were 74–148 km from the tagging site. Interestingly, the cod not only returned to the site of previous capture in the next year but did so throughout their lives (Table 4.1). Considering the broad distribution of cod, their annual migration to offshore overwintering and spawning areas, and the large area over which the fishery operates, these data imply a strong tendency to home to nearshore feeding areas. There is no indication of the specific mechanisms which guide these migrations. The most obvious ecological basis is the migration of a key prey species, capelin (*Mallotus villosus*), to spawn on specific beaches in late spring (Jangaard, 1974).

Some fishes, especially in northern areas, do not spawn annually and migrate from feeding areas to overwintering areas. One such example is the anadromous Dolly Varden (*Salvelinus malma*: Armstrong, 1984). Dolly Varden spawn in streams associated with

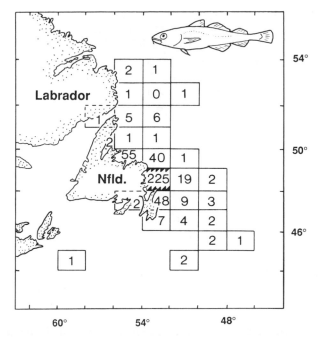

Figure 4.3 Homing of Atlantic cod (*Gadus morhua*) to feeding grounds. Cod were tagged at Bonavista, Newfoundland (area hash-marked) in 1954 and recovered in the surrounding areas from 1955 to 1966 (from Templeman, 1974).

lakes or in systems without lakes. In the spring there is a seaward migration of juveniles going to sea for the first time and of adults. In the autumn, all will return to freshwater and overwinter in lakes as cold temperatures and low water levels make many streams unsuitable in winter. Dolly Varden spawned in streams associated with lakes home to those systems to overwinter and, for some, to spawn as well. In the autumn, those originating in streams not associated with lakes have more complex patterns. Those which will not spawn that autumn enter streams until they find one with a lake and they overwinter there. Spawners return to their natal stream to spawn but then leave and overwinter in lakes elsewhere.

4.2.3 Spawning migrations: daily

The best-known migration patterns are those from feeding areas to spawning grounds. While the most famous are the long-distance annual migrations, there are also daily spawning migrations. Daily

Table 4.1 Relationship between years at large and net distance between tagging and recapture locations for Atlantic cod off Newfoundland. Numbers shown are percentages of fish caught at least 1 year after release, excluding nine very distant recoveries. A total of 18 822 cod were tagged at 13 locations (see Fig. 4.3 for an example of the detailed results of one such location; from Templeman, 1974)

Distance (km)	Number of years at large		
	1	2–3	4–17
0–74	55.2	46.7	43.8
76–148	21.4	19.6	20.6
150–296	14.6	18.7	20.2
298–444	6.3	9.2	6.9
446–926	2.5	5.7	8.5
Sample size	1848	1659	635

spawning is principally a phenomenon of tropical species, though by no means all tropical fishes spawn daily. Some coral reef fishes move from the substratum to spawn in the water column above their territories. Myrberg *et al.* (1988) recently documented a striking daily migration by surgeonfish (chiefly *Acanthurus nigrofuscus*) in the Red Sea (Fig. 4.4). These fishes migrate up to 2 km each evening, travelling rapidly in single file with 25–60 cm between individuals. Such trains of fish move along the shore to specific sites where they mill, then move offshore and spawn. After spawning, the trains reverse and the fish return to nocturnal shelter sites. The results of tagging experiments suggested but were not adequate to demonstrate fidelity for the two spawning sites.

4.2.4 Spawning migrations: seasonal

As was stated earlier, seasonal spawning migrations are typical of many temperate and boreal marine species. These migrations may bring demersal or pelagic fishes from broadly defined feeding areas to very discrete spawning sites. Homing probably occurs in many marine fishes but tagging studies often are not sufficiently fine-scaled to demonstrate it. Moreover, the patterns of tagging and recapture effort are seldom uniform. Tagging studies often reveal as much about the patterns of fishing as they do about fish migrations. Nevertheless, there is evidence of homing to the site of previous

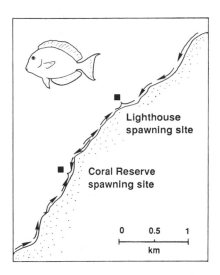

Figure 4.4 Map showing the daily migration of surgeonfish (*Acanthurus nigrofuscus*) from feeding areas to spawning sites off Eilat, Israel (from Myrberg *et al.*, 1988). Arrows indicate direction of surgeonfish movement towards the spawning sites, marked with squares.

spawning by a number of species (e.g. sea bass, *Dicentrarchus labrax*: Pawson *et al.*, 1987; Atlantic halibut, *Hippoglossus hippoglossus*: Godo and Haug, 1988). For some species, differences in meristic or morphometric characters among populations imply homing (e.g. capelin: Misra and Carscadden, 1987).

Pacific herring (*Clupea harengus pallasi*) spawn on macrophytes on specific beaches. Years of tagging have indicated that most individuals (66–96%) return to the area ('management unit') where they spawned previously (Hourston, 1982). A similar homing tendency is evident in Atlantic herring (*C. h. harengus*; Wheeler and Winters, 1984). Larval herring are too small to tag and there is no direct evidence that adults home specifically to their natal sites. However, the regular patterns of spawning time, depth and substratum at specific sites imply that a certain degree of homing has given rise to local adaptations (Iles and Sinclair, 1982). Iles and Sinclair (1982) hypothesized that the number of genetically distinct herring stocks in the Atlantic is determined by the number of geographically stable larval retention areas.

Spawning migrations of fishes to intertidal beaches are often closely synchronized to the lunar and tidal cycles. The California grunion (*Leuresthes tenuis*) spawns in the upper intertidal zone on spring high tides. Maturation is also synchronized to a lunar rhythm (Walker, 1952; Gibson, 1978). Many other fishes including surf smelt (*Hypomesus pretiosus*), four-eyed fish (*Anableps microlepis*) and capelin also display rhythmic migrations and spawning (Gibson, 1978).

Larvae retained in certain nearshore areas by current patterns may have the opportunity to learn some characteristic(s) of the areas. They home to these areas when they first mature and select a spawning site based on the presence of conspecifics or a set of substratum preferences. They may then learn some characteristic(s) of the specific spawning site but the spatial scales of the herring tag–recapture studies by Hourston (1982) and Wheeler and Winters (1984) do not permit determination of homing to specific sites.

It is not clear what mechanisms might guide herring homing and there have been no experimental studies to provide insights. In some cases herring might have the opportunity to learn the odours of nearby rivers, or beaches themselves may have distinct odours. However, homing must be guided at least in part by mechanisms not directly related to the site, as the herring may travel considerable distances from the spawning sites to feed. If the herring imprint on their natal area, information must be acquired at hatching or during the period of retention by current patterns early in life. The eggs are demersal and adhesive but there is no parental care and larvae drift from the spawning site after hatching.

Seasonal spawning migrations are much easier to study in freshwater or anadromous fishes. Shad (*Alosa sapidissima*) and salmon will be discussed in detail in sections 4.3 and 4.4, respectively. However, there are many other fishes with documented migratory patterns. Lowe-McConnell (1975) reviewed the evidence for homing to spawning sites by *Tilapia variabilis* in Lake Victoria. There are also many species of tropical riverine fishes which migrate long distances, typically from upriver spawning sites to downriver feeding areas (e.g. *Prochilodus platensis*, *P. scrofa*, *Salminus maxillosus* and some large catfish: Lowe-McConnell, 1975; Northcote, 1984; Petrere Junior, 1985).

In higher latitudes, homing to reproductive sites has been indicated for riverine fishes such as the Colorado River squawfish (*Ptychocheilus lucius*: Tyus, 1985) and inconnu (*Stenodus leucichthys*: Alt, 1977). The squawfish migration is particularly interesting from the viewpoint of homing mechanisms because fish converge on the spawning grounds from both downstream and upstream locations, travelling at least 160 km. For those moving downstream, at least the initial movements cannot be guided by odours or other information emanating from the spawning site.

Homing has also been reported for fishes migrating within lakes (e.g. muskellunge, *Esox masquinongy*: Crossman, 1990) and from lacustrine feeding areas to spawning streams (e.g. roach, *Rutilus rutilus*: l'Abee-Lund and Vollestad, 1987; white suckers, *Catostomus commersoni*: Olson and Scidmore, 1963). In the case of white suckers, olfaction seems to be the basis for homing, as blinded suckers homed but anosmic ones did not (Werner, 1979).

Olson *et al.* (1978) reviewed the results of many tagging studies with walleye (*Stizostedion vitreum*). These fish generally spawn in river inlets to lakes or in shoals in the lakes themselves. The newly hatched larval walleye drift for several days with bottom currents, then are transported by wind-driven currents in the upper water column for at least another week. Immature fish seem to disperse widely and regular movement patterns were not discerned but adults tagged on spawning grounds tended to be recovered there in subsequent years. It therefore appears that while adults home to the site of previous spawning, they do not have the opportunity to learn the features of their natal site and do not necessarily return to it. Presumably, first-time spawners are attracted to the presence of conspecifics or have innate attraction to the physical characteristics of spawning areas. Walleye also seem to have home feeding areas to which they return after spawning, though these tend to be larger and more loosely defined than the spawning grounds.

Fishes generally home to well-defined sites and tangible habitats. However, many species spawn in pelagic environments defined only by oceanographic features. The specific geographic location of these features may vary from year to year with variations in oceanic temperature and current patterns. For example, McCleave *et al.* (1987) reviewed the evidence that American and European eels (*Anguilla rostrata* and *A. anguilla*) converge from their respective freshwater rearing locations to overlapping (but not coincident) spawning areas in the region of the Sargasso Sea.

The life cycle of eels is an intriguing example of migration from several perspectives. First, how do maturing eels locate (i.e. home to) the Sargasso Sea? Tests in experimental arenas indicate compass orientation capabilities but have not conclusively demonstrated the mechanism(s) (Miles, 1968; Tesch, 1974; Karlsson, 1985). Ultrasonic telemetry (e.g. Tesch, 1989) results indicate that adult eels may swim as deep as 700 m but eels have been tracked at sea for only relatively brief periods. No adult American or European eels have been caught during their migration in the open ocean; the location of the spawning grounds has been inferred from the capture of larval eels and the prevailing current patterns. Kleckner and McCleave (1988) hypothesized that eel spawning grounds are defined by an oceanic front.

Larval eels are probably passively transported north by the Gulf Stream but some active mechanism may be involved in their departure from this current into coastal waters. In coastal waters, eels apparently use olfactory information (Creutzberg, 1959) in conjunction with selective tidal stream transport (McCleave and Wippelhauser, 1987) to ascend rivers. This transport mechanism requires ascent into the water column on flooding (landward) tides and descent to the relatively slack water near the bottom to prevent advection out to sea on ebbing tides. The ascent of rivers by eels is not a homing phenomenon, in contrast to the upriver migrations of anadromous fishes. North American eels seem to be one population that does not mate assortatively with respect to origin. Thus larval eels entering rivers throughout the range do not represent locally adapted populations but rather are derived from a common gene pool (Helfman *et al.*, 1987).

In summary, fishes inhabiting a variety of habitats make regular movements between feeding and breeding areas. There is a distinct tendency for adults to return to the site of previous spawning rather than other sites occupied by conspecifics. In some cases homing to feeding areas is also indicated. For most of these fishes, little if anything is known about the mechanisms underlying their homing. We hypothesize that fishes tend to home to the area as precisely as they can. Many species have eggs and/or larval forms which drift from the spawning substratum or are spawned in the water column. These fishes will probably not return to the exact site where they were spawned, though they may return to the area in which they were retained as larvae (e.g. herring). Species in which the young are associated with the spawning area for a longer period of time may be expected to home to it more precisely than species with larval dispersal. Homing to the more-or-less natal site permits the evolution of population-specific specializations in spawning time, temperature tolerance, fecundity and so forth and these have been documented for many fishes.

4.2.5 Homing movements within the home range

Many fishes apparently do not migrate or at least confine their movements for significant periods of time to actively defended territories or commonly-used home ranges. Establishment of a stable home range often follows a period or larval dispersal. One might assume that species with limited movements would either be unable or disinclined to return to their home range if displaced far beyond it. However, experimental displacements of such fishes generally reveal a strong tendency to return home and an ability to do so if the distance is not too great.

While migrants are generally large (Roff, 1988), these non-migrants may be either large or small. Moreover, non-migrants are generally substratum-oriented and many lack swim bladders or are weak swimmers. Thus the distances over which they home may not seem impressive in comparison to upriver migrations of hundreds or thousands of kilometres by squawfish, inconnu, salmon or South American characoids. However, in terms of body length or expended energy, they may be quite substantial. This section provides a brief review of some representative displacement experiments to give a perspective on these kinds of homing movements. Studies of intertidal fishes and centrarchids (freshwater basses) will be discussed in detail in section 4.3.

Most displacement studies involve either electronic tracking (radio or ultrasonic) or conventional tagging and are often undermined by one or more problems. First, when ultrasonic or radiotelemetry is employed to follow individual fish, the sample size is generally limited, owing to the cost of the transmitters and the labour needed to follow the fish. Second, while telemetry can reveal the behaviour of the experimental subject after displacement (i.e. whether or not it homed and, if so, by what route), it is not always clear what the fish's familiar area was prior to displacement. In the case of fish in a small lake, exploratory movements outside the home range or seasonal changes in distribution may give the fish familiarity with the whole lake. Homing in such cases may merely require piloting within an extended familiar area. Third, continuous tracks are usually limited to a few days, owing to fatigue of the personnel. In small bodies of water where the fish can be relocated, transmitter life generally limits the length of the study to a matter of weeks or months.

Mark–recapture studies using conventional tags are not only weakened by lack of information on the familiar area of the fish but are also biased by geographically non-uniform recapture effort. When tags are not recovered, it is not possible to determine whether the fish died, homed and was not recaptured or failed to home and was not recaptured.

In spite of the ambiguities in interpreting homing by displaced fish, some interesting findings have been reported. Several studies have shown that fish displaced from one part of a lake to another tend to return to the site of capture (e.g. carp, *Cyprinus carpio*: Schwartz, 1987; flathead catfish, *Pylodictis olivaris*: Hart and Summerfelt, 1973; Fig. 4.5). Similar evidence has been reported for English sole (*Parophrys vetulus*) in Puget Sound, Washington

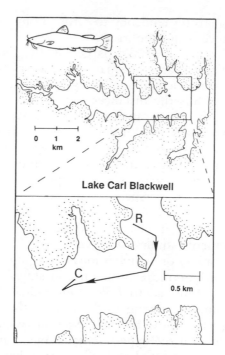

Figure 4.5 Upper panel: map of Lake Carl Blackwell, Oklahoma, where flathead catfish (*Pylodictis olivaris*) homing studies were carried out. Lower panel: track of a flathead catfish that returned to its capture site 'C' in 1.7 h after displacement to release site 'R' (fish no. 226 from Hart and Summerfelt, 1973).

(Day, 1976). Thompson (1983) displaced blennies (*Forsterygion varium*) from their territories on subtidal rocky reefs around New Zealand. These small (7–10 cm) fish are bottom-dwelling and maintain 1.5–2.0 m² territories year-round. The fish had been monitored for at least 6 months prior to displacement. In light of their size, limited swimming capacity and small familiar areas, homing by 8 of 10 male blennies displaced 700 m in 4–6 days was an impressive feat. Upon release, the blennies first sought shelter for up to an hour but then invariably moved in the homeward direction in a series of short darts from one refuge to another. One blenny took only 30 min to return to its territory from 100 m away. No information was available on the mechanisms of homing or its significance for these small fishes.

While a number of studies have displaced fishes from or within their home ranges, Green and Fisher (1977) carried these

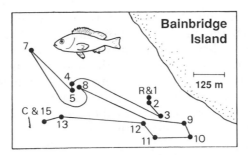

Figure 4.6 Movements of a quillback rockfish (*Sebastes maliger*) homing after displaced from its home reef off Bainbridge Island, Puget Sound, Washington. Numbers represent locations of the fish on days following release; lines connecting them do not imply actual paths between locations. 'C' and 'R' represent the capture and release sites, respectively (fish no. 6329 from Matthews, 1990b).

experiments further in their studies with the radiated shanny (*Ulvaria subbifurcata*). After determining that the home range of the adult shanny is less than 3 m² and that they would home after displacements of up to 270 m, they tested the directional preferences of displaced shanny underwater in radially symmetrical arenas. Their tests showed very strong homeward orientation and further experiments indicated that a combination of vision and olfaction seemed to be involved (though no other sensory systems were examined; Goff and Green, 1978).

Research on rockfishes (*Sebastes* spp.) provides some insights into the significance of homing by non-migratory fishes. Tagging studies have revealed that several species associated with rocky reefs tend to remain at the site of capture (yellowtail, *S. flavidus*: Carlson and Haight, 1972; copper, *S. caurinus*; quillback, *S. maliger*; brown, *S. auriculatus*: Matthews, 1990a). When displaced as far as 22.5 km (Carlson and Haight, 1972), rockfishes may return to their former habitat, even to the very crevice of rock in many cases (Matthews, 1990a). Ultrasonic tracking by Matthews (1990b) indicated that displaced rockfishes initially moved back and forth over the new site for several days, then returned quite directly to their former site (Fig. 4.6). However, it is also well known that artificial reefs are quickly colonized by adult fishes of many species, including rockfishes (Buckley and Huckle, 1985; Matthews, 1985).

This combination of exploration and homing is consistent with the results of displacements between reefs with different characteristics. In general, rockfishes from high relief reefs return there if displaced but those moved from low relief reefs to high relief reefs often remain at the high relief reef (Matthews, 1990a). Moreover, rockfishes tend to leave low relief reefs in winter, perhaps responding to storms and the loss of structure when kelp dies back (Matthews, 1990c). It seems that rockfishes on high relief ('high quality') sites move very little and return to these sites if displaced. Those on lower quality reefs periodically make exploratory movements and return home unless they find a reef superior to their own.

4.3 CASE HISTORIES

While the homing migrations of a wide variety of fishes have been described, detailed examination of the mechanisms and adaptive significance of these phenomena generally have been restricted to a very few economically important or easily studied species. Such in-depth studies have been critical, however, for our general understanding of the sensory systems involved in guiding homing fishes and the integration of these systems during homing. This section examines the homing patterns and guidance mechanisms that may be involved in the homing of six well-studied fishes with very different life histories and movement patterns. Using this 'case history' approach, we hope to illustrate (1) the importance of homing for finding and exploiting habitats needed for feeding, spawning and refuge, (2) the diversity and sensitivity of sensory systems involved in homing, and (3) some of the techniques used by researchers in studying fish homing.

4.3.1 Daily feeding migrations

(a) Scalloped hammerhead sharks (Sphyrna lewini)

Like many other foraging species, the scalloped hammerhead shark has distinct diel activity periods. During the day the sharks form small groups in the vicinity of seamounts in the Gulf of California and at night individuals leave the seamount and make extensive migrations into the surrounding pelagic environment, returning the next morning. While it appears that foraging is the primary purpose of these nightly migrations, it is less clear why the sharks remain at the seamounts during the day. The seamounts may function as refuge areas for the sharks to conserve energy or to facilitate social interactions (Klimley and Nelson, 1984). It has also been proposed, however, that the seamounts serve as centrally located orientation landmarks which are used by the sharks as reference points for their

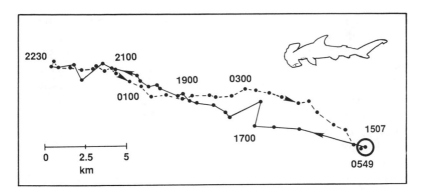

Figure 4.7 Horizontal track of a scalloped hammerhead shark (*Sphyrna lewini*) during its nocturnal movement away from El Bajo Espiritu Santo seamount in the Gulf of California (solid line) and its return to the seamount the next morning (dashed line). The shark was tagged at the area marked with the open circle at 1507 h (from A.P. Klimley, unpublished).

nightly feeding movements (Klimley and Nelson, 1984; Klimley *et al.*, 1988).

To study the sharks' diel movement patterns, Klimley *et al.* (1988) tagged 18 sharks with ultrasonic transmitters and monitored their movements with automated recording devices. They found that many sharks not only return to the same seamount each morning after nightly foraging trips but actually return to the specific site where they were tagged. This high degree of homing fidelity for specific regions of the seamount suggests that orientation to these sites is not only important for the shark but also that they must possess sensitive mechanisms to locate these sites. To return to the seamount from distant (*ca* 20 km; Fig. 4.7) foraging excursions, the sharks must orient at night in a midwater pelagic environment with few obvious guidance clues and strong water currents.

One intriguing proposal is that the sharks detect and orient to the electrical fields produced by water current eddies at the seamount moving through the earth's magnetic field (Kalmijn, 1978, 1981, 1982; Klimley *et al.*, 1988). Alternatively, local anomalies in the magnetic field associated with the seamount could provide guidance information (Vacquier and Uyeda, 1967; Klimley *et al.*, 1988).

These hypotheses are plausible, given the extraordinary sensitivity of the Ampullae of Lorenzini to electric fields (Kalmijn, 1978). Behavioural studies have demonstrated that sharks and rays are able to utilize their electroreceptive capabilities both for detection of prey and for orientation with respect to the earth's magnetic field (Kalmijn, 1981). The movement of sea-water through the earth's magnetic field induces electric fields well within the detection limit of elasmobranch fishes. Electric fields induced by the movement of the fish with respect to the earth's magnetic field also allow them to detect their compass heading (Kalmijn, 1982). However, the rapid water currents in the pelagic zone where the sharks forage at night would not permit them to home with the use of only a compass. The weak magnetic anomalies around the seamounts could serve as an additional reference system, if the sharks are able to detect them (Klimley et al., 1988).

(b) Grunts, Haemulidae

Like the hammerhead sharks, juvenile French and white grunts (*Haemulon flavolineatum* and *H. plumieri*) make precise diel movements between feeding and refuge areas. However, the habitats and mechanisms are quite different. During the day, the grunts form heterospecific aggregations closely associated with the protection of shallow Caribbean coral reefs (Ogden and Ehrlich, 1977). As light levels fall after sunset, the grunts assemble at the edge of the reef and migrate together along very precise routes into the surrounding grassbeds where they disperse and feed (Fig. 4.8). At dawn, the grunts reassemble and return to the patch reefs along the same routes used for emigration. While such twilight migrations between refuge and feeding areas are common in coral reef communities (Hobson, 1972, 1973), the precision of the grunts' diel activities, both in time and space, is remarkable. The 100–200 m routes across sand-bottomed seagrass beds have no obvious landmarks to human observers (Ogden and Ehrlich, 1977). As post-larval grunts (which are diurnal planktivores) shift to the nocturnal, benthic foraging pattern of migrants, they seem to learn the routes from experienced migrants. Thus the routes are passed on from one generation of grunts to another and maintained for many years (Helfman et al., 1982).

In addition to the spatial orientation of the migrations, they are noteworthy for their temporal precision. Both the dawn and dusk migrations occur during very specific light intensities. McFarland et al. (1979) and Helfman et al. (1982) proposed that the migratory timing minimizes predation during the twilight periods of heavy predation. As the light levels change at dawn and dusk, the grunts undergo a series of physiological adaptations including changes in body coloration and migration of the retinal pigments to facilitate

Figure 4.8 Photograph of juvenile white and French grunts (*Haemulon plumieri* and *H. flavolineatum*) on their twilight migration from a patch reef to seagrass feeding areas off St Croix, Virgin Islands, USA (photograph by T.P. Quinn).

vision at low light levels (McFarland *et al.*, 1979). McFarland *et al.* theorized that these adaptations give the grunts a visual advantage over their predators during the specific light intensities associated with their migrations. Apparently, these light levels are sufficient for the grunts to orient and school but too low for visual predators to be effective.

The adaptive advantage of the spatial precision of outward and return migrations is not fully clear. One possibility is that the specific routes allow for the orderly partitioning of the food resources of the surrounding grassbeds (Ogden and Ehrlich, 1977). The routes cross halos of bare sand around the patch reefs and seem to offer no special physical refuge from predation. Indeed, the highly predictable nature of the migration allows predators to assemble in preparation for the grunts' migration. Apparently the advantages of migration in a tightly regulated pattern outweigh the risks associated with predictability. Therefore, the departure points and routes may be important for focusing the schools but may not have inherently special features (Helfman and Schultz, 1984).

Displacement experiments have yielded different results, depending upon when and how far the grunts were taken from their home site. Grunts have returned after displacement beyond their normal home range (as far as 2.8 km: Ogden and Ehrlich, 1977). Shorter displacements suggested that grunts may recognize some features of their routes because the fish returned directly home as soon as they encountered their route. However, grunts captured during morning and evening migrations and displaced to nearby areas (*ca* 100 m) did not behave as though the grassbeds were familiar to them. Rather than compensate for their displacement, they generally headed in the direction that they had been going before being caught, even if that direction took them away from their normal feeding area (Quinn and Ogden, 1984). It is not clear what mechanism(s) might orient these migrations. The grunts leave after the sun has set and return before it rises. They might still use solar information by using patterns of light polarization or brightness or they may use a magnetic compass.

Grunts and other coral reef fishes with diel migrations are excellent subjects for orientation research. Their arrival and departure times are usually highly predictable, the spatial aspects of their movements are very stereotyped and they generally migrate throughout the year. Moreover, these fishes are often small enough to facilitate laboratory as well as field experiments. Besides grunts, few species have been studied in any detail and opportunities for further research are abundant.

4.3.2 Seasonal reproductive migrations

(a) American shad (Alosa sapidissima)

Unlike grunts, which make very precise, short distance daily migrations, the homing patterns of many other fishes involve much greater distances and longer periods of time. The American shad is an anadromous clupeid which spawns in rivers along the eastern coast of North America from the St Johns River in Florida to the St Lawrence River in Canada. Transplanted populations of shad also exist in rivers on the west coast of North America. At specific times of the spring/summer, juvenile shad emigrate from their natal rivers and join the large body of immature and adult shad moving along the coast. Maturing shad will subsequently separate from this group and return to their natal river to spawn (Walburg and Nichols, 1967; Melvin *et al.*, 1986).

Leggett and Whitney (1972) emphasized the role of temperature in the migratory behaviour of shad. As river temperatures fall below 15.5°C, juvenile shad move downstream into the ocean to feed. The timing of these movements (progressively later from the north to south), allows the juveniles to join the shad already in the ocean as they migrate south in the autumn. During the period of ocean residence, shad from the entire coast congregate in waters with temperatures from about 13 to 18°C. Thus, during the autumn and winter, the centre of concentration moves southward. In the spring, the shad reverse direction and move northward as water temperatures rise. As they near their natal river, mature shad leave the group and migrate upstream to spawn.

Because shad generally remain near the coast during their oceanic migrations, these temperature-linked movements may often be sufficient to place the adults in the vicinity of their natal river, where odors from the river may be detectable (Dodson and Leggett, 1973). However, Leggett and Whitney (1972) hypothesized that the shad on the Pacific coast may be forced offshore during certain periods of the year by cool upwelling water, yet large numbers return to the Columbia River each year. Furthermore, Dadswell *et al.* (1987) concluded that, on the basis of recent tagging studies, shad are often located outside the 'preferred' temperatures along the Atlantic coast. These two pieces of information imply that temperature-regulated migration may not be sufficient to direct shad to the vicinity of their natal stream at the appropriate time for spawning.

The question of what mechanisms allow shad to recognize their natal river and ascend it from coastal waters has been studied most closely in Long Island Sound for shad homing to the Connecticut River. Dodson and Leggett (1973) used ultrasonic transmitters and found that the shad tended to swim into (i.e. against) both flood and ebb tides. However, the shad swam faster and oriented more precisely into ebbing currents than they did into flooding currents. This resulted in net movement towards the mouth of the river.

While this counter-current orientation explains the process by which shad approach their home river, it does not indicate what clues associated with the river elicit the responses. To examine this question, Dodson and Leggett (1974) occluded the olfactory organs of shad captured in Long Island Sound, tagged the fish and released them. Significantly fewer anosmic shad were recovered in the Connecticut River than controls, indicating that olfaction was involved. When such anosmic fish were tracked they displayed much of the normal counter-current behaviour but there was no longer a tendency to swim faster into ebbing tides, hence no westerly displacement was achieved. Dodson and Leggett (1974) concluded that olfactory information is needed to provide a

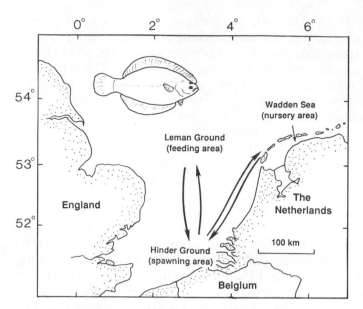

Figure 4.9 Map of the North Sea showing some general features of plaice (*Pleuronecles platessa*) migration (from Arnold, 1981).

necessary river-ward bias in the counter-current orientation, enabling shad to migrate through the reversing odour field of the estuary.

(b) *Plaice* (Pleuronectes platessa)

The extensive spawning migrations of anadromous fishes such as shad and salmon are well known but many marine fishes also make extensive seasonal spawning migrations. One particularly well-studied species is the plaice, a flatfish which lives on the sandy-bottomed European continental shelf (reviewed by Arnold, 1981). Mature plaice of the Southern Bight stock in the North Sea migrate each winter from feeding grounds centred near Leman Ground to their spawning area 200–300 km south at the Hinder Ground (Fig. 4.9). Peak spawning occurs in January when females produce large numbers of pelagic eggs. After spawning, the spent adults return to their northerly feeding grounds while the developing embryos and larvae drift northeast with the residual currents until they metamorphose into bottom-oriented juveniles (usually along the coast of Holland). These juveniles then migrate inshore to their

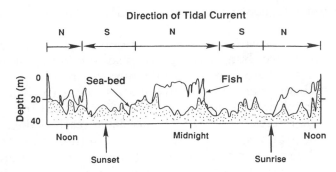

Figure 4.10 Vertical movements of a plaice over the course of two tidal changes (fish no. 7 from Arnold, 1981).

intertidal nursery areas located in the rich mudflats of the Wadden Sea. After 3–4 years the plaice begin to mature and migrate south along the eastern side of the Southern Bight to the spawning areas at Hinder Ground, converging on the same area as adults that had spawned previously.

This pattern of discrete spawning, nursery and feeding areas is a common feature in the life histories of many marine fishes in boreal and temperate waters (Arnold, 1981). In the case of North Sea plaice, every migratory phase is intimately linked to the semidiurnal tidal currents which prevail on the European continental shelf. Greer Walker *et al.* (1978) fitted adult plaice with transponding acoustic transmitters and monitored their horizontal and vertical movements during reversing tidal cycles in the Southern Bight. The plaice demonstrated a semidiurnal pattern of vertical movements, termed selective tidal stream transport (Fig. 4.10). Migrating plaice came off the bottom during high slack water and remained in midwater for 5–6 hours during the northerly-flowing ebb tide before returning to the bottom during low slack water. During the flooding tides the fish appeared to hold their position on the bottom. This pattern of vertical movement allowed the fish to take advantage of the tidal velocities that favoured their migration and avoid those that opposed migration.

Based on these studies, Harden Jones *et al.* (1979) predicted that maturing plaice migrating towards their spawning area from Leman Ground would swim up in the water column primarily during southerly tides and spent plaice returning to their feeding areas would be found in midwater during northerly (ebb) tides. Paired midwater trawls conducted during consecutive tides along the migratory route confirmed this prediction (Harden Jones *et al.*, 1979).

Selective tidal stream transport is not only used by maturing and spent adult plaice but larval plaice also selectively move into midwater to facilitate transport to the nursery areas of the Wadden Sea (Creutzberg *et al.*, 1978). Juveniles returning to their nursery grounds after their first winter offshore also use tidal transport (de Veen, 1978). Interestingly, when plaice have reached their feeding or spawning grounds, the semidiurnal pattern (high tides separated by about 12.25 h) of vertical movement is replaced by a diurnal one; plaice remain on the bottom during the day and move into midwater at night. These vertical movements are not coupled to the tidal cycle and there is no net horizontal movement.

In areas with strong tidal currents, the energetic savings associated with selective tidal stream transport are significant. Weihs (1978) estimated that such behaviour could conserve up to 40% of the energy that would be necessary if the plaice swam continuously while migrating. The effectiveness of this strategy is further illustrated by the number of fishes utilizing tidal stream transport: sole, *Solea solea* (de Veen, 1978; Greer Walker *et al.*, 1980); cod, *Gadus morhua* (Harden Jones, 1977); eels, *Anguilla anguilla* and *A. rostrata* (Arnold, 1981; McCleave and Kleckner, 1982); dogfish, *Scyliorhinus canicula* (Greer Walker *et al.*, 1980) and flounder, *Platichthys flesus* (de Veen, 1978).

While it is apparent that selective tidal stream transport is an effective and common migratory strategy, a number of questions remain regarding the sensory mechanisms involved in the temporal and spatial control of the behaviour.

(1) What cues trigger the transition from diurnal to semidiurnal vertical movements? Plaice are able to detect changes in hydrostatic pressure which could indicate tidal state (Gibson *et al.*, 1978; Gibson, 1984) but there is no direct evidence that plaice actually use this information.

(2) Once initiated, how is the pattern of semidiurnal vertical movements maintained? While on the seabed, plaice presumably are able to detect water current direction by direct mechanical stimulation. However, plaice in midwater at night would have no obvious cues to indicate when the current is reversing and they should descend to the bottom. Laboratory studies have demonstrated that juvenile plaice display free-running circa-tidal activity rhythms (Gibson, 1976). Greer Walker *et al.* (1978) proposed that the endogenous rhythm triggers exploratory excursions to the bottom to obtain the rheotactic information needed to determine tidal flow.

(3) Upon reaching the spawning or feeding grounds, what signals the plaice to discontinue the semidiurnal vertical movements? Harden Jones (1979) hypothesized that ground water seepage could provide site-specific information for migrants but there is no evidence to support this at present.

(4) Do plaice need navigational abilities to successfully migrate between nursery, feeding and spawning grounds? Most models of tidal transport (e.g. Arnold and Cook, 1984) assume no absolute sense of direction but indirect evidence indicates that a directional component may also be involved. There are apparently four stocks of plaice in the North Sea, each with distinct spawning grounds (Arnold, 1981). Tagging studies have demonstrated that juveniles recruit into the same stock as their parents (Harden Jones, 1968; Lockwood and Lucassen, 1984). Because they appear to overlap on the feeding grounds (Arnold, 1981), plaice of different stocks must have some mechanism by which they segregate to their spawning grounds. This segregation could involve one or a combination of the following mechanisms: (1) movement to different areas within the feeding grounds from which tidal streams would allow them to reach their spawning grounds, (2) directionally oriented swimming in conjunction with tidal stream transport to adjust their migration *en route*, and (3) clues emanating from the spawning grounds which the plaice detect and respond to according to their site of origin.

There is little direct evidence to shed light on the question of directed migration. Harden Jones (1981) reported that adult plaice maintain relatively constant compass headings for extended periods in midwater. Moreover, some fish appeared to swim actively in one direction while the tide carried them in another direction. These observations are consistent with the hypothesis that plaice migrations involve at least some directional component in addition to timed tidally-related vertical movements.

A computer model by Arnold and Cook (1984) interpolated the speed and direction of tidal currents throughout the North Sea and permits, by simulation, examination of different migratory patterns. The initial results of this model indicated that exclusive reliance on tidal transport could explain observed movement patterns but only with three stipulations. First, to return to the appropriate feeding grounds after spawning, the point of departure is critical. Second, for certain migrations, plaice would need to switch tides in mid-migration. Third, depending on their location on the feeding grounds, some plaice may need to initiate migration at either high or low slack water to reach the same spawning grounds. It seems unlikely that plaice rely exclusively on tidal stream transport but rather use it for the energetic savings which it permits as they travel in directions determined at least in part by other mechanisms.

4.3.3 Homing movements within the home range

(a) Centrarchids

As mentioned earlier, many fishes are relatively sedentary during most of their lives, restricting their movements to a home range or defended territory (Geking, 1953). Individuals may occasionally make excursions from the home range or move to a distinctly new area (e.g. for spawning or overwintering) but they generally do not undergo extensive migrations such as those described for shad and plaice. Despite the restricted nature of their movements under most circumstances, such fishes often display the ability to home when displaced from their familiar areas.

The family Centrarchidae, including such North American species as largemouth bass (*Micropterus salmoides*), smallmouth bass (*M. dolomieui*) and the sunfishes (*Lepomis* spp.), provides excellent examples of fishes displaying home range and homing abilities. While the species differ in specific habitats and behaviour, centrarchids generally live in structured littoral areas of lakes, ponds or slow-moving stretches of rivers. In the spring, males establish territories and build nests. After spawning, the males guard the eggs and newly hatched fry. During the remainder of the year, both sexes occupy and forage within nearshore home ranges, though they may overwinter in deeper waters offshore.

Numerous tag–release–recapture studies established the tendency of these fishes to maintain home ranges: smallmouth bass (Pflug and Pauley, 1983), largemouth bass (Lewis and Flickinger, 1967), longear sunfish (*L. megalotis*) and green sunfish (*L. cyanellus*; Gerking, 1953) and bluegill sunfish (*L. macrochirus*; Gunning and Shoop, 1963). These findings have been supported by work with ultrasonic and radio transmitters (e.g. Winter, 1977; Savitz *et al.*, 1983; Mesing and Wicker, 1986). Such studies with conventional and electronic tags have revealed the general seasonal movement patterns and also indicate that some fish leave their home ranges and move considerable distances. There is also some evidence that largemouth bass (Lewis and Flickinger, 1967) and green sunfish (Hasler and Wisby, 1958) may return to the same spawning site each spring after overwintering offshore. However, this does not seem to be true of rock bass (*Ambloplites rupestris*; Storr *et al.*, 1983).

While they generally remain within their home ranges,

experimentally displaced centrarchids often return to their place of capture (Parker and Hasler, 1959; Pflug and Pauley, 1983; Mesing and Wicker, 1986; Gerber and Haynes, 1988). In general, shorter displacements often result in higher proportions of fish returning and more rapid homing. However, it is unclear whether delays in homing or the establishment of new home areas are related to the inability of the fish to find their original site or a lack of motivation to do so. For example, a fish might initiate homeward movement but encounter an area along the way that it somehow judged adequate or superior to the original site, as seems to be the case with rockfishes.

On the basis of Griffin's (1952) classification of homing behaviour in birds, Hasler et al. (1958) described three types of homing orientation that might be employed by fish displaced from their home range: (1) the fish might move randomly in unfamiliar areas until they encounter their home range, (2) the fish might maintain directional orientation in unfamiliar areas until they encounter familiar landmarks, or (3) the fish might be able to determine the correct homeward direction without reference to landmarks. Displacement experiments in which the fish were marked with floats that could be seen at the surface yielded mixed results. Green sunfish often returned directly to the capture site from open water (Hasler and Wisby, 1958) but studies with centrarchids in a larger lake revealed more indirect movements and searching along the shoreline (Parker and Hasler, 1959).

In spite of the equivocal results from field tests, laboratory experiments have shown that centrarchids (and other pond-dwelling fishes) are capable of using the sun's position as a compass (i.e. type II) orientation. Hasler et al. (1958) first demonstrated sun orientation in fishes by training a bluegill sunfish to seek shelter in a hiding box in a circular arena. After training, the fish repeatedly chose the appropriate hiding box when the sun was visible but appeared disoriented on overcast days. When an artificial light was substituted for the sun, the fish moved to the hiding box at the same angle with respect to the light that it would have moved with respect to the sun at that time of day. Subsequent studies with largemouth bass (Loyacano et al., 1977) and bluegill sunfish (Goodyear and Bennett, 1979) indicated that these fishes use a sun compass to orient their movements perpendicular to the shoreline (so-called y-axis orientation; Fig. 4.11). This orientation seems to facilitate movement to onshore or offshore refuges from predators.

In summary, centrarchids and other lacustrine fishes probably have familiar areas which exceed the limits of their home ranges, as they would be generally defined. Familiarity with these areas may

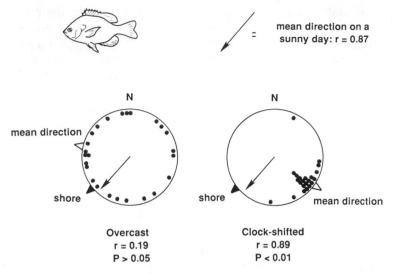

Figure 4.11 Orientation of juvenile bluegill sunfish (*Lepomis macrochirus*) in circular arenas. The left diagram compares the shoreward orientation of sunfish on sunny days (shown with an arrow) to the orientation on overcast days. The right diagram compares shoreward orientation to the orientation of clock-shifted sunfish (exposed to a photoperiod artificially advanced by 6 h; from Goodyear and Bennett, 1979).

be acquired during occasional, lengthy excursions or seasonal movements between habitats. It is also clear that many such fishes can use the sun's position to orient short-range movements, typically perpendicular to the shoreline. Evidence that the sun compass is involved in homing, however, is not conclusive (Hasler *et al.*, 1958). Finally, there seems to be no evidence of true navigation in these fishes, though it is difficult to conduct convincing experiments on homing mechanisms without knowing the size of the subject's familiar area. Centrarchids are widely distributed and easily studied and hence are good candidates for further studies on homing.

(b) Intertidal fishes

Unlike the centrarchids, which occupy relatively stable habitats in lakes and streams, intertidal fishes live in a dynamic environment, characterized by surging currents and dramatic changes in water level, temperature and salinity. The homing behaviour of several such hearty species has been studied, particularly the tidepool sculpin (*Oligocottus maculosus*), an inhabitant of the rocky intertidal habitat of the North Pacific Ocean. After a pelagic larval stage, juvenile tidepool sculpins move into the intertidal zone. Juveniles often move on- and offshore with the tides and show little fidelity to particular areas (Green, 1971; Craik, 1981). However, as they mature these small (*ca* 8 cm) fish begin to establish home areas consisting of several neighbouring tidepools or a single pool (Green, 1971; Khoo, 1974). Movements from the home pool seem to be more restricted in sites exposed to wave action than in more protected sites.

Green (1971) demonstrated homing in tidepool sculpins by marking them with embroidery beads and displacing them up to 100 m. Of the 117 sculpins displaced, 100 were subsequently recaptured in their home pools and only three were recovered in non-home pools. Most sculpins left the release site during the next high tide and homing was usually completed within one or two tidal cycles. Homing probably enables sculpins to explore areas beyond the home pool during high tides and return to their familiar shelter when water levels drop. Such ability would also facilitate their return if they are driven from their home pool by storms.

Khoo (1974) examined the sensory mechanisms underlying homing in tidepool sculpins by displacing anosmic, blinded and control sculpins. The homing of anosmic fish was significantly reduced and that of blinded fish less so, compared to the performance of controls. Preliminary laboratory tests indicated that sculpins could distinguish between home and non-home pool water, supporting the hypothesis that olfaction is the primary homing mechanism (Khoo, 1974; Craik, 1981). Interestingly, juvenile sculpins required both vision and olfaction to home (Craik, 1978). While olfaction may be essential for the final stages of homing, it seems unlikely that sculpins can determine the direction of their pool from odors at distances of over 100 m in the turbulent, rocky intertidal areas in which they live.

There are three obvious hypotheses for homing by sculpins. First, their familiar areas may be large and the displacements require nothing more than piloting within their home range. Second, they may move randomly until they encounter familiar odours or visual landmarks. Finally, they may possess some form of true navigation system (Green, 1971). Craik's (1981) finding that juveniles move more widely than adults tends to support the hypothesis that sculpins explore a large area before establishing a home pool. Thereafter, they may use their spatial map to facilitate excursions in search of food or high-quality pools to colonize or to permit them to return home after accidental displacement by storms.

4.4 PATTERNS AND MECHANISMS OF HOMING IN SALMON

Atlantic salmon (*Salmo salar*) and Pacific salmon (*Oncorhynchus* spp.) are the classic examples of homing in fishes. However, homing prevails throughout the family Salmonidae in species with very different life history patterns. Salmonids may be iteroparous or semelparous, (i.e. repeat spawners or species that die after spawning, respectively). They may reside exclusively in freshwater (in rivers, lakes or both) or be anadromous. This suggests that the adaptive significance of homing is not linked to one specific life history pattern or habitat and that the mechanisms underlying homing are flexible. In this section the Pacific salmon will be used as an example to illustrate the variety of sensory systems and behaviour patterns involved in the complex series of habitat changes exhibited by salmon.

Pacific salmon typically spawn in the autumn in streams or, less often, on beaches of lakes. Females dig gravel depressions (known as redds) into which the eggs are deposited and fertilized by one or more males. In the case of most species, all individuals die shortly after spawning. In some, such as steelhead or rainbow trout (*O. mykiss*), the fish may survive to spawn again but provide no parental care beyond redd construction and some limited redd defence. The embryos hatch but remain buried in the gravel until the yolk sac has been absorbed. At this time (usually spring) they emerge and remain in the stream (e.g. steelhead; chinook, *O. tshawytscha*; coho, *O. kisutch*; cherry, *O. masu*; cutthroat trout, *O. clarki*), migrate directly to sea (pink, *O. gorbuscha*; chum, *O. keta*) or migrate to a lake (sockeye, *O. nerka*).

4.4.1 Orientation mechanisms of juvenile salmon

While salmon are known for their homing migrations from oceanic feeding grounds back to their natal streams, our conception of the orientation mechanisms guiding these movements relies heavily on studies with juvenile salmon. Studies have described the migratory patterns and orientation mechanisms of juvenile salmon, documenting sensory systems and responses to stimuli which might be involved in oceanic migration. Prior to entering seawater, juvenile salmon often display highly oriented movement patterns. The most complex and most thoroughly studied migration patterns are those of sockeye salmon. Juveniles generally reside in lakes for the first year or two of their lives. Because the incubation sites are usually in the inlets or outlets of these lakes, emerging fry must make the appropriate migration (downstream or upstream, respectively) to reach the lake. In some cases, migration from the incubation environment to the nursery lake requires sequential downstream and upstream migration. Tests of juvenile sockeye salmon in experimental arenas revealed that those from populations spawning below lakes display upstream orientation and those from populations spawning above lakes display downstream orientation (Raleigh, 1967). Brannon (1972) provided evidence for genetic control over such rheotactic responses when he showed that the responses of fry of mixed parentage (upstream × downstream populations) were intermediate between those of their parents. Subsequent studies with other species have revealed that genetic control of rheotactic behaviour is a common trait in salmonid populations (cutthroat trout: Raleigh and Chapman, 1971; rainbow trout: Kelso *et al* ., 1981; chinook salmon: Taylor, 1988).

The lake-finding behaviour of newly emerged sockeye fry is based not only on innate rheotactic responses but also on differential expression of rheotaxis depending on whether or not the water current carries an appropriate odour. Brannon (1972) and Bodznick (1978a) presented suggestive but circumstantial evidence that sockeye fry can somehow discriminate lake water from non-lake sources and are more attracted to lake water. However, sockeye fry also tend to prefer their incubation water source over novel waters, indicating olfactory learning at this stage (Brannon, 1972).

While sockeye fry have innate and learned responses to water flow and odours that apparently facilitate migration to a lake in an unfamiliar river system, Brannon (1972) also documented directional preferences in the absence of rheotactic or olfactory clues. These preferences are apparently controlled by a combination of celestial and magnetic compasses (Quinn, 1980). For example, Chilko River fry migrate upstream towards Chilko Lake during the day. This migration is in a southerly direction and their subsequent migration in the lake is to the south (Fig. 4.12). Fry were captured as they migrated towards the lake and were tested in four-armed arenas (Fig. 4.12). Tests were conducted with celestial information available (on sunny days) and with celestial information not available (on overcast days and under translucent plastic covers). Tests were also conducted under the ambient magnetic field and with the horizontal component rotated 90° counter-clockwise. The salmon moved to the south or south-southeast under all test conditions except when the magnetic field was altered and sky cues were not available. In these tests their orientation was shifted counter-clockwise 100–120° (Fig. 4.12).

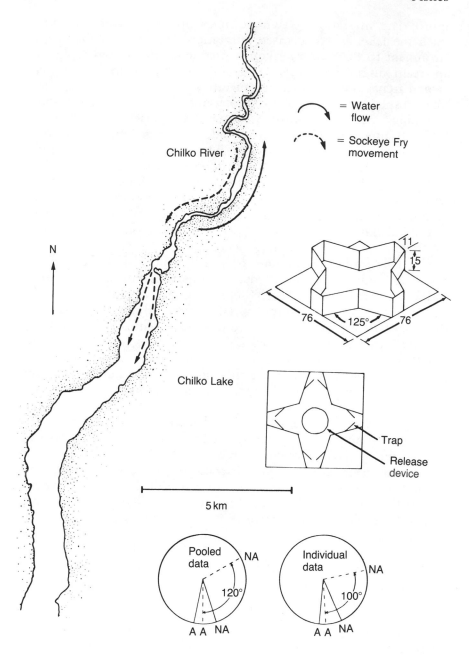

The failure of fry to orient to the magnetic field unless the sky cues were obscured by clouds or covers indicated that the magnetic field served as a backup orientation system. The magnetic field might have provided the basis for learning the celestial orientation system in this case, as the fry had probably emerged from the gravel as much as 7–10 days prior to testing. Such reliance on the magnetic field as the reference for learning celestial rotation patterns has been established in many avian orientation studies (Chapter 7) and indeed it seems that celestial orientation systems are generally learned, not innate (but see Pardi and Scapini, 1983). The only reason to doubt this scenario is the indication that newly emerged (i.e. first day) sockeye fry from Weaver Creek also seemed to rely on the sun or some celestial cue rather than the magnetic field (Quinn, 1982a). However, this population normally migrates at night and their daytime orientation was very weak, hence this result must be considered provisional.

Tests conducted with a night-migrating population from the Cedar River indicated that the magnetic field played a dominant role, as orientation shifted counter-clockwise with a shift in the magnetic field, even when there was a view of a starry sky. During night-time tests, Weaver Creek fry showed generally weak northerly orientation but this direction was strongly preferred when the moon was present (Brannon *et al.*, 1981). Three explanations were presented. First, the moon may have provided directional information. This is possible but it is difficult to understand how fry could show appropriate responses to the complex lunar pattern within hours of emergence from their gravel incubation environment. Second, the responses to the moon may have been merely phototactic. This seems unlikely, as northerly orientation of the fry brought them into the region of the tanks illuminated by moonlight and they avoided the shaded southerly region. Moreover, the fry orientation did not shift with the changing shadow patterns during

Figure 4.12 Map of the Chilko River – Chilko Lake system, British Columbia, and the experimental apparatus used to test the directional orientation of sockeye salmon fry in the absence of olfactory and rheotactic clues (dimensions in cm). The lower panel displays a summary of tests conducted during the day with sky cues available (A; clear days) and not available (NA; overcast days and translucent covers) in the ambient magnetic field (solid line) and the field shifted 90° counter-clockwise (dashed lines). The data were analysed counting each test of about 30 fish as a data point ('pooled data') and counting each fish as a data point ('individual data'). Composite figure from Quinn (1980, 1982a).

the night, as would be expected from a phototactic or fixed-angle response. Finally, the presence of the moonlight may have stimulated northerly orientation even though the moon provided no directional information. There appears to be no other evidence of lunar compass orientation in fishes and the sockeye results are inconclusive.

While the Weaver Creek fry showed only weak northerly orientation when tested in water from the creek, their orientation strengthened dramatically when they were tested in water from Harrison River, which drains Harrison Lake. Thus there may be a link between the apparent attraction of fry to lake water and the expression of compass orientation appropriate for migration in their nursery lake.

The compass directional responses of sockeye fry probably facilitate distribution in the nursery lakes. This would serve to reduce predation on the fry and intraspecific competition for food. After usually 1 or 2 years in the lake, the sockeye (now termed smolts) migrate to the lake's outlet and downstream to the ocean. As with the newly emerged fry, the migrations of sockeye smolts in lakes are rapid and a combination of celestial (sun compass and polarized light) and magnetic cues are apparently used (Groot, 1965; Dill, 1971; Quinn and Brannon, 1982).

4.4.2 Homing by adults

After varying periods of freshwater residence, salmon generally spend 1–4 years at sea, returning in the autumn to spawn and die. The initial marine migrations of the species fall into four categories. Sockeye, chum and pink salmon from North America migrate north along the coast until they are in the Gulf of Alaska, generally in their first autumn (Hartt and Dell, 1986). At this time they move offshore until they mature and return to the coastal waters to ascend their home river. While coho and chinook may migrate north and out into open water, they tend to migrate more slowly and often confine their movements to coastal waters (Fisher and Pearcy, 1987; Pearcy and Fisher, 1988). The movements of steelhead are not as well known as those of the other species but they seem to move directly offshore upon entry into the ocean and migrate far west before returning to their home stream. Anadromous cutthroat trout apparently spend only the summer at sea and do not move far from their river of origin, though the movements of this species are very poorly documented.

While at sea the net movements (i.e. point to point) of immature salmon are relatively slow. Tag–release–recapture studies indicate displacements of about 20 km/day (Hartt, 1966). These movements

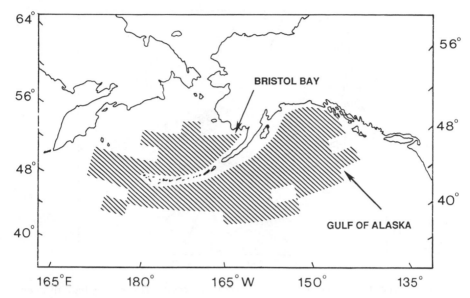

Figure 4.13 Distribution of sockeye salmon tagged at sea and recovered in Bristol Bay, Alaska, as maturing adults that season, illustrating the broad range of these populations and the variety of directions in which they must swim to return home (from Quinn and Groot, 1984).

seem to take the salmon in counter-clockwise gyral patterns, coincident with the dominant ocean currents. However, in the spring and summer of the year in which they will spawn, the movements become much more rapid as homing begins. From the open ocean to their natal stream, salmon travel through very different environments. While these environments grade into each other, for purposes of discussion they are separated into (1) open ocean, (2) coastal and estuarine waters, and (3) rivers.

4.4.3 Homing in the open ocean

It is difficult to generalize about salmon migration patterns owing to differences among species and populations over the geographical range. However, Neave (1964), Royce *et al.* (1968) and Quinn (1982b) summarized several characteristics of salmon migrations at sea.

1. Salmon converge from broadly defined feeding areas on specific coastal regions near their river of origin. Thus members of the population do not experience a common environment or use a common route when homing (Fig. 4.13).

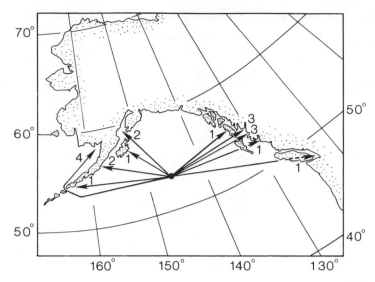

Figure 4.14 Recovery locations of adult sockeye salmon that had all been captured on a single long-line set on 27 April, 1962 in the Gulf of Alaska, tagged and released (from Neave, 1964).

2. While foraging at sea, species and populations have overlapping (but not coinciding) distributions. Thus different species and populations may experience the same environment and yet their paths diverge as they home to their respective rivers (Fig. 4.14).

3. The movements are rapid, often covering 40–60 km/day, though species and populations differ in this regard (Hiramatsu and Ishida, 1989). These travel rates greatly exceed the water current velocities over most of the range of salmon and homing salmon may swim with, against or across current on their homeward journeys.

4. The return migration does not necessarily follow the outward path, either in pelagic or coastal waters (Groot and Cooke, 1987).

5. Salmon appear at specific coastal locations at specific times. While there are interannual variations in time of return, populations nevertheless migrate on relatively fixed schedules (Blackbourn, 1987).

There has been controversy in the literature for over five decades concerning the mechanisms of homing by salmon at sea. While additional data are being collected which refine our understanding of the movements of immature and maturing (i.e. homing) salmon

at sea, the basic facts have been documented since the mid-1960s
(e.g. Neave, 1964; Hartt, 1966; Royce *et al.*, 1968). There is no
experimental evidence which directly addresses the question: what
are the mechanisms guiding salmon homing from the open ocean to
the coastal waters near their home river? Consequently, the controversy surrounding the mechanisms has focused on interpretation
of tagging data, data on timing of return to coastal waters, and
simulation models.

Early models by Saila and Shappy (1963) and Patten (1964)
demonstrated the power of simulations for testing the hypothesis
that salmon movements must be highly oriented to move from the
open ocean to coastal waters near their natal river. These models
(and a recent one by Jamon, 1990) indicated that little orientation is
needed to explain the oceanic homing patterns. In the models,
a large number of simulated salmon were 'released' at sea and
allowed to 'migrate' with various degrees of homeward orientation.
Swimming speed was simulated by step length and salmon not
returning home within set periods of time were considered to have
been lost. Those encountering North America were allowed to
move along the coast until they neared the home river (when
olfaction was presumed to take over as the control mechanism) or
they exceeded the allotted time period.

While the simulations by Saila and Shappy (1963) and Patten
(1964) encouraged careful analysis of the data on salmon movements at sea, they were based on assumptions regarding swimming
speed, endurance and recovery proportion which are now known
to be incorrect (Quinn and Groot, 1984). Jamon's (1990) model
incorporated some updated information but repeated a fundamental
error of the earlier models. The models were judged to have accurately simulated salmon homing success if the combination of model
parameters resulted in about 10% of the salmon homing. This
figure was taken to approximate the proportion of adult salmon
which locate the vicinity of their natal stream because only about
10% of the tags placed on salmon at sea are returned from coastal
areas and rivers. However, this does not take into consideration tag
loss, natural mortality, tag-induced mortality, non-reporting by
fishermen and failure of salmon escaping the fisheries to be noticed
on spawning grounds. In general, it should be assumed that virtually
all salmon will return to the vicinity of their natal stream if they
survive. Only models which can explain the rapid movement of
salmon from the open ocean to the mouth of their natal river are
given credence.

It has been hypothesized that species and populations of salmon
display seasonally changing preferences for environmental conditions (e.g. temperature and salinity) and this optimization of

physiological conditions plays a role in homing (Leggett, 1977). The distributions of salmon at sea do show some relationship with sea surface temperature. In general, as the central North Pacific Ocean warms in summer, the distribution of salmon shifts northward. In winter, a more southerly distribution is evident (Royce et al., 1968). While salmon feeding distributions are correlated to some extent with temperature, this does not necessarily indicate that temperature is a guidance mechanism. Salmon may actually be responding to changes in prey density which are correlated with temperature patterns.

Indirect evidence for the importance of temperature in ocean distribution and homing by salmon comes from the studies of Blackbourn (1987) and Groot and Quinn (1987) on sockeye salmon returning to the Fraser River, British Columbia. Blackbourn (1987) summarized evidence that sockeye salmon runs to the Fraser River tributaries are generally later after warm springs (sea surface temperatures in the Gulf of Alaska) than after cold springs. Burgner (1980) had previously documented the reverse pattern for sockeye salmon returning to Bristol Bay, Alaska. In this region, warm years are associated with early returns. Blackbourn (1987) hypothesized that the feeding distributions of sockeye salmon are affected by water temperature; the distribution is farther north when the water is warm than when it is cold. He further hypothesized that both the date when salmon depart for home and their direction of travel are fixed for each population and that they do not compensate for their location (i.e. a calendar and a compass but no map). Thus the Alaskan fish return early after warm springs because they are closer to home when they initiate migration, whereas the Fraser River fish must travel south to get home and so are farther away after a warm year and return later.

The spatial patterns of return of Fraser River sockeye also support Blackbourn's (1987) hypothesis regarding time of return and temperature. Groot and Quinn (1987) summarized evidence that the region where these fish make landfall varies from year to year and that they make landfall farther north after warm springs than after cold springs. This is consistent with the idea that the salmon approach the coast on a fixed compass orientation but do not compensate for anomalies in the location from which they begin migration. There is no direct evidence of compass orientation by homing salmon at sea but experiments with juvenile sockeye salmon discussed earlier demonstrated this capability in salmon.

While there is no experimental evidence to support or refute Blackbourn's (1987) hypothesis, the distribution of Bristol Bay sockeye salmon prior to homeward migration is such that no

Table 4.2 Catches of maturing sockeye salmon entering Bristol Bay, Alaska, expressed in fish/hour/100 fathom (182.7 m) net (a measure of catch per unit of effort). Distances are offshore from Port Moller; data are from 1985 and 1987–89 (Helton, 1991)

Distance (km)	Average catch of sockeye
42.6	3.77
61.1	11.54
79.6	22.11
98.2	17.09
116.7	8.28
135.2	1.77

compass direction could get them all home. Those in the Gulf of Alaska must swim west, then north through the Aleutian passes and finally east into the bay (Fig. 4.13). Bristol Bay sockeye migrating from the western North Pacific Ocean would not migrate west at all and so would not be able to home with the same set of instructions which the fish in the Gulf of Alaska would need. This suggests that a map, perhaps based on the magnetic field, may be involved (Quinn, 1982b). Whatever their mechanism at sea, they enter the bay migrating in a discrete offshore band, not broadly dispersed or following the shoreline (Table 4.2).

The question of how the homing migration is timed has not been fully addressed. Salmon from a given population (e.g. Bristol Bay sockeye) may be so broadly distributed that some individuals might experience day lengths up to 150 min longer than others at the time when they commence homeward migration, depending on their location (Quinn, 1982b). Unlike many migratory animals, salmon are continuously moving during their years of feeding prior to homeward movement. Thus they experience varied environmental conditions depending on where they are at sea, but are able to rapidly migrate homeward from a broad range of locations.

4.4.4 Homing in coastal and estuarine waters

As the salmon migrate homeward from the open ocean towards their natal stream they move through coastal and estuarine waters.

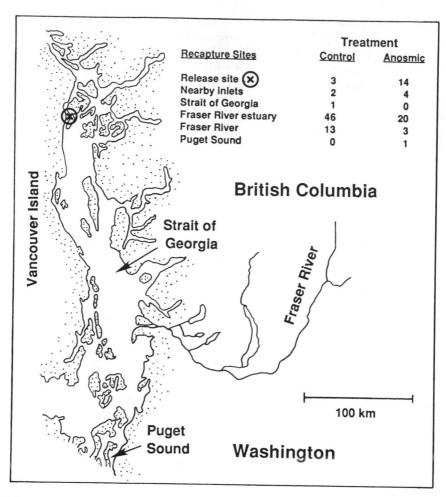

Figure 4.15 Map of the southern coast of British Columbia showing the location where control and anosmic sockeye salmon were tagged and released and the distribution of their recoveries (from Craigie, 1926).

These habitats pose special challenges for homing. The oceanic region is characterized by relatively cold water with little freshwater influence and slow, gyral current patterns. The coastal waters are generally warmer, have increasing amounts of freshwater on the surface and have rapid, reversing tidal currents.

The salmon undergo major physiological changes in coastal and estuarine waters. They mature rapidly and also make the osmoregulatory adjustments needed for the transition to freshwater. In addition to the changes in physical environment and physiology, the nearshore waters are probably also a transition zone between the homing mechanisms used in the open ocean and those used in the upriver movements.

Apparently the first experimental study of homing mechanisms in coastal waters was conducted by Craigie (1926). He captured adult sockeye salmon in Deepwater Bay on Quadra Island, between Vancouver Island and the mainland of British Columbia (Fig. 4.15). These salmon were presumably homing to the Fraser River system because it produces virtually all the sockeye salmon in this region. Approximately half the salmon (254) were tagged and released as controls and another 259 had the olfactory nerves severed before being released. Craigie's results (summarized in Fig. 4.15) have been interpreted to indicate an effect of handling and operation on behaviour because more anosmic fish failed to leave the release site. The results also provided experimental evidence that olfaction was involved in homing, as more of the controls reached and ascended the river than did the anosmic fish. However, the generally homeward movements of the anosmic salmon indicated that other mechanisms might also be operating.

Coastal and estuarine regions pose problems for the study of orientation because the salmon homing through these waters may be heading for different rivers and be on different schedules. Some populations remain in estuaries for days or weeks before ascending their natal river whereas others move rapidly upriver. Consequently, descriptive studies of salmon movements often reveal considerable variation in movement pattern. The coastal movements have been most closely studied for Fraser River sockeye salmon. Tagging work by Verhoeven and Davidoff (1962) demonstrated average travel rates of 25–35 km/day, in contrast to the 40–60 km/day often accomplished by these fish in the open ocean (Groot and Quinn, 1987).

Further information on movement patterns is available from studies using ultrasonic transmitters (Stasko *et al.*, 1976; Quinn *et al.*, 1989a). While telemetry reveals considerable variation in the movements of individual salmon, several patterns emerge. (1) The salmon generally swim at relatively efficient speeds of about 1 body length/s (Brett, 1983; Quinn, 1988). (2) They tend to swim in the upper part of the water column, though there seem to be interspecific differences. Moreover, their depth of travel may change from day to night or when oceanographic conditions change. (3) Salmon tend to swim actively during the day, though close to the river's mouth some appear to drift with the tidal currents. (4) The salmon do not use the tidal currents to facilitate their migration. They could do so by either vertical or horizontal movements into areas of slack water when the currents are not favourable. (5) Salmon do not merely follow shorelines on their homeward migration but often maintain a constant direction in open water for many hours.

Adult Fraser River sockeye salmon do not usually return home along the same route that they travelled on their seaward migration as juveniles (Groot and Cooke, 1987). They must therefore have some set of rules governing their behaviour which permit travel through structurally complex, unfamiliar regions with rapid, reversing currents. These waters seem to be the interface between compass mechanisms and olfaction. The behaviour of adult pink salmon in experimental arenas led Churmasov and Stepanov (1977) and Stepanov et al. (1979) to hypothesize that sun compass orientation was used on the homeward migration. Quinn et al. (1989a) reported that the Fraser River sockeye salmon generally swam roughly east-southeast until they encountered an island or other land mass. They did not follow the shoreline but rather reversed and moved in the opposite direction for several hours and later resumed homeward orientation.

A simulation model indicated that variations on this pattern of reversing orientation produced movement patterns which resembled those of tracked salmon (Pascual and Quinn, 1991). However, the simulation resulted in an unrealistic number of salmon becoming trapped in the complex inlets and islands between Vancouver Island and the mainland. It was concluded that some other mechanism in addition to compass orientation (such as avoidance of freshwater from non-natal sources) was needed to explain homing through coastal waters.

Assuming that the odours of the home river remain with the lighter freshwater on the surface, there will be an odour gradient associated with the current shear at the halocline (region of steepest salinity gradients). Research on Atlantic salmon led Westerberg (1984) and Doving et al. (1985) to hypothesize that salmon in coastal waters swim up and down through this interface between fresh- and salt-water, comparing the odour concentrations. They further hypothesized that salmon detect the small-scale current shears in this layer of water and infer the direction of their home river from the differential movements of the water masses. Doving et al. (1985) reported that the vertical movements of an anosmic salmon differed from those of controls. While this study was hampered by small sample size and the natural variability of salmon behaviour in such regions, the combined use of vertical stratification of odours and currents to orient in coastal and estuarine waters remains a viable hypothesis. Salmon often swim below or above the halocline for hours but perhaps they use olfactory information to set a course and then maintain it using some compass mechanism(s).

The failure of salmon to make use of the rapid tidal currents in these regions to facilitate migration implies that they are unable to detect the movements of the water in which they are swimming without a fixed visual reference. They are generally too far from shore and from the bottom to see any structure. Quinn *et al.* (1989a) reported that sockeye salmon tended to swim into the current more often than with the current but the relationship was weak. It is difficult to record water current directions and speeds at the location and depth where the fish are swimming. However, such data are necessary for an understanding of fish orientation to currents in open water. Ideally, information on the vertical profile of currents would be combined with information on the compass heading of the fish to indicate the orientation patterns and energetics of swimming in currents.

4.4.5 Homing in rivers

As salmon enter the mouth of their natal river they begin the final, freshwater, phase of their homing migration. In some cases this migration may take them almost 3000 km upstream to their spawning site, though other populations may spawn just above or even within the river's estuary. The orientation challenges facing these migrating adults can be formidable. As they migrate upstream, salmon must ascend the appropriate channels and bypass myriad non-natal tributaries, many of which may contain conspecifics. For species that spend many years at sea, the challenge may be compounded by changes in the composition of the river from natural flooding events or human activities. In some river systems salmon may also have to migrate through lakes with little water current available for orientation.

In spite of the challenges, salmon can be remarkably accurate in their ability to return to their natal site. For example, coho and chinook salmon reared and released from two hatcheries (Issaquah Creek and the University of Washington) demonstrate extraordinary homing fidelity (Fig. 4.16 and Table 4.3). Even though the Issaquah Creek fish must swim within a few hundred meters of the University of Washington hatchery, only 0.14% of the marked Issaquah Creek coho salmon and 0.18% of the chinook entered the University of Washington facility. Even fewer University of Washington salmon bypassed their home site and entered the hatchery on

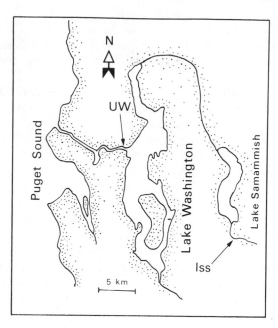

Figure 4.16 Map of the Lake Washington watershed showing the locations of the University of Washington (UW) and Issaquah Creek (Iss) hatcheries (from Quinn *et al.*, 1989b).

Table 4.3 Homing of coho and chinook salmon to the University of Washington (UW) and Issaquah Creek (Iss) hatcheries and straying between these sites, 1980–85 (from Quinn *et al.*, 1989b and unpublished data)

Species	Origin	Return site	
		UW	*Iss*
Coho salmon	UW	3712	2
	Iss	1	737
Chinook salmon	UW	5863	3
	Iss	1	542

Issaquah Creek (Table 4.3). While many other populations also show great fidelity for their home river (e.g. Quinn and Fresh, 1984; Quinn *et al.*, 1987), there are cases of straying (i.e. non-homing) levels of 10–20% (Quinn *et al.*, 1991).

Straying has generally been viewed as a failure of the sensory systems underlying home stream recognition. However, the variations in homing within and between species suggest both mechanistic and evolutionary explanations for straying. Some rivers may be either more difficult to detect or more difficult to distinguish from other rivers, hence more straying takes place. Alternatively, straying may be an alternative strategy to homing. The relative benefits of these two strategies might depend on the other life history traits of the species and the stability of the spawning environment (Quinn, 1984).

The mechanisms underlying salmon homing during riverine migration have been intensively studied. The pioneering experiment by Craigie (1926) notwithstanding, Hasler and Wisby (1951) were the first to formally demonstrate that fish can discriminate between the waters of different streams on the basis of odours. Based on their experiments (with a minnow, *Hyborhynchus notatus*, not salmon), they hypothesized that juvenile salmon learn odours associated with their 'parent stream' prior to seaward migration and later use these odours to locate the stream during the final phase of homing. The hypothesis has several components. First, streams must differ in chemical characteristics that are stable over seasons and years. Second, these differences must be distinguishable by salmon. Third, the salmon must learn the differences prior to seaward migration, remember them without reinforcement during the period of ocean residence, and respond to them as adults. This type of unconditioned, irreversible learning taking place during a particular developmental period resembles the phenomenon of offspring learning the appearance of their parents ('imprinting'), hence the term imprinting has been applied to olfactory learning in juvenile salmon.

As an initial test of their hypothesis, Wisby and Hasler (1954) captured migrating adult coho salmon that had ascended Issaquah Creek, Washington or its East Fork. They occluded the nares of half of the fish and released both the anosmic salmon and untreated controls below the confluence of the streams. Controls generally repeated their initial choice of stream, though the fidelity for the smaller East Fork was less than for the larger main creek (Table 4.4). Anosmic salmon showed much less fidelity for their stream of capture and the distribution of recaptures suggested a tendency to ascend the larger stream in the absence of olfactory cues.

A number of subsequent studies demonstrating the reduced homing ability of olfactory-occluded salmonids have provided further evidence for the role of olfaction in the final stages of homing (Hasler and Scholz, 1983). For example, coho salmon displaced from the University of Washington hatchery generally return if they are not caught by fishermen whereas anosmic coho do not return (18 of 20 controls vs 0 of 20 anosmics in 1988;

Table 4.4 Evidence for the importance of olfaction in home-stream selection by coho salmon (*Oncorhynchus kisutch*). Adult salmon were captured in the main channel of Issaquah Creek (Iss) or its East Fork and were tagged and released below the confluence. Controls were untreated, anosmic fish had their nares plugged with cotton and vaseline (Wisby and Hasler, 1954)

	Recapture site			
	Controls		Anosmic	
Capture site	Iss	E. Fork	Iss	E. Fork
Issaquah Creek	46	0	39	12
East Fork	8	19	16	3

Dittman, unpublished data). Brett and Groot (1963) questioned the validity of such sensory-impairment studies because motivation to home may be reduced. However, blinded salmon generally home (Hiyama *et al.*, 1966; Groves *et al.*, 1968), indicating that without olfaction, salmon cannot make the final selection of home stream.

Laboratory studies involving the introduction of natural stream water into experimental arenas demonstrated that adult salmon increased their locomotor activity in the presence of home water but not non-home water (Idler *et al.*, 1961; Fagerlund *et al.*, 1963). Electrophysiological recordings of summated olfactory bulbar activity (electro-olfactograms or EOGs) also indicated that home waters elicited strong responses in adult salmon (e.g. Hara *et al.*, 1965; Ueda *et al.*, 1967). While EOGs have demonstrated that salmon can discriminate between the odours of different streams, their usefulness as a measure of home-stream recognition is compromised by the fact that salmon may respond as strongly to non-home water as to home water (Oshima *et al.*, 1969a; Bodznick, 1975).

The second component of Hasler and Wisby's (1951) olfactory imprinting hypothesis is that adult salmon rely on long-term odour memories of the parent stream, acquired as juveniles, for orientation during the homeward migration. While the ability to distinguish water from different streams has been repeatedly demonstrated, the studies described above were not designed to address the issue of olfactory memory. Salmon for the olfactory occlusion and EOG studies were generally captured in the home river prior to the

Table 4.5 Homing of adult coho salmon (*Oncorhynchus kisutch*) to tributaries of Lake Michigan. Juvenile coho that had been exposed to the synthetic odorants morpholine (MOR) or phenylethyl alcohol (PEA) and unexposed controls were differentially marked and released into the lake. When the salmon were expected to mature, MOR and PEA were metered into selected rivers; these and other rivers were surveyed for returning salmon (Scholz *et al.*, 1976)

| Return sites | Experimental group | | |
	MOR	PEA	Controls
Unscented[a]	14	9	154
MOR-scented	659	20	76
PEA-scented	9	333	55

[a] Sixteen sites near the MOR- and PEA-scented rivers, including the site where the juvenile salmon had been released, were surveyed

experiment, hence it was not clear whether the responses of the fish were based on long-term memory or recent exposure to river odours (Brett and Groot, 1963; Oshima *et al.*, 1969b). To address this ambiguity, Hasler and his colleagues designed and conducted an elegant series of experiments demonstrating that odours learned as juveniles are used by homing adults (Cooper *et al.*, 1976; Scholz *et al.*, 1976; reviewed by Hasler and Scholz, 1983).

Hasler and Wisby (1951) had proposed that juvenile salmon exposed to a synthetic chemical might subsequently be attracted to an unfamiliar stream as adults if it contained the chemical. Scholz *et al.* (1976) carried out such an experiment by rearing juvenile coho salmon in a hatchery and exposing some of them to morpholine or phenethyl alcohol (PEA) and rearing some without exposure to either synthetic odorant. The salmon were differentially marked and released into Lake Michigan. During the spawning migration 1.5 years later, morpholine and PEA were metered into separate rivers and these and other rivers were monitored for marked salmon. The pooled results from 1974 and 1975 indicated that 96.6% of the adults that had been exposed to morpholine as juveniles were recovered in the morpholine-scented river and 92.0% of the PEA-exposed fish were recovered in the PEA-scented river (Table 4.5). Controls (with no prior exposure to the odorants) showed no strong preference for any of the streams.

The experiment by Scholz *et al.* (1976) demonstrated the reliance on learned odours in homing but did not indicate the linkage between olfaction and locomotion needed for successful upstream migration in a river system containing both home and non-home odours. Johnsen and Hasler (1980) found that the presence of learned artificial odorants elicited positive rheotaxis (upstream swimming) but in the absence of such odorants the salmon swam downstream. Based on these experiments, Johnsen (1982) developed a simple behavioural control model for home-stream selection consisting of positive rheotaxis released by the presence of the learned odour, negative rheotaxis in its absence and zig-zagging behaviour at low concentration such as might occur at the confluence of home and non-home rivers. It is not clear whether this model is compatible with Fretwell's (1989) finding that adult salmon distinguish 100% home water from home water diluted only 10% with non-home water.

While Johnsen's model may explain steady upstream migration in typical dendritic river systems, it does not account for the populations of salmon that enter freshwater and migrate a substantial distance upriver but then take up residence for many months prior to spawning. These holding areas are often quite far below the spawning areas and may even be on other rivers (e.g. spring chinook salmon: Berman and Quinn, 1991). The relationship between olfactory responses and the endocrine events associated with initial homing behaviour, holding, and final homing to the spawning site may be a fruitful area for further investigation. A second homing pattern not fully explained by the rheotaxis model is the behaviour of salmon returning to small tributaries or islands in large lakes. There is insufficient information on the movements of adult salmon in lakes to determine whether some compass mechanisms may also be involved.

Hasler and Wisby's hypothesis that juvenile salmon learn the chemical characteristics of their home stream and use these odours to guide their homing as adults is now commonly accepted. Although there is considerable evidence to justify this acceptance, it is important to note that the imprinting and homing process may be much more complex than is suggested by studies using artificial odorants. Considering the intensity with which salmon homing has been studied, it is remarkable how much of the process (particularly in natural systems) remains a mystery. The following sections describe several such areas of uncertainty and examine some new approaches to the phenomena of imprinting and homing.

4.4.6 Timing of imprinting

Hasler and Scholz (1983) proposed that the process of imprinting and homing is intimately linked to hormone levels at different life stages. Their studies with artificial odorants suggested that olfactory learning (imprinting) occurs during a sensitive period associated with the increased levels of the hormone thyroxine that characterize the period of transition from freshwater to sea-water (Dickhoff and Sullivan, 1987). This transition from freshwater residents, termed parr, to seaward migrants, termed smolts, is commonly called smolt transformation or smolting. Parr injected with thyroid stimulating hormone learned and retained the memory of odours whereas uninjected parr did not (Hasler and Scholz, 1983). The recent demonstration of a thyroid hormone-sensitive period in Atlantic salmon is consistent with these results (Morin et al., 1989a,b). Furthermore, Nevitt (1990) reported that after exposure to PEA during the smolt period, the olfactory receptor neurons of coho salmon were sensitized to this odorant. The hypothesis of imprinting at the smolt stage would explain why salmon reared at one site but moved to a second site prior to or during smolt transformation tend to return to the second site, not the first (e.g. Donaldson and Allen, 1957). It is therefore widely accepted that imprinting occurs at a single time and place.

In a natural river system, however, with the opportunity to experience a variety of water sources in addition to the natal site, the imprinting process may be more complex than one might infer from studies in hatcheries with artificial odours. Given the highly variable patterns of freshwater residence and anadromy exhibited by salmonid species, the olfactory learning process must be spatially and temporally flexible. For example, most sockeye salmon fry emerging from gravel nests in streams migrate immediately downstream to a lake and reside there for 1 or 2 years. As smolts, they emigrate from the lake's outlet, not the tributary where they were incubated and where they will return as adults. Coho and chinook salmon may also move extensively in freshwater prior to seaward migration (Peterson, 1982; Murray and Rosenau, 1989) yet home to their site of incubation, not smolting. The homing patterns of coho salmon that had been displaced from the rearing site for release as smolts indicated that they had learned more than one water source (Quinn et al., 1989b).

The hypothesis that salmon imprint during the incubation stage or at the time of emergence from the gravel gains support from Courtenay's (1989) evidence that juvenile coho salmon respond to odours that they have experienced shortly after or even prior to hatching. Morphological evidence is consistent with the possibility of early imprinting, as the olfactory system is apparently fully functional by the time of hatching (Brannon, 1972; Hara and Zielinski, 1989).

The paradox, thus, is that most experiments highlight the importance of olfactory learning at the smolt stage but the natural history of many salmonids seems to require learning at earlier stages. The resolution of this question may lie in the discovery that exposure to a novel water source causes a transient increase in plasma thyroid hormone levels in juvenile coho salmon (Dickhoff *et al.*, 1982; Hoffnagle and Fivizzani, 1990). If these hormone levels are associated with olfactory learning, then juvenile salmon migrating downstream might learn a series of olfactory waypoints (e.g. at the confluences of rivers) which they can retrace years later during homeward migration. Such a scenario is consistent with the sequential imprinting theory proposed by Harden Jones (1968) and Brannon (1982).

4.4.7 Natural odourants

Little is known about the nature of the odours that allow homing salmon to identify their natal stream. Hasler and Wisby (1951) proposed that salmon recognize a unique bouquet of odours emanating from the soil and vegetation along the stream. Initial attempts to characterize the components of home-stream water indicated that the stimulatory material was volatile and organic (Idler *et al.*, 1961). Subsequent studies have indicated that salmon are also able to differentiate water sources on the basis of inorganic components (Bodznick, 1978b). While it is undeniable that artificial odorants have been a critical tool for studying imprinting and homing, their use shifted the focus of research away from the chemicals to which salmon respond under natural conditions. Under natural conditions, fishes would be exposed to chemicals in complex combinations and might respond differently than if they detected the chemicals in isolation (Sandoval, 1980; Dodson and Bitterman, 1989). The ability of salmon to home in spite of changes in natural stream run-off and man-made wastes indicates a sensitive but flexible learning system.

One possible source of natural odours is the salmon themselves. Nordeng (1971, 1977) proposed that population-specific odours (pheromones) emanating from juvenile conspecifics residing in freshwater or migrating to sea guide homing adults. There are three key components to this hypothesis. Salmon populations must produce different odours, the differences must be detectable, and the salmon must rely on them for homing. Experiments have indicated that the first two conditions are met because salmon distinguish their own population from others (Groot *et al.*, 1986; Quinn and Tolson, 1986; Courtenay, 1989). However, transplanting experiments (e.g. Donaldson and Allen, 1957; Brannon and Quinn, 1990) and studies with artificial odorants (e.g. Scholz *et al.*, 1976) indicate that salmon will home to sites lacking conspecifics if other appropriate odours are present. The ability of salmon to distinguish siblings from non-siblings (Quinn and Busack, 1985; Quinn and Hara, 1986; Courtenay, 1989) suggests that pheromones may be involved in social behaviour of juveniles or perhaps mate selection of adults rather than homing.

The homing process eventually grades into the processes of spawning site selection and mate selection. The transition from migration to reproduction is not well understood but some combination of imprinted odours, chemicals from conspecifics, and perhaps innate preferences for certain substratum types may terminate homing.

4.4.8 Genetic control of homing

While the migration of juvenile salmon in freshwater and out to their marine feeding grounds seems to involve many innate responses, the great majority of experiments support the hypothesis that homing adults are responding to information that they acquired on the outward journey. However, two reports lend credence to the idea that homing has innate components as well. Bams (1976) reported that transplanted pink salmon did not home as well as locally adapted salmon and that hybrid stock (genes from local males and transplanted females) displayed intermediate levels of homing. McIsaac and Quinn (1988) reported that when adult chinook salmon from the upper Columbia River were spawned at a hatchery on the lower river, some of the progeny returned to the lower river site (as learning would dictate) but many swam past it, continuing upriver, and some even reached the ancestral site where they had never been. Neither of these two studies was able to explain the observed genetic differences in homing and the relative roles of learned and innate responses merit more attention.

4.5 SUMMARY AND CONCLUSIONS

Previous reviews of the migratory behaviour of fishes and the mechanisms guiding these migrations include those by Stasko (1971), Leggett (1977), Tesch (1980), Dodson (1988) and Arnold and Metcalfe (1989). We are impressed by the diversity of habitats through which fishes migrate and home and the diversity of ways in which these behaviour patterns fit into the lives of fishes. However, our understanding of the patterns and particularly the mechanisms of homing in fishes is derived from research on a very small number of species. While research has focused on relatively few species, some forms of migration and homing appear to be common, perhaps even the rule, in fishes.

The apparent prevalence of homing in fishes notwithstanding, the mechanisms underlying homing in open water are poorly understood. One or more species have been shown to use a sun compass, polarized light and/or electric and magnetic fields for orientation in experimental arenas. However, it is not clear how important these sources of information are and how they are integrated by migrants in their natural habitats. Species such as salmon display rapid and apparently directed oceanic migrations, leading some authors to invoke some form of navigation. Others are skeptical and find it more parsimonious to assume less oriented behaviour. To date there are only models to fuel to controversy; experimental evidence remains lacking.

Odours and rheotactic responses seem to be closely linked and are critical in the homing migrations of riverine fishes. However, the processes by which salmon (by far the most carefully studied migrants) learn the odours characteristic of their natal environment and use this information to return years later are still unclear in several key respects. We do not know what odours salmon learn, when they are learned, or how their memory is retained and replayed on the homeward journey. Indeed, the acquisition of information about the home site is almost totally unknown in any fishes except salmon.

We foresee three general trends in research on homing in fishes. First, there will continue to be descriptive studies on movement patterns of little-known species which will document more or less accurate homing. Such studies will spring both from the zoologists' interest in fish behaviour and from the needs of fisheries management. Fisheries scientists often tag fish for population estimates and to provide information on seasonal movements and population structure. Such studies often produce rich information on migration and homing patterns.

In addition to descriptive studies of homing, we anticipate increasing interest in homing in fishes from scientists interested in evolutionary biology and conservation. Reproductive homing is often associated with the separation of species into semi-isolated populations. The specializations of populations for their habitat provide excellent examples of evolution in action. Such work has so far been pursued most actively with salmonids (Ricker, 1972; Taylor, 1991) and, to a lesser extent, shad (Carscadden and Leggett, 1975). Comparisons with panmictic species such as eels (Helfman *et al.*, 1987) are particularly interesting. Moreover, the differentiation of species into genetically distinct, non-interbreeding populations poses special challenges both for rational management of fishing and conservation in general. The extent of genetic isolation of many salmon populations would qualify them as separate species by many definitions. In this context, the evolutionary aspects of straying should receive more attention.

Finally, there are three mechanistic aspects of homing which seem most intriguing. First, with the exception of fishes with known electroreceptive organs, the transduction system for the earth's magnetic field remains unknown. Several experiments have indicated that fishes can detect the magnetic field. However, neither convincing field demonstrations nor complete transduction mechanisms have been reported. Second, it is still not clear whether fishes (again, with the exception of electroreceptive species) detect and compensate for displacement by water currents when there is no fixed visual reference. Third, little is known about how the myriad orientation mechanisms are related ontogenetically and how they are integrated during migration through diverse and complex physical environments (as with diadromous fishes).

REFERENCES

Alt, K.T. (1977) Inconnu, *Stenodus leucichthys*, migration studies in Alaska 1961–1974. *J. Fish. Res. Bd Can.*, **34**, 129–33.

Armstrong, R.H. (1984) Migration of anadromous Dolly Varden charr in southeastern Alaska – a manager's nightmare, in *Biology of the Arctic Charr: Proceedings of the International Symposium on Arctic Charr* (eds L. Johnson and B. Burns), University of Manitoba Press, Winnipeg, pp. 559–70.

Arnold, G.P. (1981) Movements of fish in relation to water currents, in *Animal Migration* (ed. D.J. Aidley), Cambridge University Press, Cambridge, pp. 55–79.

Arnold, G.P. and Cook, P.H. (1984) Fish migration by selective tidal stream transport: first results with a computer simulation model for the European continental shelf, in *Mechanisms of Migration in Fishes* (eds J.D. McCleave, G.P. Arnold, J.J. Dodson and W.H. Neill), Plenum Press, New York, pp. 227–61.

Arnold, G.P. and Metcalfe, J.D. (1989) Fish migration: orientation and navigation or environmental transport? *J. Inst. Nav.*, **42**, 367–74.

Bams, R.A. (1976) Survival and propensity for homing as affected by presence or absence of locally adapted paternal genes in two transplanted populations of pink salmon (*Oncorhynchus gorbuscha*). *J. Fish. Res. Bd Can.*, **33**, 2716–25.

Berman, C.H. and Quinn, T.P. (1991) Behavioural thermoregulation and homing by spring chinook salmon, *Oncorhynchus tshawytscha* (Walbaum), in the Yakima river. *J. Fish Biol.*, **39**, 301–12.

Blackbourn, D.J. (1987) Sea surface temperature and pre-season prediction of return timing in Fraser River sockeye salmon (*Oncorhynchus nerka*). *Can. Sp. Publ. Fish. Aquat. Sci.*, **96**, 296–306.

Bodznick, D. (1975) The relationship of the olfactory EEG evoked by naturally-occurring stream waters to the homing behavior of sockeye salmon (*Oncorhynchus nerka*, Walbaum). *Comp. Biochem. Physiol.*, **52A**, 487–95.

Bodznick, D. (1978a) Water source preference and lakeward migration of sockeye salmon fry (*Oncorhynchus nerka*). *J. Comp. Physiol.*, **127**, 139–46.

Bodznick, D. (1978b) Calcium ion: an odorant for natural water discriminations and the migratory behavior of sockeye salmon. *J. Comp. Physiol.*, **127**, 157–66.

Brannon, E.L. (1972) Mechanisms controlling migration of sockeye salmon fry. *Int. Pacific Salmon Fish. Comm. Bull.*, **21**, 86 pp.

Brannon, E.L. (1982) Orientation mechanisms of homing salmonids, in *Salmon and Trout Migratory Behavior Symposium* (eds E.L. Brannon and E.O. Salo), School of Fisheries, University of Washington, Seattle, Washington, pp. 219–27.

Brannon, E.L. and Quinn, T.P. (1990) A field test of the pheromone hypothesis for homing by Pacific salmon. *J. Chem. Ecol.*, **16**, 603–9.

Brannon, E.L., Quinn, T.P., Lucchetti, G.L. and Ross, B.D. (1981) Compass orientation of sockeye salmon fry from a complex river system. *Can. J. Zool.*, **59**, 1548–53.

Bray, R.N. (1980) Influence of water currents and zooplankton densities on daily foraging movements of blacksmith, *Chromis punctipinnis*, a planktivarous reef fish. *Fish. Bull.*, **78**, 829–42.

Brett, J.R. (1983) Life energetics of sockeye salmon, *Oncorhynchus nerka*, in *Behavioral Energetics: the cost of survival in vertebrates* (eds W.P. Aspey and S.I. Lustick), Ohio State University Press, Columbus, Ohio, pp. 29–63.

Brett, J.R. and Groot, C. (1963) Some aspects of olfactory and visual responses in Pacific salmon. *J. Fish. Res. Bd Can.*, **20**, 287–303.

Buckley, R.M. and Huckle, G.J. (1985) Biological processes and ecological development on an artificial reef in Puget Sound, Washington. *Bull. Mar. Sci.*, **37**, 50–69.

Burgner, R.L. (1980) Some features of ocean migrations and timing of Pacific salmon, in *Salmonid Ecosystems of the North Pacific* (eds W.J. McNeil and D.C. Himsworth), Oregon State University Press, Corvallis, Oregon, pp. 153–64.

Carey, F.G. and Robison, B.H. (1981) Daily patterns in the activities of swordfish, *Xiphias gladius*, observed by acoustic telemetry. *Fish. Bull.*, **79**, 277–92.

Carey, F.G. and Scharold, J.V. (1990) Movements of blue sharks (*Prionace glauca*) in depth and course. *Mar. Biol.*, **106**, 329–42.

Carlson, H.R. and Haight, R.E. (1972) Evidence for a home site and homing of adult yellowtail rockfish, *Sebastes flavidus*. *J. Fish. Res. Bd Can.*, **29**, 1011–14.

Carscadden, J.E. and Leggett, W.C. (1975) Life history variations in populations of American shad, *Alosa sapidissima* (Wilson), spawning in tributaries of the St. John River, New Brunswick. *J. Fish Biol.*, **7**, 595–609.

Churmasov, A.V. and Stepanov, A.S. (1977) Sun orientation and guide-posts of the humpback salmon. *Sov. J. Mar. Biol.*, **3**, 55–63.

Clarke, T.A. and Wagner, P.J. (1976) Vertical distribution and other aspects of the ecology of certain mesopelagic fishes taken near Hawaii. *Fish. Bull.*, **74**, 635–45.

Cooper, J.C., Scholz, A.T., Horrall, R.M., Hasler, A.D. and Madison, D.M. (1976) Experimental confirmation of the olfactory hypothesis with artificially imprinted homing coho salmon (*Oncorhynchus kisutch*). *J. Fish. Res. Bd Can.*, **33**, 703–10.

Courtenay, S.C. (1989) Learning: memory of chemosensory stimuli by underyearling coho salmon (*Oncorhynchus kisutch*). PhD Dissertation, University of British Columbia, Vancouver, BC.

Craigie, E.H. (1926) A preliminary experiment upon the relation of the olfactory sense to the migration of the sockeye salmon (*Oncorhynchus nerka* Walbaum). *Trans R. Soc. Can.*, **20**, 215–24.

Craik, G.J.S. (1978) A further investigation of the homing behavior of the intertidal cottid, *Oligocottus maculosus* Girard. PhD Dissertation, University of British Columbia, Vancouver, BC.

Craik, G.J.S. (1981) The effects of age and length on homing performance in the intertidal cottid, *Oligocottus maculosus* Girard. *Can. J. Zool.*, **59**, 598–604.

Creutzberg, F. (1959) Discrimination between ebb and flood tide in migrating elvers (*Anguilla vulgaris* Turt.) by means of olfactory perception. *Nature*, **184**, 1961–2.

Creutzberg, F., Eltink, A.T.G.W. and van Noort, G.J. (1978) The migration of plaice larvae *Pleuronectes platessa* into the western Wadden Sea, in *Physiology and Behaviour of Marine Organisms* (eds D.S. Mclusky and A.J. Berry), Pergamon Press, Oxford, pp. 243–51.

Crossman, E.J. (1990) Reproductive homing in muskellunge, *Esox masquinongy*. *Can. J. Fish. Aquat. Sci.*, **47**, 1803–12.

Dadswell, M.J., Melvin, G.D., Williams, P.J. and Themelis, D.E. (1987) Influences of origin, life history, and chance on the Atlantic coast migration of American shad. *Am. Fish. Soc. Symp.*, **1**, 313–30.

Day, D.E. (1976) Homing behavior and population stratification in central Puget Sound English sole (*Parophrys vetulus*). *J. Fish. Res. Bd Can.*, **33**, 278–82.

Dickhoff, W.W., Darling, D.S. and Gorbman, A. (1982) Thyroid function during smoltification of salmonid fish. *Gunma Symposia on Endocrinology*, **19**, 45–61.

Dickhoff, W.W. and Sullivan, C. (1987) Involvement of the thyroid gland in smoltification, with special reference to metabolic and developmental processes. *Am. Fish. Soc. Symp.*, **1**, 197–210.

Dill, P.A. (1971) Perception of polarized light by yearling sockeye salmon (*Oncorhynchus nerka*). *J. Fish. Res. Bd Can.*, **28**, 1319–22.

Dodson, J.J. (1988) The nature and role of learning in the orientation and migratory behavior of fishes. *Envir. Biol. Fishes*, **23**, 161–82.

Dodson, J.J. and Bitterman, M.E. (1989) Compound uniqueness and the interactive role of morpholine in fish chemoreception. *Biol. Behav.*, **14**, 13–27.

Dodson, J.J. and Leggett, W.C. (1973) Behavior of adult American shad (*Alosa sapidissima*) homing to the Connecticut River from Long Island Sound. *J. Fish. Res. Bd Can.*, **30**, 1847–60.

Dodson, J.J. and Leggett, W.C. (1974) Role of olfaction and vision in the behavior of American shad (*Alosa sapidissima*) homing to the Connecticut River from Long Island Sound. *J. Fish. Res. Bd Can.*, **31**, 1607–19.

Donaldson, L.R. and Allen, G.H. (1957) Return of silver salmon, *Oncorhynchus kisutch* (*Walbaum*) to point of release. *Trans Am. Fish. Soc.*, **87**, 13–22.

Doving, K.B., Westerberg, H. and Johnsen, P.B. (1985) Role of olfaction in the behavioral and neuronal responses of Atlantic salmon, *Salmo salar*, to hydrographic stratification. *Can. J. Fish. Aquat. Sci.*, **42**, 1658–67.

Dubin, R.E. and Baker, J.D. (1982) Two types of cover-seeking behavior at sunset by the princess parrotfish, *Scarus taeniopterus*, at Barbados, West Indies. *Bull. Mar. Sci.*, **32**, 572–83.

Fagerlund, U.H.M., McBride, J.R., Smith, M. and Tomlinson, N. (1963) Olfactory perception in migrating salmon. III. Stimulants for adult sockeye salmon (*Oncorhynchus nerka*) in home stream waters. *J. Fish. Res. Bd Can.*, **20**, 1457–63.

Fisher, J.P. and Pearcy, W.C. (1987) Movements of coho, *Oncorhynchus kisutch*, and chinook, *O. tshawytscha*, salmon tagged at sea off Oregon, Washington and Vancouver Island during the summers 1982–85. *Fish. Bull.*, **85**, 819–26.

Fretwell, M.R. (1989) Homing behavior of adult sockeye salmon in response to a hydroelectric diversion of homestream waters at Seton Creek. *Int. Pacific Salmon Fish. Comm. Bull.*, **15**, 38 pp.

Gerber, G.P. and Haynes, J.M. (1988) Movements and behavior of smallmouth bass, *Micropterus dolomieui*, and rock bass, *Ambloplites rupestris*, in Southcentral Lake Ontario and two tributaries. *J. Freshwater Ecol.*, **4**, 425–40.

Gerking, S.D. (1953) Evidence for the concept of home range and territory in stream fishes. *Ecology*, **34**, 347–65.

Gibson R.N. (1976) Comparative studies on the rhythms of juvenile flatfish, in *Biological Rhythms in the Marine Environment* (ed. P.J. DeCoursey), University of South Carolina Press, Columbia, pp. 199–213.

Gibson, R.N. (1978) Lunar and tidal rhythms in fish, in *Rhythmic Activity of Fishes* (ed. J.E. Thorpe), Academic Press, London, pp. 201–13.

Gibson, R.N. (1984) Hydrostatic pressure and the rhythmic behavior of intertidal fishes. *Trans Am. Fish. Soc.*, **113**, 479–83.

Gibson, R.N., Blaxter, J.H.S. and de Groot, S.J. (1978) Developmental changes in the activity rhythms of the plaice (*Pleuronectes platessa* L.), in *Rhythmic Activity of Fishes* (ed. J.E. Thorpe), Academic Press, London, pp. 169–86.

Gladfelter, W.B. (1979) Twilight migrations and foraging activities of the copper sweeper *Pempheris schomburgki* (Teleostei: Pempheridae). *Mar. Biol.*, **50**, 109–19.

Godo, O.R. and Haug, T. (1988) Tagging and recapture of Atlantic halibut (*Hippoglossus hippoglossus*) in Norwegian waters. *J. Cons. Int. Explor. Mer.*, **44**, 169–79.

Goff, G.P. and Green, J.M. (1978) Field studies of the sensory basis of homing and orientation to the home site in *Ulvaria subbifurcata* (Pisces: Stichaeidae). *Can. J. Zool.*, **56**, 2220–4.

Goodyear, C.P. and Bennett, D.H. (1979) Sun compass orientation of immature bluegill. *Trans Am. Fish. Soc.*, **108**, 555–9.

Green, J.M. (1971) High tide movements and homing behavior of the tidepool sculpin *Oligocottus maculosus*. *J. Fish. Res. Bd Can.*, **28**, 383–9.

Green, J.M. and Fisher, R. (1977) A field study of homing and orientation to the home site in *Ulvaria subbifurcata* (Pisces: Stichaeidae). *Can. J. Zool.*, **55**, 1551–6.

Greer Walker, M., Harden Jones, F.R. and Arnold, G.P. (1978) The movements of plaice (*Pleuronectes platessa* L.) tracked in the open ocean. *J. Cons. Int. Explor. Mer.*, **38**, 58–86.

Greer Walker, M., Riley, J.D. and Emerson, L. (1980) On the movements of sole (*Solea solea*) and dogfish (*Scyliorhinus canicula*) tracked off the East Anglican coast. *Neth. J. Sea. Res.*, **14**, 58–86.

Griffin, D.R. (1952) Bird navigation. *Biol. Rev.*, **27**, 359–93.

Groot, C. (1965) On the orientation of young sockeye salmon (*Oncorhynchus nerka*) during their seaward migration out of lakes. *Behaviour (Suppl.)*, **14**, 198 pp.

Groot, C. and Cooke, K. (1987) Are the migrations of juvenile and adult Fraser River sockeye salmon (*Oncorhynchus nerka*) in near-shore waters related? *Can. Sp. Publ. Fish. Aquat. Sci.*, **96**, 53–60.

Groot, C. and Quinn, T.P. (1987) The homing migration of sockeye salmon, *Oncorhynchus nerka*, to the Fraser River. *Fish. Bull.*, **85**, 455–69.

Groot, C., Quinn, T.P. and Hara, T.J. (1986) Responses of migrating adult sockeye salmon (*Oncorhynchus nerka*) to population-specific odors. *Can. J. Zool.*, **64**, 926–32.

Groves, A.B., Collins, G.B. and Trefethen, G.B. (1968) Roles of olfaction and vision in choice of spawning site by homing adult chinook salmon (*Oncorhynchus tshawytscha*). *J. Fish. Res. Bd Can.*, **25**, 867–76.

Gunning, G.E. and Shoop, C.R. (1963) Occupancy of home range by longear sunfish, *Lepomis m. megalotis* (Rafinesque), and bluegill, *Lepomis m. macrochirus* Rafinesque. *Anim. Behav.*, **11**, 325–30.

Hara, T.J., Ueda, K. and Gorbman, A. (1965) Electroencephalographic studies of homing salmon. *Science*, **149**, 884–5.

Hara, T.J. and Zielinski, B. (1989) Structural and functional development of the olfactory organ in teleosts. *Trans Am. Fish. Soc.*, **118**, 183–94.

Harden Jones, F.R. (1968) *Fish Migration*, Arnold, London.

Harden Jones, F.R. (1977) Performance and behavior on migration, in *Fisheries Mathematics* (ed. J.H. Steele), Academic Press, London, pp. 145–70.

Harden Jones, F.R. (1979) The migration of plaice (*Pleuronectes platessa*) in relation to the environment, in *Fish Behavior and its Use in the Capture and Culture of Fishes* (eds J.E. Bardach, J.J. Magnuson, R.C. May and J.M. Reinhart), International Center for Living Aquatic Resource Management, Manila, pp. 383–99.

Harden Jones, F.R. (1981) Fish migration: strategy and tactics, in *Animal Migration* (ed. D.J. Aidley), Cambridge University Press, Cambridge, pp. 135–65.

Harden Jones, F.R., Arnold, G.P., Greer Walker, M. and Scholes, P. - (1979) Selective tidal stream transport and the migration of plaice (*Pleuronectes platessa* L.) in the southern North Sea. *J. Cons. Int. Explor. Mer.*, **38**, 331–7.

Hart, L.G. and Summerfelt, R.C. (1973) Homing behavior of flathead catfish, *Pylodictis olivaris* (Rafinesque), tagged with ultrasonic transmitters. *Proc. 27th A. Conf. S.E. Assoc. Game Fish. Comm.*, pp. 520–31.

Hartt, A.C. (1966) Migrations of salmon in the North Pacific Ocean and Bering Sea as determined by seining and tagging, 1959–1960. *Int. N. Pacific Fish. Comm. Bull.*, **19**, 141 pp.

Hartt, A.C. and Dell, M.B. (1986) Early oceanic migrations and growth of juvenile Pacific salmon and steelhead trout. *Int. N. Pacific Fish. Comm. Bull.*, **46**, 105 pp.

Hasler, A.D., Horrall, R.M., Wisby, W.J. and Braemer, W. (1958) Sun-orientation and homing in fishes. *Limnol. Oceanogr.*, **3**, 353–61.

Hasler, A.D. and Scholz, A.T. (1983) *Olfactory Imprinting and Homing in Salmon*. Springer-Verlag, Berlin, 134 pp.

Hasler, A.D. and Wisby, W.J. (1951) Discrimination of stream odors by fishes and relation to parent stream behavior. *Am. Nat.*, **85**, 223–38.

Hasler, A.D. and Wisby, W.J. (1958) The return of displaced largemouth bass and green sunfish to a 'home' area. *Ecology*, **39**, 289–93.

Hawryshyn, C.W., Arnold, M.G., Bowering, E. and Cole, R.L. (1990) Spatial orientation of rainbow trout to plane-polarized light: the ontogeny of E-vector discrimination and spectral sensitivity characteristics. *J. Comp. Physiol.*, **166**, 565–74.

Hawryshyn, C.W. and Beauchamp, R.D. (1985) Ultraviolet photo-sensitivity in goldfish: an independent UV retinal mechanism. *Vision Res.*, **25**, 11–20.

Helfman, G.S. (1981) Twilight activities and temporal structure in a freshwater fish community. *Can. J. Fish. Aquat. Sci.*, **38**, 1405–20.

Helfman, G.S., Facey, D.E., Hales, L.S. Jr and Bozeman, E.L. Jr (1987) Reproductive ecology of the American eel. *Am. Fish. Soc. Symp.*, **1**, 42–56.

Helfman, G.S. Meyer, J.L. and McFarland, W.N. (1982) The ontogeny of twilight migration patterns in the grunts (Pisces: Haemulidae). *Anim. Behav.*, **30**, 317–26.

Helfman, G.S. and Schultz, E.T. (1984) Social transmission of behavioral traditions in a coral reef fish. *Anim. Behav.*, **32**, 379–84.

Helton, D.R. (1991) An analysis of the Port Moller offshore test fishing forecast of sockeye and chum salmon runs to Bristol Bay, Alaska. MS Thesis, University of Washington, Seattle.

Hiramatsu, K. and Ishida, Y. (1989) Random movement and orientation in pink salmon (*Oncorhynchus gorbuscha*) migrations. *Can. J. Fish. Aquat. Sci.*, **46**, 1062–6.

Hiyama, Y., Taniuchi, T., Suyama, K., *et al.* (1966) A preliminary experiment on the return of tagged chum salmon to the Otsuchi River, Japan. *Jpn. Soc. Sci. Fish.*, **33**, 18–19.

Hobson. E.S. (1972) Activity of Hawaiian reef fishes during evening and morning transitions between daylight and darkness. *Fish. Bull.*, **70**, 715–40.

Hobson, E.S. (1973) Diel feeding migrations in tropical reef fishes. *Helgol. Wiss. Meeresunters.*, **24**, 361–70.

Hobson, E.S. and Chess, J.R. (1973) Feeding oriented movements of the atherinid fish *Pranesus pinguis* at Majuro Atoll, Marshall Islands. *Fish. Bull.*, **71**, 777–86.

Hoffnagle, T.L. and Fivizzani, A.J. (1990) Stimulation of plasma thyroxine levels by novel water chemistry during smoltification in chinook salmon (*Oncorhynchus tshawytscha*). *Can. J. Fish. Aquat. Sci.*, **47**, 1513–17.

Holland, K.N., Brill, R.W. and Chang, R.K.C. (1990a) Horizontal and vertical movements of yellowfin and bigeye tuna associated with fish aggregating devices. *Fish. Bull.*, **88**, 493–507.

Holland, K.N., Brill, R.W. and Chang, R.K.C. (1990b) Horizontal and vertical movements of Pacific blue marlin captured and released using sportfishing gear. *Fish. Bull.*, **88**, 397–402.

Hourston, A.S. (1992) Homing by Canada's west coast herring to management units and divisions as indicated by tag recoveries. *Can. J. Fish. Aquat. Sci.*, **39**, 1414–22.

Idler, D.R., McBride, J.R., Jonas, R.E.E. and Tomlinson, N. (1961) Olfactory perception in migrating salmon. II. Studies on a laboratory bio-assay for homestream water and mammalian repellent. *Can. J. Biochem. Physiol.*, **39**, 1575–83.

Iles, T.D. and Sinclair, M. (1982) Atlantic herring: stock discreteness and abundance. *Science*, **215**, 627–33.

Jamon, M. (1990) A reassessment of the random hypothesis in the ocean migration of Pacific salmon. *J. Theor. Biol.*, **143**, 197–213.

Jangaard, P. (1974) The capelin (*Mallotus villosus*) biology, distribution, exploitation, utilization, and composition. *Bull. Fish. Res. Bd Can.*, **186**, 70 pp.

Johnsen, P.B. (1982) A behavioral control model for homestream selection in migratory salmonids, in *Salmon and Trout Migratory Behavior Symposium* (eds E.L. Brannon and E.O. Salo), School of Fisheries, University of Washington, Seattle, pp. 266–73.

Johnsen, P.B. and Hasler, A.D. (1908) The use of chemical cues in the upstream migration of coho salmon, *Oncorhynchus kisutch, Walbaum. J. Fish Biol.*, **17**, 67–73.

Kalmijn, A.J. (1978) Electric and magnetic sensory world of sharks, skates and rays, in *Sensory Biology of Sharks, Skates and Rays* (eds E.S. Hodgson and R.R. Mathewson), U.S. Government Printing Office, Washington, DC, pp. 507–28.

Kalmijn, A.J. (1981) Biophysics of geomagnetic field detection. *Trans Inst. Elec. Electron. Engrs* (*Magnetics*), **17**, 1113–24.

Kalmijn, A.J. (1982) Electric and magnetic detection in elasmobranch fishes. *Science*, **218**, 916–18.

Karlsson, L. (1985) Behavioural responses of European silver eels (*Anguilla anguilla*) to the geomagnetic field. *Helg. Wiss. Meeresunters.*, **39**, 71–81.

Kelso, B.W., Northcote, T.G. and Wehrhahn, C.F. (1981) Genetic and environmental aspects of the response to water current by rainbow trout (*Salmo gairdneri*) originating from inlet and outlet streams of two lakes. *Can. J. Zool.*, **59**, 2177–85.

Khoo, H.W. (1974) Sensory basis of homing in the intertidal fish *Oligocottus maculosus* Girard. *Can. J. Zool.*, **52**, 1023–9.

Kleckner, R.C. and McCleave, J.D. (1988) The northern limit of spawning by Atlantic eels (*Anguilla* spp.) in the Sargasso Sea in relation to thermal fronts and surface water masses. *J. Mar. Res.*, **46**, 647–67.

Klimley, A.P., Butler, S.B., Nelson, D.R. and Stull, A.T. (1988) Diel movements of scalloped hammerhead sharks, *Sphyrna lewini* Griffith and Smith, to and from a seamount in the Gulf of California. *J. Fish Biol.*, **33**, 751–61.

Klimley, A.P. and Nelson, D.R. (1984) Diel movement patterns of the scalloped hammerhead shark (*Sphyrna lewini*) in relation to El Bajo Esiritu Santo: a refuging central-position social system. *Behav. Ecol. Sociobiol.*, **15**, 45–54.

L'Abee-Lund, J.H. and Vollestad, L.A. (1987) Homing precision of roach *Rutilus rutilus* in Lake Arungen, Norway. *Envir. Biol. Fishes*, **13**, 235–9.

Leggett, W.C. (1977) The ecology of fish migrations. *A. Rev. Ecol. Syst.*, **8**, 285–308.

Leggett, W.C. and Whitney, R.R. (1972) Water temperature and the migrations of American shad. *Fish. Bull.*, **70**, 659–70.

Lewis, W.M. and Flickinger, S. (1967) Home range tendency of the largemouth bass, *Micropterus salmoides*. *Ecology*, **48**, 1020–3.

Lockwood, S.J. and Lucassen, W. (1984) The recruitment of juvenile plaice (*Pleuronectes platessa* L.) to their parent spawning stock. *J. Cons. Int. Explor. Mer.*, **41**, 268–75.

Lowe-McConnell, R.H. (1975) *Fish Communities in Tropical Freshwaters*. Longman, London.

Loyacano, H.A. Jr, Chappell, J.A. and Gauthreaux, S.A. (1977) Sun-compass orientation in juvenile largemouth bass *Micropterus salmoides*. *Trans Am. Fish. Soc.*, **106**, 77–9.

Matthews, K.R. (1985) Species similarity and movement of fishes on natural and artificial reefs in Monterey Bay, California. *Bull. Mar. Sci.*, **37**, 252–70.

Matthews, K.R. (1990a) An experimental study of the habitat preferences and movement patterns of copper, quillback and brown rockfishes (*Sebastes* spp.). *Envir. Biol. Fishes*, **29**, 161–78.

Matthews, K.R. (1990b) A telemetric study of the home ranges and homing routes of copper and quillback rockfishes on shallow rocky reefs. *Can. J. Zool.*, **68**, 2243–50.

Matthews, K.R. (1990c) A comparative study of habitat use by young-of-the-year, subadult, and adult rockfishes on four habitat types in central Puget Sound. *Fish. Bull.*, **88**, 223–39.

McCleave, J.D. (1967) Homing and orientation of cutthroat trout (*Salmo clarki*) in Yellowstone Lake, with special reference to olfaction and vision. *J. Fish. Res. Bd Can.*, **24**, 2011–44.

McCleave, J.D. and Kleckner, R.C. (1982) Selective tidal stream transport in the estuarine migration of glass eels of the American eel (*Anguilla rostrata*). *J. Cons. Int. Explor. Mer.*, **40**, 262–71.

McCleave, J.D., Kleckner, R.C. and Castonguay, M. (1987) Reproductive sympatry of American and European eels and implications for migration and taxonomy. *Am. Fish. Soc. Symp.*, **1**, 286–97.

McCleave, J.D. and Wippelhauser, G.S. (1987) Behavioral aspects of selective tidal stream transport in juvenile American eels. *Am. Fish. Soc. Symp.*, **1**, 138–50.

McDowall, R.M. (1988) *Diadromy in Fishes: migrations between freshwater and marine environments*, Timber Press, Portland, 308 pp.

McFarland, W.N., Ogden, J.C. and Lythgoe, J.N. (1979) The influence of light on the twilight migrations of grunts. *Envir. Biol. Fish.*, **4**, 9–22.

McIsaac, D.O. and Quinn, T.P. (1988) Evidence for a hereditary component in homing behavior of chinook salmon (*Oncorhynchus tshawytscha*). *Can. J. Fish. Aquat. Sci.*, **45**, 2201–5.

Melvin, G.D., Dadswell, M.J. and Martin, J.D. (1986) Fidelity of American shad, *Alosa sapidissima* (Clupeidae), to its river of previous spawning. *Can. J. Fish. Aquat. Sci.*, **43**, 640–6.

Mesing, C.L. and Wicker, A.M. (1986) Home range, spawning

migrations, and homing of radio-tagged Florida largemouth bass in two central Florida lakes. *Trans Am. Fish. Soc.*, **115**, 286–95.

Miles, S.G. (1968) Laboratory experiments on the orientation of the adult American eel, *Anguilla rostrata*. *J. Fish. Res. Bd Can.*, **25**, 2143–55.

Miller, J.M., Crowder, L.B. and Moser, M.L. (1985) Migration and utilization of estuarine nurseries by juvenile fishes: an evolutionary perspective, in *Migration: mechanisms and adaptive significance* (ed. M.A. Rankin), *Contr. Mar. Sci. (Suppl.)*, **27**, 338–52.

Misra, R.K. and Carscadden, J.E. (1987) A multivariate analysis of morphometrics to detect differences in populations of capelin (*Mallotus villosus*). *J. Cons. Int. Explor. Mer.*, **43**, 99–106.

Morin, P.-P., Dodson, J.J. and Dore, F.Y. (1989a) Cardiac responses to a natural odorant as evidence of a sensitive period for olfactory imprinting in young Atlantic salmon, *Salmo salar*. *Can. J. Fish. Aquat. Sci.*, **46**, 122–30.

Morin, P.-P., Dodson, J.J. and Dore, F.Y. (1989b) Thyroid activity concomitant with olfactory learning and heart rate changes in Atlantic salmon, *Salmo salar*. *Can. J. Fish. Aquat. Sci.*, **46**, 131–6.

Murray, C.P. and Rosenau, M.L. (1989) Rearing of juvenile chinook salmon (*Oncorhynchus tshawytscha*) in nonnatal tributaries of the lower Fraser River, British Columbia. *Trans Am. Fish. Soc.*, **118**, 284–9.

Myrberg, A.A., Montgomery, W.L. and Fishelson, L. (1988) The reproductive behavior of *Acanthurus nigrofuscus* (Forskal) and other surgeonfishes (Fam. Acanthuridae) off Eilat, Israel (Gulf of Aqaba, Red Sea). *Ethology*, **79**, 31–61.

Neave, F. (1964) Ocean migrations of Pacific salmon. *J. Fish. Res. Bd Can.*, **21**, 1227–44.

Nevitt, G.A. (1990) Properties of olfactory receptor cells of coho salmon (*Oncorhynchus Kisutch*). PhD Dissertation, University of Washington, Seattle.

Neilson, J.D. and Perry, R.I. (1990) Diel vertical migrations of marine fishes: an obligate or facultative process? *Adv. Mar. Biol.*, **26**, 115–68.

Nordeng, H. (1971) Is the local orientation of anadramous fishes determined by pheromones? *Nature*, **233**, 411–13.

Nordeng, H. (1977) A pheromone hypothesis for homeward migration in anadramous salmonids. *Oikos*, **28**, 155–9.

Northcote, T.G. (1984) Mechanisms of fish migration in rivers, in *Mechanisms of Migration in Fishes* (eds J.D. McCleave, G.P. Arnold, J.J. Dodson and W.H. Neill), Plenum Press, New York, pp. 317–56.

Ogden, J.C. and Buckman, N.S. (1973) Movements, foraging groups, and diurnal migrations of the striped parrotfish *Scarus croicensis* Bloch (Scaridae). *Ecology*, **54**, 589–96.

Ogden, J.C. and Ehrlich, P.R. (1977) The behavior of heterotypic resting schools of juvenile grunts (Pomadasyidae). *Mar. Biol.*, **42**, 273–80.

Olson, D.E., Schupp, D.H. and Macins, V. (1978) A hypothesis of homing behavior of walleyes as related to observed patterns of passive and active movement. *Am. Fish. Soc. Spec. Publ.*, **11**, 52–7.

Olson, D.E. and Scidmore, W.J. (1963) Homing tendency of spawning white suckers in Many Point Lake, Minnesota. *Trans Am. Fish. Soc.*, **92**, 13–16.

Oshima, K., Hahn, W.E. and Gorbman, A. (1969a) Olfactory discrimination of natural waters by salmon. *J. Fish. Res. Bd Can.*, **26**, 2111–21.

Oshima, K., Hahn, W.E. and Gorbman, A. (1969b) Electroencephalographic olfactory responses in adult salmon to waters traversed in the homing migration. *J. Fish. Res. Bd Can.*, **26**, 2123–33.

Pardi, L. and Scapini, F. (1983) Inheritance of solar direction finding in sandhoppers: mass-crossing experiments. *J. Comp. Physiol.*, **151**, 435–40.

Parker, R.A. and Hasler, A.D. (1959) Movement of some displaced centrachids. *Copeia*, **1959**, 11–13.

Pascual, M.A. and Quinn, T.P. (1991) Evaluation of alternative models of the coastal migration of adult Fraser River sockeye salmon (*Oncorhynchus nerka*). *Can. J. Fish. Aquat. Sci.*, **48**, 799–810.

Patten, B.C. (1964) The rational decision process in salmon migration. *J. Cons. Int. Explor. Mer.*, **28**, 410–17.

Pawson, M.G., Kelley, D.F. and Pickett, G.D. (1987) The distribution and migrations of bass, *Dicentrarchus labrax* L., in waters around England and Wales as shown by tagging. *J. Mar. Biol. Assoc. U.K.*, **67**, 183–217.

Pearcy, W.G. and Fisher, J.P. (1988) Migrations of coho salmon, *Oncorhynchus kisutch*, during their first summer in the ocean. *Fish. Bull.*, **86**, 173–95.

Peterson, N.P. (1982) Immigration of juvenile coho salmon (*Oncorhynchus kisutch*) into riverine ponds. *Can. J. Fish. Aquat. Sci.*, **39**, 1308–10.

Petrere Junior, M. (1985) Migraciones de peces de agua dulce en America Latina: algunos comentarios. *FAO: COPESCAL Doc. Ocas.*, **1**, 17 pp.

Pflug, D.E. and Pauley, G.B. (1983) The movement and homing of smallmouth bass, *Micropterus dolomieui*, in Lake Sammamish, Washington, *Cal. Fish Game*, **69**, 207–16.

Quinn, T.P. (1980) Evidence for celestial and magnetic compass orientation in lake migrating sockeye salmon fry. *J. Comp. Physiol.*, **137**, 243–8.

Quinn, T.P. (1982a) Intra-specific differences in sockeye salmon fry compass orientation mechanisms, in *Salmon and Trout Migratory Behavior Symposium* (eds E.L. Brannon and E.O. Salo), School of Fisheries, University of Washington, Seattle, pp. 79–85.

Quinn, T.P. (1982b) A model for salmon navigation on the high seas, in *Salmon and Trout Migratory Behavior Symposium* (eds E.L. Brannon and E.O. Salo), School of Fisheries, University of Washington, Seattle, pp. 229–37.

Quinn, T.P. (1984) Homing and straying in Pacific salmon, in *Mechanisms of Migration in Fishes* (eds J.D. McCleave, G.P. Arnold, J.J. Dodson and W.H. Neill), Plenum Press, New York, pp. 357–62.

Quinn, T.P. (1988) Estimated swimming speeds of migrating adult sockeye salmon. *Can. J. Zool.*, **66**, 2160–3.

Quinn, T.P. and Brannon, E.L. (1982) The use of celestial and magnetic cues by orienting sockeye salmon smolts. *J. Comp. Physiol.*, **147**, 547–52.

Quinn, T.P., Brannon, E.L. and Dittman, A.H. (1989b) Spatial aspects of imprinting and homing in coho salmon (*Oncorhynchus kisutch*). *Fish. Bull.*, **87**, 769–74.

Quinn, T.P. and Brodeur, R.D. (1991) Intra-specific variations in the movement patterns of marine animals. *Am. Zool.*, **31**, 231–41.

Quinn, T.P. and Busack, C.A. (1985) Chemosensory recognition of siblings in juvenile coho salmon (*Oncorhynchus kisutch*). *Anim. Behav.*, **33**, 51–6.

Quinn, T.P. and Fresh, K. (1984) Homing and straying in chinook salmon (*Oncorhynchus kisutch*) from Cowlitz River Hatchery, Washington. *Can. J. Fish. Aquat. Sci.*, **41**, 1078–82.

Quinn, T.P. and Groot, C. (1984) Pacific salmon (*Oncorhynchus*) migrations: orientation versus random movement. *Can. J. Fish. Aquat. Sci.*, **41**, 1319–24.

Quinn, T.P. and Hara, T.J. (1986) Sibling recognition and olfactory sensitivity in juvenile coho salmon (*Oncorhynchus kisutch*). *Can. J. Zool.*, **64**, 921–5.

Quinn, T.P., Miller, B.S. and Wingert, R.C. (1980) Depth distribution and seasonal and diel movements of ratfish, *Hydrolagus colliei*, in Puget Sound, Washington. *Fish. Bull.*, **78**, 816–21.

Quinn, T.P., Nemeth, R.S. and McIsaac, D.O. (1991) Patterns of homing and straying by fall chinook salmon in the lower Columbia River. *Trans Am. Fish. Soc.*, **120**, 150–6.

Quinn, T.P. and Ogden, J.C. (1984) Field evidence of compass orientation in migrating juvenile grunts (*Haemulidae*). *J. Exp. Mar. Biol. Ecol.*, **81**, 181–92.

Quinn, T.P., terHart, B.A. and Groot, C. (1989a) Migratory orientation and vertical movements of homing adult sockeye salmon, *Oncorhynchus nerka*, in coastal waters. *Anim. Behav.*, **37**, 587–99.

Quinn, T.P. and Tolson, G.M. (1986) Evidence of chemically mediated population recognition in coho salmon (*Oncorhynchus kisutch*). *Can. J. Zool.*, **64**, 84–7.

Quinn, T.P., Wood, C.C., Margolis, L., Riddell, B.E. and Hyatt, K.D. (1987) Homing in wild sockeye salmon (*Oncorhynchus nerka*) populations as inferred from differences in parasite prevalence and allozyme allele frequencies. *Can. J. Fish. Aquat. Sci.*, **44**, 1963–71.

Raleigh, R.F. (1967) Genetic control in the lakeward migrations of sockeye salmon (*Oncorhynchus nerka*) fry. *J. Fish. Res. Bd Can.*, **24**, 2613–22.

Raleigh, R.F. and Chapman, D.W. (1971) Genetic control in lakeward migration of cutthroat trout fry. *Trans Am. Fish. Soc.*, **100**, 33–40.

Reese, E.S. (1989) Orientation behavior of butterflyfishes (family

Chaetodontidae) on coral reefs: spatial learning of route specific landmarks and cognitive maps. *Envir. Biol. Fishes*, **25**, 79–86.

Ricker, W.E. (1972) Hereditary and environment factors affecting certain salmonid populations, in *The Stock Concept in Pacific Salmon* (eds R.C. Simon and P.A. Larkin), H.R. Macmillan Lectures in Fisheries, University of British Columbia, Vancouver, pp. 19–160.

Robblee, M.B. and Zieman, J.C. (1984) Diel variation in the fish fauna of a tropical seagrass feeding ground. *Bull. Mar. Sci.*, **34**, 335–45.

Roff, D.A. (1988) The evolution of migration and some life history parameters in marine fishes. *Envir. Biol. Fishes*, **22**, 133–46.

Royce, W.F., Smith, L.S. and Hartt, A.C. (1968) Models of oceanic migrations of Pacific salmon and comments on guidance mechanisms. *Fish. Bull.*, **66**, 441–62.

Saila, S.B. and Shappy, R.A. (1963) Random movement and orientation in salmon migration. *J. Cons. Int. Explor. Mer.*, **28**, 153–66.

Sandoval, W.A. (1980) Odor detection by coho salmon (*Oncorhynchus kisutch*): a laboratory bioassay and genetic basis. MSc Thesis, Oregon State University, Corvallis, Oregon.

Savitz, J., Fish, D.A. and Weszely, R. (1983) Effects of forage on home range size of largemouth bass. *Trans Am. Fish. Soc.*, **112**, 772–6.

Scholz, A.T., Horrall, R.M., Cooper, J.C. and Hasler, A.D. (1976) Imprinting to chemical cues: the basis for home stream selection in salmon. *Science*, **192**, 1247–9.

Schwartz, F.J. (1987) Homing behavior of tagged and displaced carp, *Cyprinus carpio*, in Pymatuning Lake, Pennsylvania/Ohio. *Ohio J. Sci.*, **87**, 15–22.

Sciarrotta, T.C. and Nelson, D.R. (1977) Diel behavior of the blue shark, *Prionace glauca*, near Santa Catalina Island, California. *Fish. Bull.*, **75**, 519–28.

Stasko, A.B. (1971) Review of field studies on fish orientation. *Ann. N.Y. Acad. Sci.*, **188**, 12–29.

Stasko, A.B., Horrall, R.M. and Hasler, A.D. (1976) Coastal movements of adult Fraser River sockeye salmon (*Oncorhynchus nerka*) observed by ultrasonic tracking. *Trans Am. Fish. Soc.*, **105**, 64–71.

Stepanov, A.S., Churmasov, A.V. and Cherkashin, S.A. (1979) Migration direction finding by pink salmon according to the sun. *Sov. J. Mar. Biol.*, **5**(2), 92–9.

Storr, J.F., Hadden-Carter, J. and Myers, J.M. (1983) Dispersion of rock bass along the south shore of Lake Ontario. *Trans Am. Fish. Soc.*, **112**, 618–28.

Taylor, E.B. (1988) Adaptive variation in rheotactic and agonistic behaviour in newly-emerged fry of chinook salmon (*Oncorhynchus tshawytscha*) from ocean- and stream-type populations. *Can. J. Fish. Aquat. Sci.*, **45**, 237–43.

Taylor, E.B. (1991) A review of local adaptation in Salmonidae, with particular reference to Pacific and Atlantic salmon. *Aquaculture*, **98**, 185–207.

Templeman, W. (1974) Migrations and intermingling of Atlantic cod (*Gadus morhua*) stocks of the Newfoundland area. *J. Fish. Res. Bd Can.*, **31**, 1073–92.

Tesch, F.-W. (1974) Influence of geomagnetism and salinity on the directional choice of eels. *Helg. Wiss. Meeresunters.*, **26**, 382–95.

Tesch, F.-W. (1980) Migratory performance and environmental evidence of orientation, in *Environmental Physiology of Fishes* (ed. M.A. Ali), Plenum Press, New York, pp. 589–612.

Tesch, F.-W. (1989) Changes in swimming depth and direction of silver eels (*Anguilla anguilla* L.) from the continental shelf to the deep sea. *Aquat. Living Res.*, **2**, 9–20.

Thompson, S. (1983) Homing in a territorial reef fish. *Copeia*, **1983**, 832–4.

Tyus, H.M. (1985) Homing behavior noted for Colorado squawfish. *Copeia*, **1985**, 213–15.

Ueda, K., Hara, T.J. and Gorbman, A. (1967) Electroencephalographic studies on olfactory discrimination in adult spawning salmon. *Comp. Biochem. Physiol.*, **21**, 133–43.

Vacquier, V. and Uyeda, S. (1967) Paleomagnetism of nine seamounts in western Pacific and three volcanoes in Japan. *Bull. Earthq. Res. Inst.*, **45**, 815–49.

Veen, J.F. de (1978) On selective tidal stream transport in the migration of North Sea plaice (*Pleuronectes platessa*) and other flatfish species. *Neth. J. Sea Res.*, **12**, 115–47.

Verhoeven, L.A. and Davidoff, E.B. (1962) Marine tagging of Fraser River sockeye salmon. *Int. Pacific Salmon Fish. Comm. Bull.*, **13**, 132 pp.

Walburg, C.H. and Nichols, P.R. (1967) Biology and management of the American shad and status of the fisheries, Atlantic coast of the United States, 1960. *U.S. Fish Wildl. Serv. Spec. Sci. Rep. Fish.*, **550**, 105 pp.

Walker, B.W. (1952) A guide to the grunion. *Cal. Fish Game*, **38**, 409–20.

Walker, M.M. (1984) Learned magnetic field discrimination in yellowfin tuna, *Thunnus albacares*. *J. Comp. Physiol.*, **155**, 673–9.

Weihs, D. (1978) Tidal stream transport as an efficient method for migration. *J. Cons. Int. Explor. Mer.*, **38**, 92–9.

Werner, R.G. (1979) Homing mechanism of spawning white suckers in Wolf Lake, New York. *N.Y. Fish Game J.*, **26**, 48–58.

Westerberg, H. (1984) The orientation of fish and the vertical stratification at fine- and micro-structure scales, in *Mechanisms of Migration in Fishes* (eds J.D. McCleave, G.P. Arnold, J.J. Dodson and W.H. Neill), Plenum Press, New York, pp. 179–203.

Wheeler, J.P. and Winters, G.H. (1984) Homing of Atlantic herring (*Clupea harengus harengus*) in Newfoundland waters as indicated by tagging data. *Can. J. Fish. Aquat. Sci.*, **41**, 108–17.

Winn, H.E., Salmon, M. and Roberts, N. (1964) Sun–compass orientation by parrot fishes. *Zeits. Tierpsychol.*, **21**, 798–812.

Winter, J.D. (1977) Summer home range movements and habitat use by four largemouth bass in Mary Lake, Minnesota. *Trans Am. Fish. Soc.*, **106**, 323–30.

Wisby, W.J. and Hasler, A.D. (1954) Effect of occlusion on migrating silver salmon (*Oncorhynchus kisutch*). *J. Fish. Res. Bd Can.*, **11**, 472–8.

Yuen, H.S.H. (1970) Behavior of skipjack tuna, *Katsuwonus pelamis*, as determined by tracking with ultrasonic devices. *J. Fish. Res. Bd Can.*, **27**, 2071–9.

Zijlstra, J.J. (1988) Fish migrations between coastal and offshore areas, in *Coastal–offshore ecosystem interactions* (ed. B.-O. Jansson), Springer-Verlag, Berlin, pp. 257–72.

Chapter 5
Amphibians
Ulrich Sinsch

5.1 INTRODUCTION

The spectacular mass migrations of many Palaearctic amphibians towards their spawning sites have long drawn the attention of naturalists to their ability for directed movements over large distances. Reports of common toads (*Bufo bufo*) returning to former breeding ponds which had been inhabited years before (Heusser, 1956) illustrate the extraordinary site fidelity of some species. The phenomenon of homing in Amphibia has been well documented since the last century (Dürigen, 1897), but the identification of directional cues guiding the spontaneous migrations and of the perceptual systems dates from the past three decades. Recent research focuses mainly on the environmental constraints of homing and the specific role of each cue in the orientation system, whereas the study of the neural mechanisms processing and evaluating directional information has been neglected.

Amphibians migrate, if important habitat resources are spatially separated. Homing to specific sites within the habitat has been observed in adults of about 50 species of terrestrial urodeles and anurans, whereas data on aquatic species and caecilians are still lacking (reviewed by Sinsch, 1991). Generally, ponds (resources: mates, water), home ranges (resources: food, shelter) and hibernation sites (resource: shelter) are subject to return migrations. However, the presence and the extent of site fidelity varies considerably among different species, sometimes even among populations of the same species. Therefore, prior to the analysis of the sensory aspects of homing two basic problems will be addressed:

Animal Homing. Edited by Floriano Papi. Published in 1992 by Chapman & Hall, 2–6 Boundary Row, London SEI 8HN. ISBN 0 412 36390 9.

1. Why do some migratory amphibians home and others do not?
2. Is homing behaviour restricted to the area of previous migratory experience?

5.1.1 Homing behaviour or random search?

Amphibians which rely on spatially separated resources perform periodic migrations between the sites of reproduction, nutrition, shelter, and hibernation. These resources are found by two basic strategies: random search, and homing to a resource known from previous excursions. Both strategies imply costs in terms of the investment of time and energy for locomotion and the increased risk of predation. The random-search strategy is adopted by juvenile amphibians which lack information about the location of resources. They disperse over the adjacent areas around their natal place and become residents at favourable sites. In contrast, adults possess local knowledge obtained during the juvenile stage and can choose between the two strategies. If return migrations cause lower costs than random search, selection pressure is expected to act in favour of homing behaviour.

Homing on direct paths minimizes the investment of time, if the resource is scarce or patchily distributed. In contrast, random search for new sites seems appropriate, if the resource is common or its location is unknown. In fact, there is a slight tendency in *Bufo bufo* and *B. woodhousii fowleri* (Fowler's toad) for the frequency of individuals exhibiting breeding site fidelity to the natal pond to decrease with the number of suitable ponds within the migratory range of each population (Table 5.1). The breaking of fidelity to the natal pond in favour of an artificial, less distant pond as evidenced in common toads (Schlupp *et al.*, 1990) also indicates the attempt of amphibians to decrease the time needed for the breeding migration. The spatial distribution of the resource is not the only factor influencing the number of homing individuals per population; the specific quality (e.g. size, number of conspecific competitors) of the resource certainly has an effect too.

The spatial range of homeward orientation is also a measure of the extent of site fidelity provided that it does not reflect the inability to relocate the home site. Two examples will illustrate this point. Non-reproductive common toads establish summer home ranges 50–3000 m from the breeding pond (Heusser, 1968). Suitable home ranges are patchy but abundant resources, and toads usually return to their home areas following foraging trips of up to 150 m distance. However, the home site fidelity is limited in time to one season and also in space (Sinsch, 1987a). Toads with extensive

Table 5.1 Frequency of individuals exhibiting breeding site fidelity and of potential breeding sites within the migratory range of five toad populations

Species	No. of breeding ponds	Homing individuals (%)	Reference
Bufo bufo	2	97	Sinsch (unpublished)
	4	95	Heusser (1969)
	6	60	Jungfer (1943)
Bufo woodhousii	7	73	Breden (1987)
fowleri	>10	24	Ferguson (1963)
Bufo viridis	15	66	Sauer (1988)
Bufo calamita	15	52	Sauer (1988)

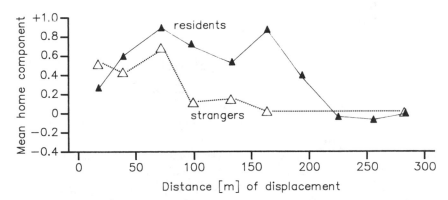

Figure 5.1 Initial orientation of common toads passively displaced from their summer home range. Residents: 23 toads originating from the local population with previous migratory experience in the release area. Strangers: 25 toads originating from a 10 km distant population displaced to the same area as the residents and allowed to establish new home ranges (modified from Sinsch, 1987a).

local knowledge cease to return to the home range at displacement distance of more than 200 m; toads with little previous migratory experience in the area fail to return home even at distances of 100 m (Fig. 5.1). Outside this small area around the previous home range they prefer to establish a new home range. In the Manitoba toad (*Bufo hemiophrys*) which inhabits the rather uniform margins of ponds during summer, home site fidelity is completely lacking

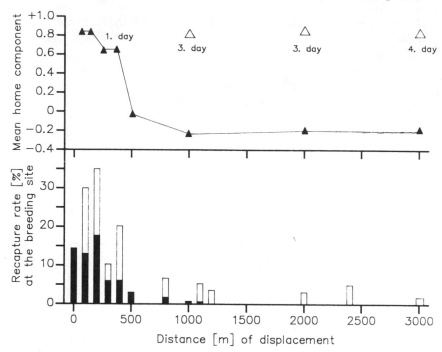

Figure 5.2 Orientation towards the breeding site and homing success of passively displaced common toads. The paths of these toads were recorded continuously (mechanical tracking device) during the first 4 days following displacement. Filled triangles represent the home component at the first day, open triangles that at the third and fourth day following release (Sinsch, 1987b; unpublished). Filled columns show the proportion of displaced tagged toads which homed during the same breeding season, open columns the total of toads recaptured 1 year later (Heusser, 1969).

(Breckenridge and Tester, 1961). On the other hand, the same toads show a remarkable, long-lasting fidelity to hibernation sites which are rare and patchy resources.

The costs of homing behaviour (energy expenditure, risk of predation) increase with the distance of migration. In fact, the frequency of homing in common toads following displacement abruptly drops at distances of more than 400 m from the breeding site. However, most individuals are able to orient towards home even at a displacement distance of 3000 m (Fig. 5.2). This finding indicates that the orientation cues used at short distances do not provide the necessary information at larger distances. The shift to an alternative and better suited sensory modality may be responsible for the delayed homeward orientation. The number of toads which return to the breeding site from distant release sites remains low even if assessed one year following displacement. Information on the further fate of the non–homing toads is still lacking. In conclusion, a distance of about 400 m probably marks the turning point where the costs of immediate homing outweigh the benefit in most individuals. This may explain why female common toads which prefer more distant summer home ranges than males do not migrate every year to the breeding site, although the ovarian cycle is completed in 3 months (Heusser, 1969). The occurrence and the spatial range of homing behaviour are subject to the following constraints:

1. the costs of random search for a new site providing the resource needed must exceed the costs of a straight return migration to a known site;
2. the frequency with which individuals exhibit site fidelity in a population and the spatial range of homing increase with the scarcity of sites providing the resource;
3. the motivation to home decreases with increasing distance of displacement.

5.1.2 Navigation or orientation within familiar areas?

Under natural conditions, an active outward journey precedes any kind of homing behaviour. Therefore, in the field, homing always takes place in an area experienced previously. The only exceptions are individuals which have been accidentally dislodged by the current of a stream or by a predator. However, passively displaced common toads which could not perceive acoustic, magnetic, mechanical, olfactory and visual cues during the outward journey still orient correctly towards home (Buck, 1988). This finding does not rule out the possibility that route-based information contributes to homing, but it is certainly not essential. However, among amphibians, only the California newt (*Taricha torosa*) has been shown to use kinesthetic orientation (Endler, 1970).

The essential criterion for navigation, i.e. homeward orientation based on a map-and-compass mechanism (Papi, 1990) must now be addressed. At present it is difficult to distinguish pilotage, steering from one familiar point to another, from navigation in Amphibia because almost all experimental displacements have been performed within the area of familiarity where either kind of orientation may work. The only definite evidence for navigation is therefore homeward orientation of passively displaced amphibians from unfamiliar release sites (Able, 1981). Unfortunately, the previous migratory experience of the displaced individuals is unknown in all studies on homing. Estimates of the migratory range of adults do not solve the problem because juveniles are distinctly more wide-ranging than adults (Breden, 1987). The largest natural migratory range of an amphibian is that of juvenile cane toads (*Bufo marinus*) with a maximum distance of 35 km from the natal pond (Freeland and Martin, 1985). Release sites outside this range can be considered as unfamiliar areas for any species of amphibian. However, in none of the few displacements outside this range were individuals oriented homewards. Yet, the migratory range of most amphibians is undoubtedly far smaller than 35 km (Sinsch, 1991). Curiously, among Amphibia the rather sedentary newt *Taricha rivularis* holds the record for long-distance homeward orientation at 12.8 km (Twitty *et al.*, 1967). These newts, and common toads (Sinsch, 1987b), are the most likely candidates for homing from unfamiliar sites. However, there is no indication that amphibians can navigate from release sites far outside their migratory range towards familiar areas. At the present state of knowledge we have to assume that homing orientation requires previous migratory experience to establish a local map which permits little or no extrapolation in unfamiliar areas.

5.2 SENSORY BASIS OF ORIENTATION BEHAVIOUR

The directional cues and the perceptual systems related to homing behaviour have long remained enigmatic because the concept of monosensory orientation predominated among herpetologists. Moreover, deprivation of certain sensory information often affects the initial orientation following displacement but does not prevent successful homing. In fact, it is almost impossible to impair homing in amphibians displaced to moderately distant release sites. These findings finally led to the conclusion that amphibians possess a multisensory orientation system utilizing a variety of directional cues (Ferguson, 1971). The following account presents the classes of cues known to be used for homing and the sensory systems involved in their perception.

5.2.1 Acoustic orientation

Potentially, the vocalization of conspecifics, and sounds originating from other animals or of abiotic origin may serve as directional cues, if they are related in an unequivocal way to the goal of a migrating animal. In fact, many amphibians have been shown to discriminate between conspecific and other calls, and to locate phonotactically the origin of conspecific calls (Rand, 1988). On the other hand, it remains unclear whether amphibians use sounds other than conspecific signals as directional references.

Vocalization is a common feature of many anuran, urodele, and caecilian species, but only the calls of anurans reach sound pressure levels which are audible at more than a few metres distance. Therefore, the main function of amphibian vocalization is apparently that of short-distance communication, e.g. mate recognition, male–male competition, and territorial defence (Wells, 1988). However, there are also notable exceptions which use calls for long-distance phonotaxis.

(a) Acoustic cues

Natterjack toads (*Bufo calamita*) produce advertisement calls which are still audible 1000 m distant from the chorus (Arak, 1983). The maintenance of high sound pressure levels implies great energetic investment as evidenced by increased oxygen consumption and partially anaerobic metabolism in calling frogs. Therefore, only anurans which do not breed at geographically fixed sites use this strategy to attract females. Nevertheless, even in the absence of guiding calls, displaced male natterjacks remain able to locate their former calling site (Sinsch, 1992).

Surprisingly little is known about the sound localization behaviour of amphibians. The available information is limited to a few anuran species tested in phonotaxis experiments at short distances. Frogs approach a conspecific sound source in a zig-zag course which causes a frequent alternation of the leading ear. Successive comparisons of the binaural cues at decreasing distances, sometimes accompanied with head scanning movements, are the basic features of sound localization (Rheinländer and Klump, 1988). Thus far, the frog's ear is thought to act as a pressure-gradient receiver that is highly directional, in particular for low-frequency sounds (Eggermont, 1988). The strategy of frogs employing phonotactic, long-distance orientation has not yet been studied in detail, but the above-described behaviour is probably also effective at greater distances.

(b) Sound receptors

In anurans the primary sound receptors are the external ears, often including a tympanic membrane. The vibrations caused by airborne sound waves above 1000 Hz are transmitted via the columnella to the papilla basilaris of the inner ear. A second pathway for frequencies below 1000 Hz consists of the opercular complex and the papilla amphibiorum of the inner ear (Lewis and Lombard, 1988). While the papilla amphibiorum is present in all amphibians studied so far, the opercular complex seems restricted to species with terrestrial habitats (McCormick, 1988). Neural audiograms prove the existence of at least two frequency ranges of extraordinary auditory sensitivity (Walkowiak et al., 1981). The upper range coincides closely with the frequency range of the conspecific vocalizations (tympanic–columnellar–papilla basilaris pathway). The lower range of sensitivity covers frequencies typically below 500 Hz and is the same in most species (opercular–papilla amphibiorum pathway). This common pathway is rarely used for conspecific communication, but is probably involved in the sound localization mechanism of anuran amphibians.

5.2.2 Magnetic field orientation

Orientation based on magnetic information was first shown in the cave salamander, *Eurycea lucifuga* (Phillips, 1977). Evidence of this ability has been extended to another newt and four toad species, by the use of either deprivation experiments or controlled alterations of magnetic parameters. The effects of these treatments range from complete disorientation to predictable false directional choice. Generally, artificial magnetic fields affect the initial orientation but do not impair homing.

(a) Cues from the earth's magnetic field

The experimental manipulation of inclination (vertical component) and declination (horizontal component) of the magnetic field alters the orientation behaviour of salamanders (Phillips, 1986) and toads (Buck, 1988). The inclination of the earth's magnetic field serves as a reference for simple compass orientation in eastern red-spotted newts (*Notophthalmus viridescens*) and in common toads (*Bufo bufo*). However, the inversion of the magnetic vertical component had no effect on the directional choice of presumably homing newts (Phillips, 1986), whereas the alteration of the horizontal component did have an effect (Phillips, 1987). Phillips' interpretation of these findings is that there are two types of magnetic compasses, one for simple compass orientation relying on axial sensitivity and another one for homing by responding to polar sensitivity. Yet, this conclusion is based on tests on only a few individuals and the

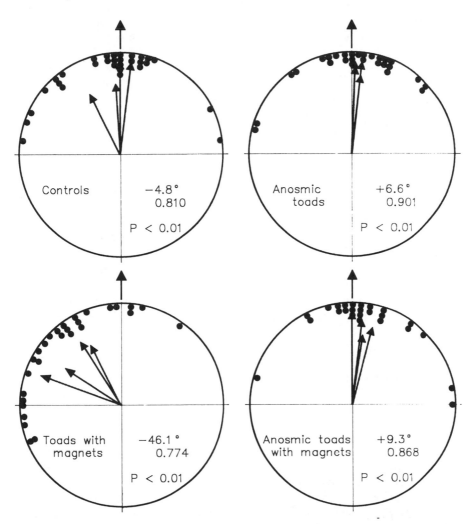

Figure 5.3 The initial orientation of passively displaced Andean toads towards their breeding site. Toads were equipped with bar magnets or brass bars of the same size, and/or made anosmic by covering their olfactory epithelia with cyanoacrylate glue. The outer arrows indicate the breeding site direction; the inner arrows represent the mean vectors of four individual releases (radius of circle, $r = 1$). The direction and length of the second-order mean vector and its significance are also given. Each dot in the periphery of the circle represents the bearing of one toad (modified from Sinsch, 1988).

experiment has not been repeated in the same or any other species. It is unknown whether amphibians can detect other parameters of the earth's magnetic field such as the intensity.

(b) Receptors

The sensory basis for the detection of directional information about the earth's magnetic field has not been identified so far. However, the skin melanophores of the clawed toad (*Xenopus laevis*) respond directly to weak magnetic fields, even if isolated (Leucht, 1987). Since melanophores are not connected with the central nervous system, these sites sensitive to magnetic fields are probably not the receptors of directional information. On the other hand, anosmic Andean toads (*Bufo spinulosus*) with magnets glued on the head oriented correctly, whereas intact toads with magnets deviated significantly from the home direction (Fig. 5.3). These findings suggest that parts of the olfactory system may be involved in detecting magnetic fields. Nevertheless, neurophysiological evidence for the location of magnetic receptors is not available at present.

5.2.3 Orientation based on mechanical stimuli

Amphibians respond to a variety of vibratory stimuli with directed movements. However, mechanical cues are used only for short-distance orientation to fixed reference points within a range of a few metres. Therefore, these cues and their receptors are mentioned only briefly. Toads use water surface-waves for communication in the course of their mating behaviour (Walkowiak and Münz, 1985). The receptors involved are the lateral line organs (Elephandt, 1984). Seismic signals provide information which helps to locate approaching predators and they are perceived by the sensory epithelia of the saccula and lagena of the inner ear (Lewis and Narins, 1985; Lewis and Lombard, 1988). Finally, rheotaxis permits aquatic salamanders to orient in the darkness of the cave habitat (Durand and Parzefall, 1987).

5.2.4 Olfactory orientation

Many amphibians use odours as directional cues for short-distance orientation towards mates or prey, or for long-distance homeward orientation. Evidence is based on deprivation experiments and on the choice of specific odours which the subjects had previously learned to recognize. The effect of anosmia ranges from complete impairment of homing in some salamanders, to reduced initial orientation in other salamanders and toads, and to normal homeward orientation in another toad (Sinsch, 1991). The importance of olfactory cues for initial orientation of toads seems to be positively related to the size of their migratory range (Sinsch, 1990).

(a) Olfactory cues

The chemical nature of the odours involved in amphibian homing behaviour is completely unknown. During the breeding migration the guiding odours originate, at least partly, from the breeding pond itself, as demonstrated in some anurans (Grubb, 1973; Sinsch, 1987b). Homing salamanders seem to use local odour patterns (Madison, 1972).

(b) Olfactory receptors

The olfactory system of amphibians is well developed and consists of two receptor organs with parallel but distinct neural channels (Duellman and Trueb, 1986). Two areas of sensory epithelia are located in the cavum principale, one on the roof and the walls, another on the floor. The first one gives rise to the dorsal division of the olfactory nerve, the second one to its ventral division. The third area of sensory epithelia forms part of the vomeronasal organ (= Jacobson's organ) situated in an accessory chamber and gives rise to the vomeronasal nerve.

5.2.5 Visual orientation

The importance of visual stimuli for the homing behaviour of amphibians has long been recognized (Buytendijk, 1918). A variety of visual cues is utilized from the larval to the adult stage to orient in space and in time. Evidence is based on the effects of false or absent visual information on the initial orientation of many amphibians after displacement. However, in most species even blinding does not impair successful homing. Visual cues are more important for the orientation of species with a small migratory range than for the orientation of wide-ranging species (Sinsch, 1990).

(a) Visual cues

There are two classes of visual cues: fixed landmarks such as shore lines or forest silhouettes, and periodically 'moving' celestial cues such as the sun, moon, stars or polarization patterns. The specific role of each cue in homeward orientation apparently differs among the species and also depends on the class of cue. At short distances, common toads use landmarks to steer straight course in familiar areas (Heusser, 1969). Celestial cues require compensation for their apparent movement before they can be used as directional reference points.

(b) Light receptors

In Amphibia the perception of visual orientation cues is not restricted to the lateral eyes but also takes place in the pineal complex (Adler, 1976). The interaction of both sensory systems has been demonstrated in frogs and salamanders: eyeless individuals and those with a covered pineal complex orient as well as normal individuals, whereas eyeless amphibians with a covered pineal complex orient randomly (Taylor and Ferguson, 1970; Taylor, 1972). The lateral eyes of most amphibians possess muscles of accommodation and a retina containing four kinds of receptors (Duellman and Trueb, 1986): green rods (maximum of absorption: 432 nm), red rods (502 nm), principal cones (580 nm) and accessory cones (502 nm). The principal visual pigment of the terrestrial species is rhodopsin, that of larval and aquatic amphibians porphyropsin. The sensitivity of the visual system is great and can reach that of owls as evidenced in tailed frogs, which hunt successfully at an ambient illumination of 10^{-5} lux (Hailman, 1982). The pineal complex of anurans consists of the intracranial pineal body (epiphysis cerebri) and the frontal organ, whereas the second is lacking in urodeles and caecilians (Adler, 1976). Both the pineal body and the frontal organ contain cells with photoreceptive ultrastructures which are sensitive to visible and ultraviolet light but not to infra-red wavelengths.

5.2.6 Direct, route-based or map-based orientation?

Amphibians do not only derive directional information from a surprising variety of cues, but also apply a number of strategies in order to home after active or passive displacement (Table 5.2). A simple method is the direct orientation towards the target demonstrated in many urodeles and anurans (e.g. Jungfer, 1943; Arak, 1983). It is based on acoustic (breeding chorus), mechanical, or olfactory (pond odours) cues which permit direct sensory contact with the goal. The reliability of these sources of information as directional cues decreases strongly with the distance from the

Table 5.2 Survey of the sources of directional information, the perception systems, and their role in the orientation system of amphibians (classification as proposed by Papi, 1990). Details and references in the text

Source of directional information	Site of perception	Function
Acoustic cues		
Calls of conspecifics	Inner ear (papilla basilaris, papilla amphibiorum)	Direct orientation
Magnetic cues		
Inclination of earth's magnetic field	?	Compass, route-based or map-based orientation
Declination of earth's magnetic field	?	
Mechanical cues		
Water surface-waves	Lateral line system	Direct orientation
Seismic signals	Inner ear (saccula, lagena)	Direct orientation
Water currents	Lateral line system	?
Olfactory cues		
Pond odours	Nares and Jacobson's organ	Direct orientation
Other odours		Pilotage or map-based orientation
Visual cues		
Landmarks	Lateral eyes	Pilotage or map-based orientation
Celestial cues		
Polarization pattern	Lateral eyes and pineal complex	Time-compensated compass of route-based or map-based orientation
Sun		
Moon		
Stars		
Position of body	Proprioreceptors	Kinesthetic orientation

target. Hence, they are mainly employed for short-distance homing.

Active displacement makes possible the use of route-based orientation by means of fixed geographical landmarks such as forest silhouettes or local odours (e.g. Heusser, 1969) during route reversal. If amphibians are passively displaced, they may employ the same cues for pilotage towards home (e.g. Sinsch, 1987b).

The most complex method of homing is map-based orientation which requires firstly the localization of the release site relative to home (map step) and then the determination of the direction leading to home (compass step). So far, there has been no un-equivocal evidence for map-based orientation in Amphibia. How-ever, blindfolded Andean toads remain disoriented, even if they are not deprived of magnetic and olfactory cues (Sinsch, 1988), and anosmic common toads cannot use magnetic or visual cues instead of olfactory ones (Sinsch, 1987b). The necessity of at least two independent cues indicates that the homing behaviour of these toads is more complex than pilotage or direct orientation which require only a single cue. Since magnetic and celestial cues serve as refer-ences for compass systems, the other necessary source of inform-ation may be related to the map step of orientation. In conclusion, it is probable that amphibians possess a mechanism of site local-ization based on release site-specific information, but further experimental analysis is needed.

There is ample evidence that amphibians use two compass systems, one based on celestial cues, and the other on the inclina-tion of the earth's magnetic field. The predictable effects of clock-shifting on the directional choice of amphibians using the celestial compass prove that they possess a circadian clock for the necessary time compensation of the cues's apparent movement (e.g. Taylor and Ferguson, 1970). Since larval and adult amphibians can derive directional information directly from the axis (e-vector) of linearly-polarized light which is visible even on overcast days, it is uncertain whether the position of the sun itself really serves as the reference of the sun compass. In nocturnal amphibians, star or lunar compasses have been demonstrated (Plasa, 1979), but their contribution to orientation is probably low because most individuals migrate on rainy and overcast nights.

5.2.7 Ranking of equivalent orientation cues

The variety of potential orientation cues and their partial inter-changeability demonstrate clearly that the orientation system of amphibians includes some redundancy in it (Table 5.2). The benefit

Table 5.3 Primary cues for long-distance orientation in toads of the genus *Bufo*. +, Disorientation following deprivation of the tested cues; −, homeward orientation despite deprivation; 1 female; 2 males

Species	Class of orientation cue				Reference
	Acoustic	Magnetic	Olfactory	Visual	
B. americanus	?	?	+	+	Dole (1972)
B. bufo	−	+	+	−	Heusser (1969), Sinsch (1987b)
B. boreas	−	?	+	−	Tracy and Dole (1969)
B. calamita	+1	+2	+2	+2	Arak (1983) Sinsch (1992)
B. spinulosus	−	+	−	+	Sinsch (1988)
B. valliceps	?	?	+	+	Grubb (1973)
B. woodhousii	+	?	+	+	Ferguson (1963), Grubb (1970)

of alternative back-up systems for every strategy of orientation is a reduction in the risk of failing to reach the target. The biology of amphibians and the intraspecific variability of their habitats partly explain the apparent contradiction to the principle of parsimony:

1. the availability of potential cues varies temporally and among different biotopes inhabited by populations of the same species;
2. the reproductive success of species with a short reproductive period depends on the time of arrival at a breeding site which may be distant;
3. the risk of dying due to dehydration and predation increases with the time spent searching for the goal.

However, there are many differences, even in closely related species, in the cues used for initial orientation (Table 5.3). The cues utilized probably provide the most reliable directional information among all the potential cues within the natural environment of each species. The migratory range, the diel activity period and the type of breeding site apparently influence the selection of these preferred or primary cues. Despite this preference, amphibians remain capable of using less reliable sources of directional information as secondary cues in the absence of the primary ones, following a considerable temporal delay. Determination of the specific rank of an orientation cue is probably not inherited, because the orientation system would lose a lot of its essential plasticity in order to adapt to changing environments. It is more reasonable to assume that at an unknown developmental stage, amphibians rank the alternative sensory input into a hierarchy of primary and secondary cues according to their availability in the habitat. Thus, the interspecific and interpopulational differences in initial orientation behaviour are probably modifications of the same basic system.

5.3 THE NEURONAL BASIS OF PROCESSING DIRECTIONAL INFORMATION

The existence of primary and secondary cues for initial orientation shows that amphibians are able to choose among all simultaneously perceived directional information. Hence, it is reasonable to assume that the sensory input is filtered and weighted in the neuronal network to select the best directional information. However, the locations of the presumed filters and the neuronal processes underlying homing behaviour are completely unknown in amphibians. Discovery of the neurophysiological basis of orientation behaviour is a major task for future research.

5.4 THE PHYLOGENETIC ORIGIN OF HOMING BEHAVIOUR

Alternative hypotheses for the evolutionary origin of orientation behaviour in amphibians are:

1. the origin dates back to the common ancestors of recent Amphibia;
2. anurans and urodeles independently developed homing behaviour.

There is convincing evidence for the monophyly of three orders of recent Amphibia (Gardiner, 1983). Fossil records indicate that the lineages of Anura on one hand, and of Caudata and Gymnophiona on the other have evolved separately since the Jurassic (Duellman and Trueb, 1986). Thus, the last common ancestor shared by all recent amphibians lived more than 200 million years ago.

There are several arguments favouring the common origin of orientation behaviour. Before separation into the recent orders the primitive amphibians began to exploit terrestrial environments. However, their skin was as water-permeable as it is in the modern amphibians. Thus, migrations on land were also constrained by the demands of water balance. It is reasonable to assume that selection

favoured those individuals which could steer straight, and therefore, short paths through unfavourable environments. The directional cues used for orientation and the perceptual system are
essentially the same in anurans and urodeles. There is no indication
of basic diversity in the orientation systems of the two recent
orders. Finally, fish and amphibians share so many features of
their orientation system and the receptors of orientation cues (see
Chapter 4), that the evolutionary origin of homing in vertebrates
probably dates back to even earlier periods.

5.5 PERSPECTIVES

Our knowledge of the homing mechanisms used by amphibians
is obviously rather fragmentary. At present, only the diffuse outline of a remarkably flexible orientation system is visible because
investigation has remained at the level of behaviour. The underlying neurophysiological level has not even been approached. In
future, a concentration on case studies of species which are either
well known from field investigations (e.g. *Taricha rivularis, Bufo
bufo*), or show homing behaviour in the laboratory (*Notophthalmus
viridescens*) appears to be most promising. The background knowledge from previous studies permits us to address, more specifically
than before, open questions such as: do amphibians navigate or
what is the neuronal basis of ranking of orientation cues?

5.6 SUMMARY AND CONCLUSIONS

Homing to the sites of reproduction, nutrition and hibernation
is a common feature of adult amphibians. The spatial range of
homeward orientation seems to be restricted almost entirely to
the area of previous migratory experience. Amphibians employ a
variety of different cues for orientation: conspecific calls, odours,
visual landmarks, celestial cues such as the position of the sun,
moon, stars and the polarization patterns in the sky, the inclination
and declination of the earth's magnetic field, seismic signals, water
surface waves and water currents. These cues serve as references,
either for pilotage or for direct, route-based or map-based orientation. The multisensory orientation system of amphibians apparently includes some redundancy, but alternative directional cues
are ranked hierarchically according to their availability in the habitat.
Therefore, the present interspecific and interpopulational differences
in initial orientation behaviour are probably modifications of
the same basic system. The perceptual systems for directional
information are well studied except for the magnetic receptors,

whereas the neuronal basis of processing and ranking of orientation cues is completely unknown. The phylogenetic origin of the orientation system appears to date back to a common ancestor of all recent Amphibia.

REFERENCES

Able, K.P. (1981) Mechanisms of orientation, navigation and homing, in *Animal Migration, Orientation and Navigation* (ed. S.A. Gauthreaux), Academic Press, New York, pp. 283–373.

Adler, K. (1976) Extraocular photoreception in amphibians. *Photochem. Photobiol.*, **23**, 275–98.

Arak, A. (1983) Sexual selection by male–male competition in natterjack toad choruses. *Nature*, **306**, 261–2.

Breckenridge, W.J. and Tester, J.R. (1961) Growth, local movements and hibernation of the Manitoba toad, *Bufo hemiophrys. Ecology*, **42**, 637–46.

Breden, F. (1987) The effect of post-metamorphic dispersal on the population genetic structure of Fowler's toad, *Bufo woodhousei fowleri. Copeia*, **1987**, 386–94.

Buck, T. (1988) Untersuchungen zur Biologie der Erdkröte *Bufo bufo* L. unter besonderer Berücksichtigung der Erscheinungsformen des Phänomens der Orientierung. Unpublished PhD Thesis, University of Hamburg, 179 pp.

Buytendijk, F.J.J. (1918) Instinct et la recherche du nid et experience chez les crapauds (*Bufo vulgaris* et *Bufo calamita*). *Arch. Neerl. Physiol.*, **2**, 1–50.

Dole, J.W. (1972) Homing and orientation of displaced toads, *Bufo americanus*, to their home sites. *Copeia*, **1972**, 151–8.

Duellman, W.E. and Trueb, L. (1986) *Biology of Amphibians*, McGraw-Hill, New York, 670 pp.

Dürigen, B. (1897) *Deutschlands Reptilien und Amphibien*, Schluss, Magdeburg.

Durand, J.P. and Parzefall, J. (1987) Comparative study of the rheotaxis in the cave salamander *Proteus anguinus* and his epigean relative *Necturus maculosus* (Proteidae, Urodela). *Behav. Processes*, **15**, 285–92.

Eggermont, J.J. (1988) Mechanisms of sound localization in anurans, in *The Evolution of the Amphibian Auditory System* (eds B. Fritzsch, M.J. Ryan, W. Wilczynski, T.E. Hetherington and W. Walkowiak), John Wiley, New York, pp. 307–36.

Elephandt, A. (1984) The role of ventral lateral line organs in water wave localization in the clawed toad (*Xenopus laevis*). *J. Comp. Physiol.*, **154A**, 773–80.

Endler, J. (1970) Kinesthetic orientation in the California newt (*Taricha torosa*). *Behaviour*, **37**, 15–23.

Ferguson, D.E. (1963) Orientation in three species of anuran amphibians, in *Animal Orientation* (ed. H. Autrum), *Ergebn. Biol.* **26**, pp. 128–34.

Ferguson, D.E. (1971) The sensory basis of orientation in amphibians. *Ann. N.Y. Acad. Sci.*, **188**, 30–6.

Freeland, W.J. and Martin, K.C. (1985) The rate of range expansion by *Bufo marinus* in Northern Australia, 1980–84. *Aust. Wildl. Res.*, **12**, 555–9.

Gardiner, B.G. (1983) Gnathostome vertebrae and the classification of the amphibia. *Zool. J. Linnaean Soc.*, **74**, 1–59.

Grubb, J.C. (1970) Orientation in post-reproductive Mexican toads, *Bufo valliceps*. *Copeia*, **1970**, 674–80.

Grubb, J.C. (1973) Olfactory orientation in *Bufo woodhousei fowleri*, *Pseudacris clarki* and *Pseudacris streckeri*. *Anim. Behav.*, **21**, 726–32.

Hailman, J.P. (1982) Extremely low ambient light levels of *Ascaphus truei*. *J. Herpet.*, **16**, 83–4.

Heusser, H. (1956) Biotopansprüche und Verhalten gegenüber natürlichen und künstlichen Umweltveränderungen bei einheimischen Amphibien. *Vierteljahreschr. Naturf. Ges. Zürich*, **101**, 189–210.

Heusser, H. (1968) Die · Lebensweise der Erdkröte (*Bufo bufo* L.): Wanderungen und Sommerquartiere. *Rev. Suis. Zool.*, **75**, 928–82.

Heusser, H. (1969) Die Lebensweise der Erdkröte (*Bufo bufo* L.): Das Orientierungsproblem. *Rev. Suis. Zool.*, **76**, 444–517.

Jungfer, W. (1943) Beiträge zur Biologie der Erdkröte (*Bufo bufo* L.) mit besonderer Berücksichtigung der Wanderung zu den Laichgewässern. *Z. f. Morphol. u. Ökol. d. Tiere*, **40**, 117–57.

Leucht, T. (1987) Magnetic effects on tail-fin melanophores of *Xenopus laevis* tadpoles *in vitro*. *Naturwissenschaften*, **74**, 441–3.

Lewis, E.R. and Lombard, R.E. (1988) The amphibian inner ear, in *The Evolution of the Amphibian Auditory System* (eds B. Fritzsch, M.J. Ryan, W. Wilczynski, T.E. Hetherington and W. Walkowiak), John Wiley, New York, pp. 93–123.

Lewis, E.R. and Narins, P.M. (1985) Do frogs communicate with seismic signals? *Science*, **227**, 188–9.

Madison, D.M. (1972) Homing orientation in salamanders: a mechanism involving chemical cues, in *Animal Orientation and Navigation* (eds S.R. Galler, L. Schmid-Koenig, G.J. Jacobs and R.E. Belleville), NASA SP 262, pp. 485–98.

McCormick, C.A. (1988) Evolution of auditory pathways in the Amphibia, in *The Evolution of the Amphibian Auditory System* (eds B. Fritzsch, M.J. Ryan, W. Wilczynski, T.E. Hetherington and W. Walkowiak), John Wiley, New York, pp. 587–612.

Papi, F. (1990) Homing phenomena: mechanisms and classification. *Ethol. Ecol. Evol.*, **2**, 3–10.

Phillips, J.B. (1977) Use of the earth's magnetic field by orienting cave salamanders (*Eurycea lucifuga*). *J. Comp. Physiol*, **121**, 273–88.

Phillips, J.B. (1986) Two magnetoreception pathways in a migratory salamander. *Science*, **233**, 765–7.

Phillips, J.B. (1987) Laboratory studies of homing orientation in the eastern red-spotted newt, *Notophthalmus viridescens*. *J. Exp. Biol.*, **131**, 215–29.

Plasa, L. (1979) Heimfindeverhalten bei *Salamandra salamandra* (L.). *Z. Tierpsychol.*, **51**, 113–25.

Rand, A.S. (1988) An overview of anuran acoustic communication, in *The Evolution of the Amphibian Auditory System* (eds B. Fritzsch, M.J. Ryan, W. Wilczynski, T.E. Hetherington and W. Walkowiak), John Wiley, New York, pp. 415–31.

Rheinländer, J. and Klump, G. (1988) Behavioural aspects of sound localization, in *The Evolution of the Amphibian Auditory System* (eds B. Fritzsch, M.J. Ryan, W. Wilczynski, T.E. Hetherington and W. Walkowiak), John Wiley, New York, pp. 297–305.

Sauer, H. (1988) Autökologische Untersuchungen der Kreuzkröte – *Bufo calamita* (Laurenti) 1768 – und der Wechselkröte – *Bufo viridis* (Laurenti) 1768 – als Grundlage für gezielte Schutzmaßnahmen. Unpublished thesis, University of Bonn, 126 pp.

Schlupp, I., Podloucky, R., Kietz, M. and Stolz, F.-M. (1990) Pilot-projekt Braken: Erste Ergebnisse zur Neubesiedlung eines Ersatzlaichgewässers durch adulte Erdkröten. *Inform. d. Naturschutz Niedersachs., Hannover*, **10**, 12–18.

Sinsch, U. (1987a) Migratory behaviour of the toad *Bufo bufo* within its home range and after displacement, in *Proc. 4th Ord. Gen. Meet. Soc. Eur. Herp.* (eds J.J. van Gelder, H. Strijbosch and P.J.M. Bergers), Nijmegen, pp. 361–4.

Sinsch, U. (1987b) Orientation behaviour of toads (*Bufo bufo*) displaced from the breeding site. *J. Comp. Physiol.*, **161A**, 715–27.

Sinsch, U. (1988) El sapo andino *Bufo spinulosus*: analisis preliminar ed su orientacion hacia sus lugares de reproduccion. *Boletin ed Lima*, **57**, 83–91.

Sinsch, U. (1990) Migration and orientation in anuran amphibians. *Ethol. Ecol Evol.*, **2**, 65–79.

Sinsch, U. (1991) Mini-review: the orientation behaviour of amphibians. *Herpet. J.* **1**, 541–4.

Sinsch, U. (1992) Sex-biased site fidelity and orientation behaviour in reproductive natterjack toads (*Bufo calamita*). *Ethol. Ecol. Evol.*, **4**, 15–32.

Taylor, D.H. (1972) Extra-optic photoreception and compass orientation in larval and adult salamanders (*Ambystoma tigrinum*). *Anim. Behav.*, **20**, 233–6.

Taylor, D.H. and Ferguson, D.E. (1970) Extraoptic celestial orientation in the southern cricket frog *Acris gryllus*. *Science*, **168**, 390–2.

Tracy, C.R. and Dole, J.W. (1969) Orientation of displaced California toads, *Bufo boreas*, to their breeding sites. *Copeia*, **1969**, 693–700.

Twitty, V.C., Grant, D. and Anderson, O. (1967) Initial homeward orientation after long-distance displacements in the newt *Taricha rivularis*. *Proc. Natl. Acad. Sci., U.S.A.*, **57**, 342.

Walkowiak, W., Capranica, R.R. and Schneider, H. (1981) A comparative study of auditory sensitivity in the genus *Bufo* (Amphibia). *Behav. Processes*, **6**, 223–37.

Walkowiak, W. and Münz, H. (1985) The significance of water surface-waves in the communication of fire-bellied toads. *Naturwissenschaften*, **72**, 49–50.

Wells, K.D. (1988) The effect of social interactions on anuran vocal behaviour, in *The Evolution of the Amphibian Auditory System* (eds B. Fritzsch, M.J. Ryan, W. Wilczynski, T.E. Hetherington and W. Walkowiak), John Wiley, New York, pp. 433–54.

Chapter 6
Reptiles
G. Chelazzi

6.1 INTRODUCTION

The approximately 6000 living species of reptiles inhabit a great variety of habitats, ranging from open seas to emerged lands. This ecological diversity is characteristic of all the main orders of this class. Chelonia, for instance, include turtles spending most of their life in open seas or coastal areas, others more or less linked to inland waters, and typical terrestrial tortoises. Many species of the two suborders of Squamata – Lacertilia (lizards and related forms) and Ophidia (snakes) – inhabit arid or semi-arid environments, but some include freshwater bodies in their habitat and others are entirely marine. On the contrary, Loricata (crocodiles, alligators and gavials) inhabit mostly swamps, lakes and rivers, and only one species is definitely linked to coastal areas.

Given their ecological diversity and related morpho-physiological heterogeneity it is not easy to consider all the living reptiles together when discussing the ecological and functional aspects of their behaviour. In particular their homing capacity in natural conditions is related to adaptive problems associated with different life strategies: there is a difference in the seasonal shuttling between overwintering dens, reproductive sites and foraging areas in garter snakes, and the return to the familiar area by a juvenile alligator after an unpredictable displacement due to flooding. The scale of the phenomenon also varies greatly: the transoceanic circuits of green turtles and sea snakes cover hundreds or thousands of miles, while the looped excursions of European yellow-green

Animal Homing. Edited by Floriano Papi.
Published in 1992 by Chapman & Hall, 2–6 Boundary Row, London SE1 8HN.
ISBN 0 412 36390 9.

racer snakes stay within a few hundred metres of their den. Thus, in the following sections the natural appearance of homing, evidence for homing ability after experimental displacement, and cues used to home will be discussed separately for each major group.

6.2 CHELONIA

Most of the studies on the homing of Chelonia refer to a few of the approximately 170 species of the suborder Cryptodira, containing the most successful and advanced turtles and tortoises, while practically nothing is known about the homing capacity of the 50-odd species comprising the more primitive Pleurodira (side-necked turtles). Regardless of their taxonomy, chelonians are considered here according to the main ecological categories: sea turtles, freshwater turtles, and land tortoises.

6.2.1 Sea turtles

The reproductive migrations of sea turtles are among the most spectacular examples of homing behaviour in animals. Extensive tagging programmes and satellite radiotracking (e.g. Carr, 1967b; Hays, 1991) have shown that many species are divided into demes inhabiting reproductive and foraging areas, hundreds or thousands of miles apart. For instance, a deme of the green turtle, *Chelonia mydas*, nests on Ascension Island in the middle of the Atlantic Ocean, about 8° south of the equator. Young hatched on the beaches of this island reach the sea and drift away with the current. After about 1 year spent in nursery areas or conducting a pelagic life, they reach the feeding grounds typical of the subadult and adult specimens on the coast of Brazil between 20° south and the equator. Every 2–3 years, the mature males and females migrate back to mate in the waters around Ascension Island where the females usually nest on their own natal beach (Carr and Hirth, 1962; Carr, 1975). This phenomenon is a very impressive example of long-range homing since many individual turtles show fidelity to a specific subsection of a given beach (Mortimer and Portier, 1989). The origin and evolution of the Brazil–Ascension Island reproductive circuit of *C. mydas* have been related by Carr and Colemann (1974) to the sea-floor spreading theory according to which, from the earliest Tertiary on, the distance between the Brazilian coast and the insular nesting sites has gradually increased due to the drifting apart of South America and Africa.

Similar reproductive circuits have been described in other demes of the same species, such as the Tortuguero–western Caribbean (Carr and Giovannoli, 1957) and Yemen–western Indian Ocean demes (Hirth and Carr, 1970). Long-range reproductive migrations have also been documented in the loggerhead turtle, *Caretta caretta*: specimens tagged at Tongaland (South Africa) were traced to the Zanzibar coast, a migration of over 2000 km (Hughes, 1974). The leatherback turtle, *Dermochelys coriacea*, also seems to perform reproductive migrations, as indicated by their recovery in extremely cold northern waters far from their nesting sites (Pritchard, 1976).

However other species, such as the hawksbill turtle (*Eretmochelys imbricata*), nest on beaches adjacent to the reefs on which they live (Carr, 1952; Bustard, 1979), or, like the olive ridley (*Lepidochelys olivacea*), seem to divide into migratory and non-migratory groups. Some turtles tagged on the beaches of Guyana were later recaptured along the coast of South America (eastern Venezuela, northern Brazil) and then back in the nesting area, while others were recaptured close to the nest area even during non-nesting periods (Pritchard, 1976).

These long circuits, often centred around a constant reproduction area, and the finding that adult females return to their original beach to nest, seem to require very precise homing mechanisms. The use of sun compass orientation during the Brazil–Ascension Island–Brazil migration of the green turtle is discussed by Carr and Colemann (1974), but no definite proof is offered.

On the contrary, much evidence is available on the use of olfactory cues in locating the original island for nesting. According to Carr (1967a), olfactory imprinting to odours of the natal beach occurs in baby turtles which, as adults, are then capable of recognizing and tracing such scents in the ocean currents. If Carr's hypothesis is right, this is one of the the most impressive examples of olfactory imprinting in animals: at the earliest green and loggerhead turtles reach maturity at the ages of 15–35 years (Balazs, 1980) which means not only that the hatchlings must learn chemical cues during a short period of a few weeks, but must also be able to discriminate these cues from many others in very low concentrations many years later.

The theoretical discussion of Koch *et al.* (1969) and the laboratory experiments of Manton (1979) support the olfactory hypothesis. In particular, Manton found that *C. mydas* can be conditioned to press a lever to obtain food upon recognition of low concentrations (10^{-5} M) of various substances, and can remember these odours for at least 1 year. Moreover, baby green turtles were 'imprinted' by Grassman (1984) to morpholine and phenylethanol (5×10^{-5} M) in

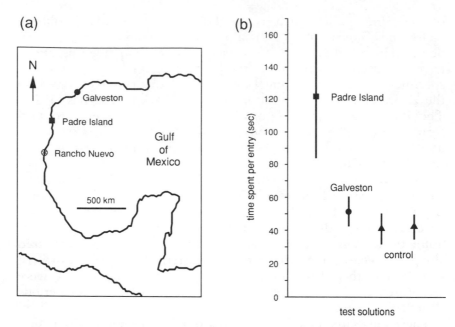

Figure 6.1 Results of an experiment performed by Grassman *et al.* (1984) to test the preference of *Lepidochelys kempi* for different sea-water solutions. (a) Eggs collected at Rancho Nuevo (Mexico) during oviposition were transported to Padre Island (Texas) and there incubated under local sand until they hatched. Hatchlings were later raised in Galveston (Texas). At the age of 4 months the ridleys were offered solutions of sand and sea-water from Padre Island, Galveston, and control solutions in a multiple choice arena. (b) The mean time spent per entry in compartments treated with the Padre Island solution was significantly higher (Tukey's test: $P < 0.01$) than that recorded in the other compartments. Vertical bars represent standard error of the mean.

their nest or in tank water. Two months later the turtles were tested in a four-choice chamber with different chemicals and significantly preferred the compartment containing the substances to which they had been exposed. A different experiment was performed by Grassman *et al.* (1984) on young *Lepidochelys kempi* which showed a preference for the sand and water collected at their hatching site (Fig. 6.1).

Owens *et al.* (1986) reviewed the available literature on the olfactory basis of homing during the breeding migrations of sea turtles. They concluded that sea turtles can orient to learned specific chemical cues; and they can orient to, and distinguish between, low concentrations of solutions prepared from natural beaches. In conclusion, though field and laboratory experiments do not prove definitely that a true chemosensory imprinting (*sensu* Lorenz, 1937) to the natal beach occurs in sea turtles, learned chemical cues seem to play an important role in navigation during their reproductive migrations.

6.2.2 Freshwater turtles

Most of the information on the homing behaviour of freshwater turtles comes from studies conducted on the families Emydidae and Trionychidae. Although Emydids are generally considered a group of freshwater turtles, the approximately 80 species, mostly from North America and Asia, are either truly aquatic (as *Clemmys marmorata*) or terrestrial (as *Terrapene carolina*) (Pritchard, 1979). The 22 species of Tryonichidae (soft-shelled turtles) inhabit marshy creeks, large swift-flowing rivers, and bayous and lakes in Africa, Asia and North America (Ernst and Barbour, 1972). Movement patterns in the field, including home range maintenance, have been studied by capture–recapture and radiotracking in freshwater turtles, since the early study of Cagle (1944) on the North American species *Chrysemys picta* and *Pseudemys scripta*. Generally speaking, Emydids maintain definite home ranges which include one or more bodies of water and more or less widely emerged areas (Gibbons, 1970).

Several studies based on releasing the turtles outside their home range suggest that Emydids are capable of external homing with various degrees of accuracy. About 28% of *Clemmys guttata* displaced up to 800 m from the capture site returned to their original pond (Ernst, 1968). In *Chrysemys picta* the homing performance varies from 50 to 20% when released at distances of 1.6 and 3.2 km, respectively (Ernst, 1970). The semi-aquatic species *Clemmys insculpta* homes well if released less than 2 km from the capture site, but its homing performance falls sharply when displaced longer distances (Carroll and Ehrenfeld, 1978). When released 24 km downstream and upstream along a river, *Graptemys pulchra* returned to the capture site in 9 and 7% of the tests (Shealy, 1976). The European pond turtle, *Emys orbicularis*, shows very stable home ranges which remain constant year after year (Lebboroni and Chelazzi, 1991) and is able to relocate its precise home range along a canal after downstream and upstream releases about 1 km away; over land it shows a homeward heading when displaced up to 1.7 km (personal observation). A good homing performance has been documented by Plummer and Shirer (1975)

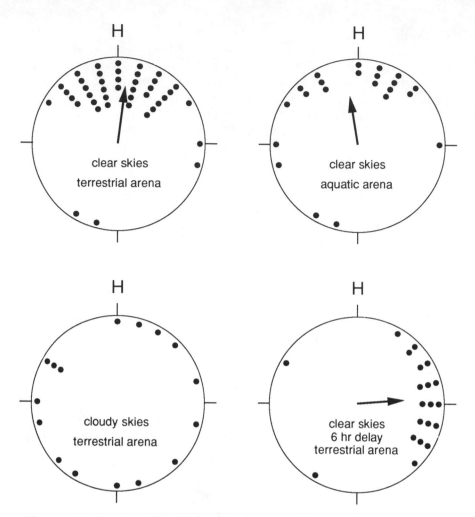

Figure 6.2 Results of experiments performed by DeRosa and Taylor (1980) on the emydid freshwater turtle, *Crysemys picta*. Tests were performed in circular arenas, both on land and in water after displacements of several kilometres from the familiar area. Each dot represents the resultant direction obtained by the headings of a turtle in the unit time. Arrows represent the resultant vectors obtained in each experiment. All directions are referred to the expected homing direction (H). In the absence of visible landmarks, turtles showed a distinct homeward orientation under clear skies, but not under cloudy skies. A clock shift 6 h slow, induced by exposing the animals to a shifted photocycle prior to the test, produced a 90° clockwise deviation in orientation, in agreement with a time-compensated sun compass mechanism. Similar results were obtained by the same authors for *Terrapene carolina* and for the soft-shelled turtle *Trionyx spinifer*. These findings support the hypothesis that these species use a two-step navigation (map and celestial compass) to home, but map cues have not yet been clarified.

in the North American *Trionyx muticus*, trailed with floating balloons and radiotelemetry.

Studies on the cues used by Emydids and Triomychids for homing are scant. Gould (1957, 1959) reports that *Terrapene carolina* and *Chrysemys picta* show accurate homing behaviour if displaced under a clear sky, but no ability to home under a completely overcast sky, and concludes that the turtles use celestial cues to orient homeward. Contradictory results were obtained by Emlen (1969) in *C. picta*: this turtle was able to home after short distance displacements (100 m) even under an overcast sky, but homing ability disappeared when released at 1 km from the original pond.

A detailed study on the homing behaviour and orientation mechanisms of *C. picta, Terrapene carolina* and *Trionyx spinifer* has been conducted by DeRosa and Taylor (1980). Turtles of the three species showed a good homeward orientation when released under sunny skies in circular arenas, both on land and in water, at distances of several kilometres from the capture site (Fig. 6.2). The headings appeared randomly distributed when the turtles were tested under cloudy skies. Moreover, when tested after a clock shift 6 h slow the three species showed a clockwise deviation of 90° with respect to the homeward direction. These findings support the hypothesis that the three species use a sun-compass orientation based on an internal clock, as part of a map and compass navigation mechanism.

The use of magnetic cues for directional and homing orientation have been tested by Mathis and Moore (1988) in the box turtle, *Terrapene carolina*. Laboratory experiments involved training the turtles to move along an east–west axis. When tested in a circular arena control turtles with dummy weights attached to the carapace oriented significantly to the training direction while specimens carrying disk magnets taped to the centre of their carapace were randomly oriented. Homeward orientation was tested in the field, releasing the animals from different directions 0.75–1.2 km from the capture site. During displacement the magnetic field around the experimental turtles was increased 15% over the natural value using Helmholtz coils generating a field strength of 0.61 G. On the contrary, control specimens were carried to the release site under normal earth field strength. The data suggest an effect of the altered magnetic field on the homeward orientation of *T. carolina*: while control specimens showed a combination of homeward and preferred (eastward) compass orientation, the experimental turtles were randomly oriented with respect to home, showing only a weak eastward orientation. However, pooling the headings recorded from different release sites with respect to the expected homeward

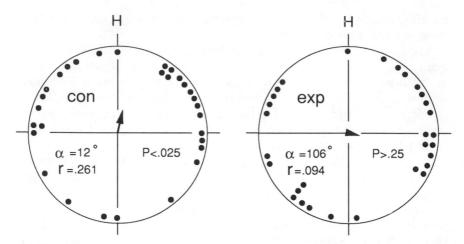

Figure 6.3 Results of field experiments conducted by Mathis and Moore (1988) to test the importance of the magnetic field in the homing of the box turtle, *Terrapene carolina*. The data relative to four releases, pooled together with respect to the homeward direction (H) give, in the authors' opinion, evidence of a weak homeward orientation in the controls (con) and a random distribution in the headings of turtles exposed to an altered magnetic field (exp). However, the two distributions do not differ according to Watson's U^2-test ($U^2_{30,31} = 0.091$; $P > 0.1$). α, Direction of resultant vectors; r, length of resultant vectors; P, probability that the distribution is taken from a randomly oriented one (V-test).

direction reveals no statistical difference between the control and experimental specimens (Fig. 6.3).

Some authors suggest the importance of other cues in the homing of Emydids. For instance, Ernst (1970) concluded that olfaction and geotaxis cannot be ruled out in the homing of *Chrysemys picta*, and Carroll and Ehrenfeld (1978) suggest olfaction as the basis for the homing capacity of *Clemmys insculpta*. However, as stressed by Owens *et al.* (1986), these conclusions have not yet been supported by adequate experimental evidence.

6.2.3 Land tortoises

Apart from a few specific studies, information on the internal and external homing and orientation of land tortoises is anecdotal and scarse. Spontaneous movements in the field have been documented in different species of the American genus *Gopherus*, in the giant turtles of the genus *Geochelone*, and in the Mediterranean species of *Testudo*.

Gopherus polyphemus and *G. flavomarginatus* perform looped excursions mostly centred around a constant burrow (Auffenberg and Weaver, 1969), and *G. agassizi* is reported by Burge (1977) to return periodically to an actively dug burrow. A few experiments have been performed on these species to test their external homing ability. Auffenberg and Weaver (1969) released *G. berlandieri* at various distances from their familiar 'loma' (broad–topped hill including the familiar area). They concluded that this species is able to home when released up to 50 m inside its home range, but not from beyond. Gourley (1974) performed orientation tests both in the field and in an arena on *G. polyphemus*: when displaced 8 km or more from the capture site the tortoises preferred compass directions not related to the home direction.

Extensive studies on the spontaneous movements of *Geochelone gigantea* at Aldabra Atoll (Seychelles) by Swingland and Lessels (1979) have revealed a behavioural polymorphism including maintenance of a definite home range, migration between different habitats and dispersive patterns. A similar plasticity in movement patterns within a natural population on the Galapagos Islands has been recorded in *Geochelone elephantopus* by Rodhouse *et al.* (1975). No quantitative tests on external homing have been conducted on these species.

Detailed information on movement patterns and homing in Testudinidae come from long-term research conducted on a central Italy population of *Testudo hermanni* (Chelazzi and Francisci, 1979). This species actively maintains a constant home range during each activity season, performing 'centrally biased' movement patterns year after year (Chelazzi and Carlà, 1986). By radiotracking tortoises displaced up to 1560 m from the capture site, well outside their home range, it was seen that they chose a significant homeward orientation soon after release (Fig. 6.4). Home range maintenance and homing behaviour after displacement were explained in terms of behavioural homeostasis, particularly related to site selection for behavioural thermoregulation (Chelazzi and Calzolai, 1986; Calzolai and Chelazzi, 1991).

Specific field tests were conducted on the same species to assess the importance of olfactory information on their external homing (Chelazzi and Delfino, 1986). Tortoises rendered anosmic by intranasal washing with a 2% solution of $ZnSO_4$ and displaced 500–1000 m from their home range wandered around the release site, while control tortoises showed a strong homeward orientation. (Fig. 6.5). This result strongly suggests the importance of olfaction

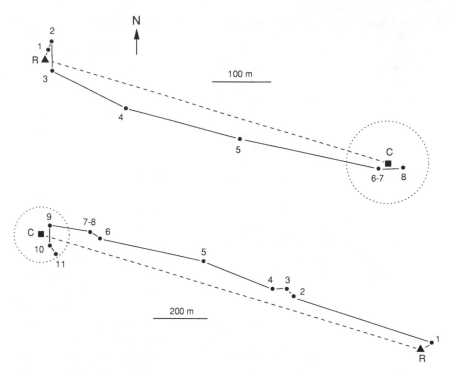

Figure 6.4 Two examples of quick homing response shown by the Mediterranean land tortoise, *Testudo hermanni*, after displacement well outside its normal home range. C indicates the capture site located at the centre of the home range (dotted circle). Animals were displaced along a linear route (dashed line) to the release site (R) in a hand–carried container which prevented vision of the landscape. Radio fixes were taken daily (numbers near dots) until the tortoises reached their home range. Immediately after release *T. hermanni* usually hides in shrubs but afterwards its mobility increases well above the usual level. Such normal activities as feeding and mate searching are suppressed during their return which is fast and along a straight route. Once in the home area, normal behaviour patterns resume and their path becomes tortuous again. (After Chelazzi and Francisci, 1979.)

in the homing of *T. hermanni*, but a non-specific effect of the treatment cannnot be excluded. Auffenberg and Weaver (1969) suggested that olfactory or visual cues are used by *Gopherus berlandieri* to home but did not perform any specific tests on homing cues. Thus the existence of olfactory navigation in land tortoises has yet to be demonstrated, as has the hypothesis that these reptiles use

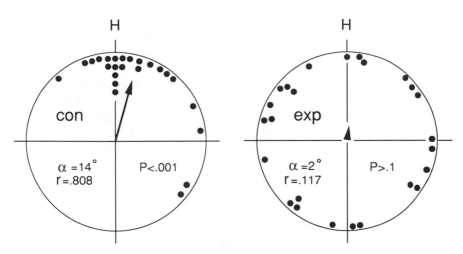

Figure 6.5 Results of an experiment conducted by Chelazzi and Delfino (1986) to test the role of olfaction in the homing of *Testudo hermanni*. Control and experimental tortoises were anaesthetized and the nasal cavities of the latter washed with a 2% solution of $ZnSO_4$. This treatment greatly altered their olfactory mucosa, as confirmed by histological control. Both groups were then displaced well outside their home range and radiotracked daily. The directions chosen by control tortoises after a week were significantly homeward oriented while the experimental animals wandered randomly around the release sites. α, direction of resultant vectors; r, length of resultant vectors; P, probability that the distribution is taken from a randomly oriented one (Rayleigh test).

two steps (map and compass) in navigating. However, the experiments of Gourley (1974) on *G. polyphemus*, based on releasing tortoises after clock shifting, support the presence of a time-compensated sun compass in this species.

6.3 LORICATA

Spontaneous movements related to habitat selection have occasionally been investigated in a few of the approximately 20 living species of Loricates. *Gavialis gangeticus* is reported by Bustard and Singh (1983) to occupy a stable home range 10.4–19.2 km long in the river. Migratory movements with periodic return to the same area have been described in crocodiles. *Crocodylus niloticus* travel long distances to reach breeding and nesting sites on the central islands of Lake Rudolf (Modha, 1967, 1968); during droughts in

Venezuela, *Crocodylus johnstoni* and *Caiman crocodilus* migrate over land to permanent water (Staton and Dixon, 1975; Gorzula, 1978).

During their terrestrial migrations these species show an excellent homing ability. In Gorzula's (1978) experiments on *C. crocodilus* more than 80% of the specimens homed successfully from distances exceeding 2 km, and Webb *et al.* (1983a, b) recorded active homing behaviour in adult *C. johnstoni* displaced 30 km upstream from capture sites. Moreover, 68% of the first and second year *C. crocodilus* and 52% of the older individuals returned to the exact location they were found in before the rainy season (Ouboter and Nanhoe, 1988). According to Webb *et al.* (1983a), the homing behaviour of *C. johnstoni* is highly adaptive in giving them the capacity to relocate specific refuge sites which tend to be used year after year by the same individuals.

Extensive studies have been conducted on the homing behaviour of young alligators. These include releasing and following individuals by resighting or radiotracking in the field after translocation, or specific orientation tests inside experimental arenas. In a first set of tests, Rodda (1984a) obtained the routes of 19 juvenile *Alligator mississipiensis* captured at night and displaced in various directions 0.9–6.6 km (1–10 home range diameters) with respect to the capture site. Homing performance was high: 10 out of 19 juveniles promptly returned, two headed in the homeward direction but stopped before reaching it, two remained at the release site. A definite homeward orientation was evident soon after release and no differences were observed in the homing behaviour of individuals displaced in different ways: more or less direct routes, view of the sky allowed or prevented. This was the first evidence that young alligators have a true navigational ability.

A larger number of releases were performed by Rodda (1984b) using young *A. mississipiensis* from various lakes in north central Florida. A total of 181 yearlings (7–14 months) and older juveniles (>14 months) were displaced 12–34 km in various directions to a 9 m diameter dodecagonal arena provided with symmetric observation points. Displacement was by boat and automobile over more or less tortuous routes, with the sky 'opaque' or 'visible'. The two age classes behaved differently: the older juveniles showed a good homeward orientation regardless of how they were transported, while the yearlings did only when displaced along a fairly linear route. When displaced along a tortuous route they chose a fixed (y-axis) orientation. Visibility of the sky during displacement did not affect orientation of the alligators, and orientation did not significantly deteriorate under an overcast sky. Moreover, a significant correlation emerged between the accuracy of homeward orientation in older juveniles and the geomagnetic dip angle (and the horizontal magnetic intensity) measured at the time of the test (Fig. 6.6). In particular, small fluctuations in the dip angle (0.05°) 'produced' a large deviation (more than 100°) from homeward orientation. These findings allowed the author to hypothesize the presence in older juveniles of a 'multi-coordinate navigation' including a geomagnetic map based on the difference between the dip angle prevailing at the time and place of capture, and those present when and where a homeward orientation is chosen.

An equally high homing performance and prompt homeward orientation was registered in 258 juveniles of the same species displaced 0.8–16 km out of their familiar area and radiotracked (Rodda, 1985). Also, the larger sample size permitted a better assessment of the effect of the displacement technique. The alligators displaced slowly (on a motor boat) showed a higher homing performance than those displaced rapidly (by car), which agrees with the hypothesis of a homing mechanism incorporating a 'displacement-route-based navigation'. However, the absence of an effect of displacement route directness argues against this. Surprisingly, the 'sky-visible' group showed a lower homing performance than the 'opaque' group: the author suggests that this could have been due to a stress effect. Finally, the finding that homing performance worsens at distances greater than 5 km is consistent with homing based on a goal-emanating odour.

In combination, Rodda's experiments on young alligators support the presence of a true navigation but his results are too contradictory to clarify the mechanisms involved, while the evidence for a magnetic map is still very weak. One obscure point is also the adaptive significance of such homing behaviour, since alligators are far more sedentary than crocodiles. Rodda (1985) argues that homing behaviour may help juveniles to return to their original site after monitoring the habitat for suitable areas free from territorial adults. Moreover, occasional 'climatic catastrophes' may force them far from their familiar areas.

Finally, a detailed investigation on the orientation of *A. mississipiensis* not in the context of homing was conducted by Murphy (1981). Juveniles from Orange Lake (Florida) were trained for 30–60 days in an outdoor pen to learn a fixed compass direction (water–land axis). Tests were performed in a land arena excluding all landmarks, but permitting a view of the sky. Under clear day skies alligators headed significantly in the expected (landward) direction. The same occurred at night with the moon visible or under moonless but clear starry skies. On the contrary, no significant departure of the headings from a uniform distribution

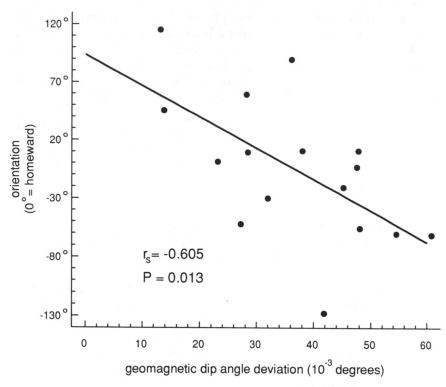

Figure 6.6 Results of experiments conducted by Rodda (1984b) on juvenile *Alligator mississipiensis*. Alligators were displaced from various lakes in Florida to a test arena 12–34 km from the capture sites in different directions. Older juveniles showed a definite homeward orientation regardless of the displacement technique (fast or slow, tortuous or straight, in darkness or in daylight with the sky visible) but the accuracy of their orientation was significantly correlated to the dip angle variations in the geomagnetic field at the test time. Each dot in the figure represents the relationship between the average orientation of the alligators and the corresponding geomagnetic value. The continuous line represents the regression line. The fact that large deviations from the homeward direction were observed when small dip angle deviations occurred suggested to the author that alligators may use the earth's magnetic field as a map cue in multicoordinate navigation and not as a compass mechanism in a two-step navigation. In other similar tests, however, no such correlation was observed. r_s, Correlation coefficient; P, probability that the two variables are not correlated.

was observed under full cloud cover at night. Clock shift tests (6 h advance) gave evidence of a time-compensated sun compass during the day but produced no expected shift in orientation under starry skies. In conclusion, the y-axis orientation experimentally tested by Murphy in *A. mississipiensis* seems based on a time-compensated sun compass, a non-time-compensated star compass, and a moon (time-compensated?) compass. Whether such compass mechanisms are used by young alligators as part of their navigation for homing remains to be investigated.

6.4 LACERTILIA

Information on movement patterns and homing of lizards come from a few studies conducted almost exclusively on Lacertids and Iguanids. The earliest study of Iguanids was performed by Noble (1934) on the North American species *Sceloporus ondulatus* which showed a distinct homing capacity within 270 m of the capture site. Later, several studies were conducted on other North American species of the same genus. Mayhew (1963) and Weintraub (1970), respectively, report successful homing in *S. orcutti* after displacements up to 150 and 215 m from the capture site, outside its familiar area. Similar results were obtained by Guyer (1978) with *S. graciosus*, and by Bissinger (1983) and Ellis-Quinn and Simon (1989) with *S. jarrovi*: these iguanids demonstrated a homing capacity after experimental displacements up to 280 m from the capture site. Other species showing a good homing capacity after releases exceeding the limits of their normal home range are *Dipsosaurus dorsalis* (Krekorian, 1977), *Uta stansburiana* (Spoecker, 1967) after displacements up to 122 m, *Phrynosoma douglassi* (Guyer, 1978) after displacements up to 148 m, and the lacertid *Takydromus tachydromoides* (Ishihara, 1969) after displacements up to 180 m. On the other hand, negative results were obtained by Rand (1967) in *Anolis lineatopus* displaced 180 m from the capture site and by Fitch (1940) in some specimens of *S. occidentalis* displaced a few hundred yards from the capture site. Neither was any homing capacity recorded by Tinkle (1967) in *U. stansburiana*.

Fewer data are available on European Lacertids. Strijbosch *et al.* (1983) displaced specimens of *Lacerta agilis* and *L. vivipara* 70 and 100 m from the capture site, recording a homing performance ranging from 81.5 to 66.7% and from 50 to 28.6%, respectively. The homing time is quite long for both species, ranging from 1 day to more than 2 months. More recently, Foà *et al.* (1990) recorded a good homing ability in *Podarcis sicula*. Their initial orientation at the release point seemed to vary with the displacement distance;

individuals released 25 and 50 m from the capture site deviated from the expected homeward direction by 29° and 41°, respectively, but in longer displacements the homeward orientation was even more accurate, deviating only 3° from the expected direction.

Although most of the above studies have documented a distinct homing capacity from sites not included in the home range of the different species of Iguanids and Lacertids, the orientation mechanisms underlying the homing behaviour of lizards are relatively unknown. The capacity to use sun-compass orientation has been demonstrated in directionally trained *Lacerta viridis* (Fischer, 1961) and *Uma notata* (Adler and Phillips, 1985), but the importance of this mechanism in the homing context remains to be demonstrated.

Recently, Ellis-Quinn and Simon (1991) conducted a detailed analysis on the importance of a sun compass in the homing of *Sceloporus jarrovi*. This iguanid is able to home from sites never visited, and clock-shifted specimens show an initial deviation with respect to the home direction consistent with the expectation of a time-compensated compass orientation. Moreover, specimens having the parietal eye and surrounding parietal scale covered with a thick layer of black paint show a significantly reduced homing performance (20%) in comparison to both controls (61%) and specimens having paint placed only alongside the parietal eye (Fig. 6.7). The important result of this study is that *S. jarrovi* cannot orient homeward using only the lateral eyes. However, this study does not clarify all the mechanisms underlying the homing capacity of this species since a sun compass alone is not sufficient to produce a goal-finding orientation from unknown sites without a map mechanism.

6.5 OPHIDIA

Several species of snakes perform spontaneous seasonal movements involving long-distance migrations between different sites (hibernacula, foraging sites, mating areas) which may involve complex orientation mechanisms (Gregory et al., 1987). For instance, the sea snake, *Pelamis platurus*, performs long-distance migrations in the Indian and Pacific Oceans, similar to the circuits of marine fish and turtles (Graham et al., 1971), but till now no experimental studies have been performed on the homing of this species.

Landreth (1973) found that in Oklahoma, *Crotalus atrox* migrates from winter dens to spring–summer activity ranges located up to 3.5 km apart, returning to the same den in autumn. Spontaneous movements between the different areas seem to be well directed,

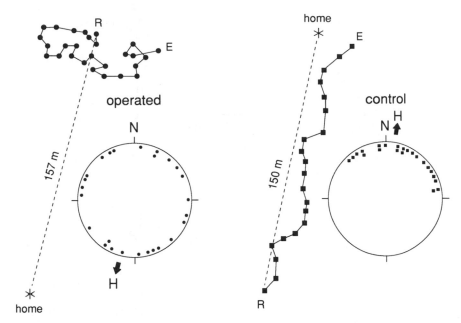

Figure 6.7 Examples of routes obtained by Ellis-Quinn and Simon (1991) in the study of the role of the parietal eye in the homing of the lizard *Sceloporus jarrovi*. The paths of an operated (dots) and a control (squares) lizard after experimental displacements (dashed lines) are shown. R and E represent the release site and the end of the daily radiotracking, respectively. Circular distributions plot the orientation of the daily steps of the two animals. While the control animal headed directly homeward, the path of the lizard whose parietal eye and surrounding parietal scale had been covered with a thick layer of black paint was definitely tortuous and not homeward oriented. These results, combined with other experiments, led the authors to conclude that the experimental lizards did not orient homeward because they could not use a time-compensated celestial compass as part of a map and compass navigation.

while snakes released in foreign habitats after translocation of 8–96 km continued to move on direct courses, usually in the same direction they were heading when captured. The North American garter snake, *Thamnophis sirtalis*, performs seasonal migrations from winter communal dens to foraging areas about 7 km apart and returns to the original sites in late autumn (Larsen, 1987).

On a smaller scale, the yellow-green racer, *Coluber viridiflavus*, performs looped excursions from the main shelter, covering trips from a few metres up to about 3 km connected to basking, feeding

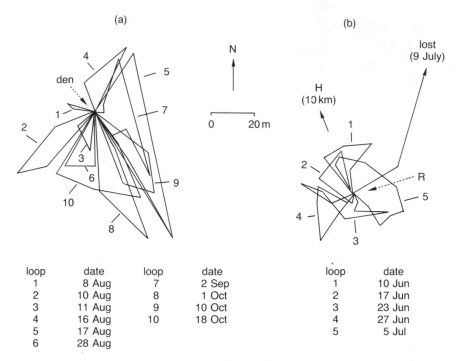

loop	date	loop	date	loop	date
1	8 Aug	7	2 Sep	1	10 Jun
2	10 Aug	8	1 Oct	2	17 Jun
3	11 Aug	9	10 Oct	3	23 Jun
4	16 Aug	10	18 Oct	4	27 Jun
5	17 Aug			5	5 Jul
6	28 Aug				

Figure 6.8 Two examples of the movement pattern recorded by Ciofi and Chelazzi (1991) by radiotracking the snake *Coluber viridiflavus* in central Italy. The resident male (a) shows typically looped excursions centred around a permanent den which is also used for overwintering. The imported male (b) performs similar loops centred around the release site (R). After about one month the radio signal from the imported male was lost when the animal suddenly left the study area, heading in a direction not very different from that of its capture site, about 10 km away. Loops longer than 15 m are reported for both snakes, connecting all the daily fixes; the departure day of each excursion is also indicated.

and mating (Fig. 6.8). During the longer trips, which may last up to 1 month, the snakes use secondary shelters but later invariably return to their main home (Ciofi and Chelazzi, 1991).

The many studies performed to assess the homing capacity of North American snakes after experimental displacements possibly out of the normal home range have given contradictory evidence. According to Imler (1945) the bullsnake, *Pituophis melanoleucus sayi*, shows poor homing capacity and the same seems true for *Heterodon platyrhinos* and *H. nasicus* (Platt, 1969). Reinert and Kodrich (1982) found no homing behaviour in *Sistrurus catenatus catenatus* from

Pennsylvania. Individuals of the worm snake, *Carphophis amoenus*, transported 800, 400, and 150 m from the capture site in Kentucky simply established a new home range where released (Barbour *et al.*, 1969). Fitch and Shirer (1971) did not find homing ability in *Natrix sipedon, Agkistrodon contortrix* and *Crotalus horridus* released up to 10 km from their place of origin in Kansas. The displaced snakes made irregular movements typical of resident individuals or moved away from the release site without heading homeward.

On the contrary, a more or less strong homing ability has been observed in other species. Hirth (1966) recorded a good homing behaviour in *Masticophis taeniatus* and *Crotalus viridis*. A total of 15 out of 30 *Coluber constrictor* homed after displacements of 100–300 m from their main den (Brown and Parker, 1976). There was no correlation between distance displaced and return time which ranged from 4 to 12 days. According to Brown and Parker, *C. constrictor*'s ability to return to the main den was significantly dependent on the direction of displacement. Most of the animals that did not return were displaced to sites from which their return brough them near other dens which conspecifics used for hibernation. Using funnel traps, Fraker (1970) studied the homing behaviour of *Natrix sipedon sipedon* by translocating the snakes up to 650 m from their familiar area. Of the displaced animals, 20% homed. Even if distance and direction of displacement were apparently unimportant, only the older snakes homed over the greater distances.

Scanter still is the evidence on orientation mechanisms involved in the homing of snakes. The visual recognition of landmarks is possibly important in short-range movements, but seems unlikely to be so in long-range travel due to the low vantage point of snakes (Gregory *et al.*, 1987). On the contrary, many studies have suggested olfaction as the primary means of goal orientation in snakes in such different functional contexts as conspecific aggregation, mate search and predation. The olfactory cues can be waterborne as in *Natrix sipedon* (Brown, 1940; Fraker, 1970), airborne as in *Diadopis punctatus* (Dundee and Miller, 1968) or in the form of scent trails.

Trail following plays an important role in the location of conspecifics by different species of the genus *Thamnophis*, especially for the male trailing of females during the mating season (e.g. Ford, 1986). Furthermore, immature individuals of *Crotalus v. viridis* trail adults to hibernacula in the autumn (Klauber, 1972; Graves *et al.*, 1986).

However, no solid evidence is available concerning the presence of olfactory navigation in snakes. Olfactive familiarity with sites around the den is considered important in the homing ability of the whipsnake, *Masticophis taeniatus*, by Brown and Parker (1976) on the basis of some evidence for homing in experimentally blinded specimens displaced 100–300 m. *Natrix sipedon* may home using airborne odours associated with their home ranges (Fraker, 1970).

Some evidence for celestial orientation in displaced snakes has been found (Gregory *et al.*, 1987). According to Landreth (1973), *Crotalus atrox* allowed to move in a circular arena show an ability to orient by solar cues. *Regina septemvittata* and *Natrix sipedon* (Newcomer *et al.*, 1974) can use celestial cues for spatial orientation. Training these snakes in an outdoor circular aquatic arena permitted assessing their directional learning. Moreover, after a 6-h shift in the light–dark cycle the snakes showed a deviation of 90° with respect to their control orientation, indicating a sun-compass mechanism based on an internal clock. However, this capacity seems to be related to a y-axis orientation to water bodies, not necessarily involved in homing. Whether snakes use polarized light is unknown, but Lawson (1985) suspected a role for it in the time-compensated solar orientation of *Thamnophis radix*. The possibility that stellar or lunar cues are used by nocturnal snakes has not yet been investigated.

6.6 SUMMARY AND CONCLUSIONS

There is much information on the homing capacity of reptiles but still not enough to reach a satisfactory understanding of the basic ecological and operational aspects of such behaviour pattern. This is due to several sources of heterogeneity, partly related to the ecological diversity of the different groups of reptiles, partly due to the different perspectives of the studies conducted on this topic, and finally to some methodological difficulties faced when testing these aspects of reptilian behaviour, both in the field and under controlled conditions in the laboratory.

In particular, field observations of both spontaneous homing and the capacity to home after experimental displacement require suitable tracking techniques. This is not a great problem for such groups as land tortoises, which are easily equipped with radio-transmitters giving information not only on the coordinates assumed by the animal in time, but also on the ethological and physio-logical variables associated to each fix (Legler, 1979). However, marking, handling and tracking most lizards and snakes is still difficult despite the miniaturization of radiotransmitters (Reinert and Kodrich, 1982; Slip and Shine, 1988; Ciofi and Chelazzi, 1991).

Obviously, difficult groups for ethological studies requiring individual marking and tracking are crocodiles and alligators, for which the use of anaesthesia still poses serious problems (Loveridge and Blake, 1987). Difficulties can also arise in tracking such reptiles as marine turtles which spend most of their life in water. In these cases radiotracking can be implemented by using a satellite repeating system which allows recording fixes in the open ocean when the 'animals emerge to breath (Stoneburner, 1982; Timko and Kolz, 1982; Byles, 1987). In the case of freshwater turtles sonar tracking can usefully substitute radiotracking (Ireland and Kanwisher, 1978).

The use of these modern techniques has revealed the presence of a distinct homing capacity in several species belonging to all the major reptile groups. The ecological context in which homing appears is quite different, ranging from the ontogenetic return migration of sea turtles and sea snakes (Baker, 1978) to the active maintenance of a constant familiar area in land tortoises (Chelazzi and Carlà, 1986) and snakes (Gregory et al., 1987). However, the mechanisms used by reptiles to home are far from being clarified. The fundamental question of whether or not the external homing observed in many groups is based on true navigation consisting of a two-phase mechanism (map and compass) has yet to be assessed. Some authors speculate on the importance of different cues in giving map information for navigation, such as the magnetic field claimed by Rodda (1984b) to be used in the map step by homing juvenile alligators. Many reports suggest the importance of olfaction in the homing of sea turtles and land tortoises, but no coherent hypothesis on the operational aspects of olfactory-based homing has yet been presented.

Much more consistent information is available on the compass mechanisms enabling reptiles to maintain a constant direction in the context of a y-axis orientation or as compass components in a two-phase navigation. The role of celestial cues in classical time-compensated orientation of freshwater turtles, lizards and snakes has been well demonstrated. On the contrary, the use of a magnetic compass, claimed for instance by Mathis and Moore (1988) in the homing of a freshwater turtle, cannot be considered as demonstrated.

REFERENCES

Adler, K. and Phillips, J.B. (1985) Orientation in a desert lizard (*Uma notata*): time-compensated compass movement and polarotaxis. *J. Comp. Physiol.*, **156**, 547–52.

Auffenberg, W. and Weaver, W.G. (1969) *Gopherus berlandieri* in southeastern Texas. *Bull. Fla. St. Mus. Biol. Sci.*, **13**, 141–203.

Baker, R.R. (1978) *The Evolutionary Ecology of Animal Migration*, Hodder and Stoughton, London, 1012 pp.

Balazs, G.H. (1980) Synopsis of biological data on the green turtle in the hawaiian islands. *NOAA-TM-NMFS-SWFC*-7, 141 pp.

Barbour, R.W., Harvey, M.J. and Hardin, J.W. (1969) Home range, movements, and activity of the eastern worm snake, *Carphophis amoenus amoenus*. *Ecology*, **50**, 470–6.

Bissinger, B.E. (1983) Homing behavior, sun-compass orientation, and thermoregulation in the lizard (*Sceloporus jarrowi*): the role of the parietal eye. PhD Thesis, The City University of New York, New York.

Brown, E.E. (1940) Life history and habits of the northern watersnake, *Natrix s. sipedon*. PhD Thesis, Cornell University, Ithaca, New York, 119 pp.

Brown, W.S. and Parker, W.S. (1976) Movement ecology of *Coluber constrictor* near communal hibernacula. *Copeia*, **1976**, 225–42.

Burge, B.L. (1977) Daily and seasonal behavior and areas utilized by the desert tortoise *Gopherus agassizi* in Southern Nevada, in *The Desert Tortoise Council, Edit. Proceedings of 1977 Symposium*, Desert Tortoise Council, Las Vegas, pp. 59–94.

Bustard, H.R. (1979) Population dynamics of sea turtles, in *Turtles: perspectives and research* (eds M. Harless and H. Morlock), John Wiley, New York, pp. 523–40.

Bustard, H.R. and Singh, L.A.K. (1983) Movement of wild gharial, *Gavialis gangeticus* (Gmelin) in the river Mahanadi, Orissa (India). *Br. J. Herpet.*, **6**, 287–91.

Byles, R. (1987) Development of a sea turtle satellite biotelemetry system. *Proc. Argos Users Conf.* Service Argos, Greenbelt, MD, pp. 199–210.

Cagle, F.R. (1944) Home range, homing and migration in turtles. *Misc. Publs. Mus. Zool. Univ. Mich.*, **61**, 1–34.

Calzolai, R. and Chelazzi, G. (1991) Habitat use in a central Italy population of *Testudo hermanni* Gmelin (Reptilia Testudinidae). *Ethol. Ecol. Evol.*, **3**, 153–66.

Carr, A. (1952) *Handbook of Turtles*. Cornell University Press, Ithaca, New York, 695 pp.

Carr, A. (1967a) *So Excellente a Fishe*, Natural History Press, New York.

Carr, A. (1967b) Adaptive aspects of the scheduled travel of *Chelonia*, in *Animal Orientation and Navigation, Proc. 27th A. Biol. Coll.* (ed. R.M. Storm), Oregon State University Press, Corvallis, pp. 35–55.

Carr, A. (1975) The Ascension Island green turtle colony. *Copeia*, **1975**, 547–55.

Carr, A. and Colemann, P.J. (1974) Seafloor spreading theory and the odyssey of the green turtle. *Nature*, **249**, 128–30.

Carr, A. and Giovannoli, L. (1957) The ecology and migrations of sea turtles, 2. Results of field work in Costa Rica, 1955. *Am. Mus. Nov.*, **1835**, 1–32.

Carr, A. and Hirth, H.F. (1962) The ecology and migration of sea turtles,

5. Comparative features of isolated green turtle colonies. *Am. Mus. Nov.*, **2091**, 1–41.

Carroll, T.E. and Ehrenfeld, D.W. (1978) Intermediate-range homing in the Wood Turtle, *Clemmys insculpta*. *Copeia*, **1978**, 117–26.

Chelazzi, G. and Calzolai, R. (1986) Thermal benefits from familiarity with the environment in a reptile. *Oecologia*, **68**, 557–8.

Chelazzi, G. and Carlà, M. (1986) Mechanisms allowing home range stability in *Testudo hermanni* Gmelin (Reptilia Testudinidae): field study and simulation. *Monitore Zoologico Italiano (Nuova Serie)*, **20**, 349–70.

Chelazzi, G. and Delfino, G. (1986) A field test on the use of olfaction in homing by *Testudo hermanni* (Reptilia Testudinidae). *J. Herpetol.*, **20**, 451–5.

Chelazzi, G. and Francisci, F. (1979) Movement patterns and homing behaviour of *Testudo hermanni* Gmelin (Reptilia Testudinidae). *Monitore Zoologico Italiano (Nuova Serie)*, **13**, 105–27.

Ciofi, C. and Chelazzi, G. (1991) Radiotracking of *Coluber viridiflavus* using external transmitters. *J. Herpetol.*, **25**, 37–40.

DeRosa, C.T. and Taylor, D.H. (1980) Homeward orientation mechanisms in three species of turtles (*Trionyx spinifer*, *Chrysemys picta*, and *Terrapene carolina*). *Behav. Ecol. Sociobiol.*, **7**, 15–23.

Dundee, H.A. and Miller, M.C. (1968) Aggregative behavior and habitat conditioning by the prairie ringneck snake, *Diadophis punctatus arnyi*. *Tulane Stud. Zool. Bot.*, **15**, 41–58.

Ellis-Quinn, B.A. and Simon, C.A. (1989) Homing behavior of the lizard *Sceloporus jarrovi*. *J. Herpetol.*, **3**, 146–52.

Ellis-Quinn, B.A. and Simon, C.A. (1991) Lizard homing behavior: the role of the parietal eye during displacement and radiotracking, and time-compensated celestial orientation in the lizard *Sceloporus jarrovi*. *Behav. Ecol. Sociobiol.*, **28**, 397–407.

Emlen, S.T. (1969) Homing ability and orientation in the painted turtle, *Chrysemys picta marginata*. *Behaviour*, **33**, 58–76.

Ernst, C.H. (1968) Homing ability of the spotted turtle, *Clemmys guttata* (Schneider). *Herpetologica*, **24**, 77–8.

Ernst, C.H. (1970) Homing ability in the painted turtle *Chrysemys picta* (Schneider). *Herpetologica*, **26**, 399–403.

Ernst, C.H. and Barbour, R.W. (1972) *Turtles of the United States*, The University Press of Kentucky.

Fischer, K. (1961) Untersuchungen zur Sonnenkompassorientierung und Laufaktivität von Smaragdeidechsen (*Lacerta viridis* Laur.). *Z. Tierpsychol.*, **18**, 450–70.

Fitch, H.S. (1940) A field study of the growth and behavior of the fence lizard. *Univ. California Pub. Zool.*, **44**, 151–72.

Fitch, H.S. and Shirer, H.W. (1971) A radiotelemetric study of spatial relationships in some common snakes. *Copeia*, **1971**, 118–28.

Foà, A., Bearzi, M. and Baldaccini, N.E. (1990) A preliminary report on the size of the home range and on the orientational capabilities in the lacertid lizard *Podarcis sicula*. *Ethol. Ecol. Evol.*, **2**, 310.

Ford, N.B. (1986) The role of pheromone trails in the sociobiology of snakes, in *Chemical Signals in Vertebrates*, Vol. 4 (eds D. Duvall, D. Müller-Schwarze and R.M. Silverstein), Plenum Press, New York, pp. 261–78.

Fraker, M.A. (1970) Home range and homing in the watersnake, *Natrix sipedon sipedon*. *Copeia*, **1970**, 665–73.

Gibbons, J.W. (1970) Terrestrial activity and the population dynamics of aquatic turtles. *Am. Midl. Nat.*, **83**, 404–14.

Gorzula, S.J. (1978) An ecological study of *Caiman crocodilus crocodilus* inhabiting savanna lagoons in the Venezuelan Guayana. *Oecologia*, **35**, 21–34.

Gould, E. (1957) Orientation in box turtles, *Terrapene c. carolina* (Linnaeus). *Biol. Bull. Mar. Biol. Lab.*, **112**, 336–48.

Gould, E. (1959) Studies on orientation of turtles. *Copeia*, **1959**, 174–6.

Gourley, E.V. (1974) Orientation of the gopher tortoise *Gopherus polyphemus*. *Anim. Behav.*, **22**, 158–69.

Graham, J.B., Rubinoff, I. and Hecht, M.K. (1971) Temperature physiology of the sea snake, *Pelamis platurus*: and index of its colonization potential in the Atlantic Ocean. *Proc. Natl. Acad. Sci.*, *U.S.A.*, **68**, 1360–3.

Grassman, M.A. (1984) The chemosensory behavior of juvenile sea turtles: implication for chemical imprinting. PhD Dissertation, Texas A&M University.

Grassman, M.A., Owens, D.W., McVey, J.P. and Marquez M.R. (1984) Olfactory-based orientation in artificially imprinted sea turtles. *Science*, **224**, 83.

Graves, B.M., Duvall, D., King, M.B., Linsted, S.L. and Gern, W.A. (1986) Initial den location by neonatal prairie rattlesnakes: functions, causes, and natural history in chemical ecology, in *Chemical Signals in Vertebrates*, Vol. 4 (eds D. Duvall, D. Müller-Schwarze and R.M. Silverstein), Plenum Press, New York, pp. 285–304.

Gregory, P.T., Macartney, J.M. and Larsen, K.W. (1987) Spatial patterns and movements, in *Snakes: ecology and evolutionary biology* (eds R.A. Seigel, J.T. Collins and S.S. Novak), Macmillan, New York, pp. 366–95.

Guyer, C. (1978) Comparative ecology of the short-horned lizard (*Phrynosoma douglassi*) and the sagebrush lizard (*Sceloporus graciosus*) in southeastern Idaho. MS Thesis, Idaho State University, 56 pp.

Hays, G. (1991) The potential for assessing nesting beach fidelity and clutch frequency for sea turtles by satellite tracking. 4th European Conference on Wildlife Telemetry, Aberdeen, UK.

Hirth, H.F. (1966) The ability of two species of snakes to return to a hibernaculum after displacement. *Southwest. Nat.*, **11**, 49–53.

Hirth, H.F. and Carr, A. (1970) The green turtle in the Gulf of Aden and Seycelles Islands. *Verhand. Konin. Nederl. Acad. Weten. Natur.*, **58**, 1–41.

Hughes, G.R. (1974) The sea turtles of South-East Africa. 2. The biology of the Tongaland loggerhead turtle *Caretta caretta L.* with comments on the leatherback turtle *Dermochelys coriacea* L. and the green turtle *Chelonla mydas* L. in the study region. *Invest. Rep. Ocean. Res. Inst.*, Durban, South Africa, **36**, 1–96.

Imler, R.H. (1945) Bullsnakes and their control on a Nebraska wildlife refuge. *J. Wildl. Mngmt.*, **9**, 265–73.

Ireland, L.C. and Kanwisher, J.W. (1978) Underwater acoustic biotelemetry: procedures for obtaining information on the behavior and physiology of free-swimming aquatic animals in their natural environments, in *The Behavior of Fish and Other Aquatic Animals* (ed. D.I. Mostofsky), Academic Press, New York, pp. 341–79.

Ishihara, S. (1969) Homing behavior of the lizard, *Tachydromus tachydromoides* (Schlegel). *Bull. Kyoto Univ. Educ. Ser. B*, **36**, 11–23.

Klauber, L.M. (1972) *Rattlesnakes*, University of California Press, Berkeley.

Koch, A.L., Carr, A. and Ehrenfeld, D.W. (1969) The problem of open sea migration: the migration of the green turtle to Ascension Island. *J. Theor. Biol.*, **22**, 163.

Krekorian, C.O. (1977) Homing in the desert iguana, *Dipsosaurus dorsalis*. *Herpetologica*, **33**, 123–7.

Larsen, K.W. (1987) Movements and behavior of migratory garter snakes, *Thamnophis sirtalis*. *Can. J. Zool.*, **65**, 2241–7.

Landreth, H.F. (1973) Orientation behavior of the rattlesnake, *Crotalus atrox*. *Copeia*, **1973**, 26–31.

Lawson, M.A. (1985) Preliminary investigations into the roles of visual and pheromonal stimuli on aspects of the behaviour of the western plains garter snake, *Thamnophis radix haydeni*. MSc Thesis, University of Regina, Saskatchewan, Canada.

Lebboroni, M. and Chelazzi, G. (1991) Activity patterns of *Emys orbicularis* L. (Chelonia Emydidae) in central Italy. *Ethol. Ecol. Evol.*, **3**, 257–68.

Legler, W.K. (1979) Telemetry, in *Turtles: perspectives and research* (eds M. Harless and H. Morlock), John Wiley, New York, pp. 61–72.

Lorenz, K.Z. (1937) The companion in the bird's world. *Auk*, **54**, 245–73.

Loveridge, J.P. and Blake, D.K. (1987) Crocodile immobilization and anaesthesia, in *Wildlife Management: crocodiles and alligators* (eds G.J.W. Webb *et al.*), Surrey Beatty and Sons, Chipping Norton, NSW, pp. 259–67.

Manton, M.L. (1979) Olfaction and behavior, in *Turtles: perspectives and research* (eds M. Harless and H. Morlock), John Wiley, New York, pp. 289–301.

Mathis, A. and Moore, F.R. (1988) Geomagnetism and the homeward orientation of the box turtle, *Terrapene carolina*. *Ethology*, **78**, 265–74.

Mayhew, W.W. (1963) Biology of the granite spiny lizard, *Sceloporus orcutti*. *Am. Midl. Nat.*, **69**, 310–27.

Modha, M.L. (1967) The ecology of the Nile crocodile (*Crocodylus niloticus* Laurenti) on Central Island, Lake Rudolf. *E. Afr. Wildl. J.*, **5**, 74–95.

Modha, M.L. (1968) Crocodile research project, Central Island, Lake Rudolf: 1967 breeding season. *E. Afr. Wildl. J.*, **6**, 148–50.

Mortimer, J.A. and Portier, K.M. (1989) Reproductive homing and internesting behavior of the Green Turtle (*Chelonia mydas*) at Ascension Island, South Atlantic Ocean. *Copeia*, **1989**, 962–77.

Murphy, P.A. (1981) Celestial compass orientation in juvenile american alligators (*Alligator mississippiensis*). *Copeia*, **1981**, 638–45.

Newcomer, T.R., Taylor, D.H. and Guttman, S.I. (1974) Celestial orientation in two species of water snakes (*Natrix sipedon* and *Regina septemvittata*). *Herpetologica*, **2**, 194–200.

Noble, G.K. (1934) Experimenting with the courtship of lizards. *Nat. Hist.*, **34**, 5–15.

Ouboter, P.E. and Nanhoe, L.M. (1988) Habitat selection and migration of *Caiman crocodilus crocodilus* in a swamp and swamp-forest habitat in northern Suriname. *J. Herpetol.*, **22**, 283–94.

Owens, D., Comuzzie, D.C. and Grassman, M. (1986) Chemoreception in the homing and orientation behavior of amphibians and reptiles, with special reference to sea turtles, in *Chemical Signals in Vertebrates*, Vol. 4. *Ecology, Evolution, and Comparative Biology* (eds D. Duvall, D. Muller-Schwarze and R.M. Silverstein), Plenum Press, New York, pp. 341–55.

Platt, D.R. (1969) Natural history of the hognose snakes *Heterodon platyrhinos* and *Heterodon nasicus*. *Univ. Kans. Publ. Mus. Nat. Hist.*, **18**, 235–420.

Plummer, M.V. and Shirer, H.W. (1975) Movement patterns in a river population of the softshell turtle *Trionyx muticus*. *Occ. Pap. Mus. Nat. Hist. Univ. Kans.*, **43**, 1–26.

Pritchard, P.C.H. (1976) Post-nesting movements of marine turtles (Chelonididae and Dermochelidae) tagged in the Guyanas. *Copeia*, **1976**, 749–54.

Pritchard, P. (1979) *Encyclopedia of Turtles.*, T.F.H. Publications, Neptune, New Jersey.

Rand, A.S. (1967) Ecology and social organization in the iguanid lizard *Anolis lineatopus*. *Proc. U.S. Nat. Mus.*, **122**, 1–79.

Reinert, H.K. and Kodrich, W.R. (1982) Movements and habitat utilization by the Massasauga, *Sistrurus catenatus catenatus*. *J. Herpetol.*, **16**, 162–71.

Rodda, G.H. (1984a) Homeward paths of displaced juvenile alligators as determined by radiotelemetry. *Behav. Ecol. Sociobiol.*, **14**, 241–6.

Rodda, G.H. (1984b) The orientation and navigation of juvenile alligators: evidence of magnetic sensitivity. *J. Comp. Physiol.*, **154**, 649–58.

Rodda, G.H. (1985) Navigation in juvenile alligators. *Z. Tierpsychol.*, **68**, 65–77.

Rodhouse P., Barling, R.W.A., Clark, W.I.C., *et al.* (1975) The feeding and ranging behaviour of the Galapagos giant tortoises (*Geochelone elephantopus*). The Cambridge and London University Galapagos Expeditions, 1972 and 1973. *J. Zool., Lond.*, **176**, 297–310.

Shealy, R.M. (1976) The natural history of the Alabama map turtle, *Graptemys pulchra* Baur, in Alabama. *Bull. Fla. State Mus., Biol. Sci.*, **21**, 47–111.

Slip, D.J. and Shine, R. (1988) Habitat use, movements, and activity patterns of free-ranging diamond pythons, *Morelia spilota spilota* (Serpentes: Boidae): a rediotelemetric study. *Aust. Wildl. Res.*, **15**, 515–31.

Spoecker, P.D. (1967) Movements and seasonal activity cycles of the lizard *Uta stansburiana stejnegeri. Am. Midl. Nat.*, **77**, 484–94.

Staton, M.A. and Dixon, J.R. (1975) Studies on the dry season biology of *Caiman crocodilus* from Venezuelan Llanos. *Mem. Soc. Cienc. Nat., La Salle*, **101**, 237–65.

Stoneburner, D.C. (1982) Sea turtle (*Caretta caretta*) migration and movements in the South Atlantic Ocean. NASA SP-457, Natl. Aeronaut. Space Admin. Washington, DC, 74 pp.

Strijbosch, H., Rooy, P.Th.J.C.v. and Voesenek, L.A.C.J. (1983) Homing behaviour of *Lacerta agilis* and *Lacerta vivipara* (Sauria, Lacertidae). *Amphibia-Reptilia*, **4**, 43–7.

Swingland, I.R. and Lessels, C.M. (1979) The natural regulation of giant tortoise populations on Aldabra Atoll: movement polymorphism, reproductive success and mortality. *J. Anim. Ecol.*, **2**, 639–54.

Timko, R.E. and Kolz, A.L. (1982) Satellite sea turtles tracking. *Mar. Fish. Rev.*, **44**, 19–24.

Tinkle, D.W. (1967) The life and demography of the side-blotched lizard, *Uta stansburiana. Misc. Publ. Mus. Zool. Univ. Michigan*, **132**, 1–182.

Webb, G.J.W., Buckworth, R. and Manolis, S.C. (1983a) *Crocodilus johnstoni* in the McKinlay River area Northern Territory Australia. 4. A demonstration of homing. *Aust. Wildl. Res.*, **10**, 403–6.

Webb, G.J.W., Manolis, S.C. and Buckworth, R. (1983b) *Crocodilus johnstoni* in the McKinlay River area Northern Territory Australia. 2. Dry season habitat selection and an estimate of the total population size. *Aust. Wildl. Res.*, **10**, 373–82.

Weintraub, J. (1970) Homing in the lizard *Sceloporus orcutti. Anim. Behav.*, **18**, 132–7.

Chapter 7
Birds

F. Papi and H.G. Wallraff

7.1 INTRODUCTION

Within the animal kingdom, birds are unrivalled at covering large distances quickly and passing over geographic barriers. They use this power to reach food supplies, go where environment conditions are most suitable, and escape predators. Many bird species are migratory and regularly shuttle between winter and summer quarters, which may be far apart. The marked site fidelity of most species both when breeding and in winter, often even at stop-over sites, makes birds an ideal object of homing studies.

While in flight, a bird might survey a large area and memorize a great many landmarks, which might be enough by themselves to permit orientation with a pilotage mechanism. This, however, is not sufficient to explain orientation during many migratory journeys. Birds that migrate night and day over monotonous landscapes or large tracts of water, such as those which fly from the northern Mediterranean countries to the African savannahs south of the Sahara, can hardly rely on landmarks to stay on course. Compass orientation is widely used by migrants and may explain such performances, but cannot account for the capacity for true navigation actually displayed by birds. When displaced by storms or experimenters to an unfamiliar area, birds are often capable of position-fixing with respect to their home or intended goal. How they manage to do this and which cues they rely on has been one of the main aims of research work on bird navigation over the last 40 years.

Animal Homing. Edited by Floriano Papi. Published in 1992 by Chapman & Hall. 2–6 Boundary Row, London SEI 8HN. ISBN 0 412 36300 9.

7.2 HOMING UNDER NATURAL CONDITIONS

7.2.1 Foraging flights

A bird incubating its eggs or feeding its young has to leave its nest
to search for food and then find its nest again. When such flights
cover only a few dozens or hundreds of metres, as is the case in
many species, nest finding is then merely one special case of
finding a goal in an environment consisting of countless familiar
landmarks. This kind of homing is classified as pilotage and
is usually discussed nowadays under the heading 'cognitive map'
(p. 13).

There are, however, gradual transitions between flying within a
territory that includes the nest and the feeding area, and flying to
and fro between a restricted nesting territory and feeding grounds a
few kilometres apart. So too there are gradual transitions between
flights of a few kilometres, such as those of starlings or hawks, and
flights over dozens or even hundreds of kilometres, such as those of
vultures or, in response to particular meteorological conditions,
swifts (e.g. Koskimies, 1950; Lack, 1958). Very long foraging
flights are performed by sea-birds. A wandering albatross (*Diomedea
exulans*) has been followed by satellite tracking over an extended
tour of more than 15 200 km above the open ocean, lasting 33 days,
after which it returned to its nest site on a small island (Fig. 1.1e). It
would be impossible to explain this homing performance on the
basis of a 'cognitive map' made up of conventional patterns of
landmarks.

7.2.2 Migrations

A well known and much admired case of homing is that of the
stork as it returns every spring to its nest in Europe after spending
the winter in Africa. The periodic migrations of birds, involving
spectacular performances, impressive mass movements and beauti-
ful scenes, are beset by problems of homing (Baker, 1978, 1984;
Able, 1980; Alerstam, 1990; Berthold, 1991a). There is no basic
difference with respect to foraging flights, however. It should be
borne in mind that migration too takes place on a sliding scale from
only a few kilometres (e.g. from mountains to lower regions) to
long-distance flights, such as the non-stop journeys of land birds
over oceans (Fig. 7.1). The maximum distance of some 30 000 km
per year is covered by the Arctic tern (*Sterna paradisea*).

There is, however, a cardinal problem in bird migration. The
young birds born in any given year travel to population-specific

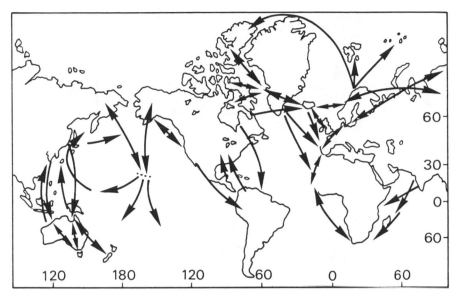

Figure 7.1 The transoceanic migrations of land birds are among the most wonderful homing performances known. The main routes are shown here. Spring migration is indicated by northward pointing arrows, and autumn migration by southward pointing arrows (From Williams and Williams, 1990.)

wintering grounds, often thousands of kilometres from their birth place. Many passerines migrate at night, the young birds often flying separately from experienced adult birds. Can an animal 'home' to a goal that has never so far been its home? This is a substantial problem. The real question is, whether it is necessary to assume that the inexperienced bird is in any way programmed with some knowledge about the geographical position of its goal area. An alternative hypothesis might be that it is merely programmed to fly in a certain compass direction and to stop flying when it has covered a certain distance.

To decide between these alternatives, Perdeck (1958) conducted an experiment which has become a classic. Of the huge numbers of starlings (*Sturnus vulgaris*) which migrate through Holland in autumn, more than 11 000 were caught and transported to Switzerland, where they were released. This was a displacement of about 600 km perpendicular to the normal migration route, which usually ends in northern France and southern England. The distribution of recovery sites in the subsequent winter months revealed differences in orientation behaviour dependent on the age

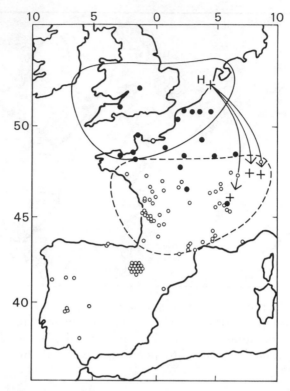

Figure 7.2 Starlings (*Sturnus vulgaris*) were caught in Holland (H) during their autumn migration in a NNE–SSW direction and released from three sites in Switzerland (crosses). Their approximate wintering range is bounded by a solid line, and a correspondingly displaced range by a dashed line. The distribution of recovery sites in the following winter shows that adults (filled circles) tended towards the normal winter range, whereas young birds (open circles) kept the normal compass direction. (Modified from Perdeck, 1958.)

of the birds (Fig. 7.2). Young starlings migrating for the first time continued to fly the normal compass course so that they arrived in an abnormal area dislocated by approximately the direction and distance of displacement. Older starlings, which had already spent at least one winter in the population-specific area, decided on an abnormal compass course which led them back to the normal wintering area.

Thus, true goal orientation was performed only by those individuals that had previously established a certain home area. Usually the young birds come to a similar final result by following

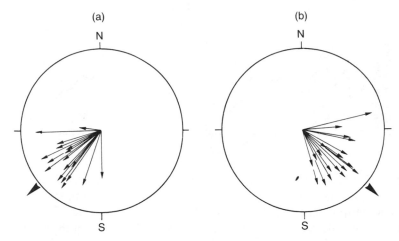

Figure 7.3 Young blackcaps (*Sylvia atricapilla*) from different populations orient in different directions according to their migratory route. Birds were taken out of their nests shortly after opening their eyes; they were then hand-raised in a laboratory, and tested outdoors in Emlen funnels (see Fig. 7.10). The arrows inside the circles are mean vectors that have been calculated for each individual bird; the arrows outside the circles indicate the overall mean direction. (a) Blackcaps from southwest Germany migrating towards France, Iberia and northwest Africa; (b) blackcaps from easternmost Austria migrating towards the eastern Mediterranean area and further. The data were recorded in September–October. (Modified from Helbig *et al.*, 1989.)

an appropriate direction over an appropriate distance. The spatial distance to be flown is programmed in terms of time (Berthold, 1991b). Both parts of the programme, direction as well as duration of migration, are specific for each population and are genetically determined (Figs 7.3 and 7.4). They are the end-products of a long phylogenetic process during which migratory routes have been adapted to the specific geomorphological and ecological conditions of the continents. Long-distance routes are often more complicated than the broad-front pattern of the starling. In part, they include temporal shifts of the intended direction. To what degree such shifts also depend on external signals has not yet been clarified fully (cf. Wallraff, 1991).

The bearing-and-distance orientation of young birds is a case of vectorial orientation. Since intended direction and distance have been developed phylogenetically, their behaviour is called 'genetically-based orientation' (p. 11). The time-and-direction

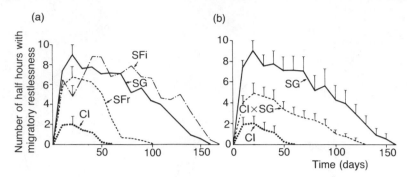

Figure 7.4 The average amount and duration of nocturnal activity in some migratory birds is related to the length of the migratory route. (a) The temporal pattern of nocturnal migratory restlessness is compared in four populations of blackcap (*Sylvia atricapilla*; SFi, southern Finland; SG, southern Germany; SFr, southern France; CI, Canary Islands). The data were averaged for all the birds (24–26 in each group), including those that showed no nocturnal activity. (b) The activity of two parental populations is compared with that of their (experimentally bred) hybrids. (Modified from Berthold and Querner, 1981.)

programme may be sufficient to guide a bird towards some point within a large area, as for instance on a flight from a native site in Europe to the Sahel zone in Africa, or from a winter quarter in Zambia to Scandinavia. Yet one would hardly expect it to be capable of guiding a stork or swallow back to the individual village where it had bred the year before. Too many accidents can happen in between. Birds are displaced by winds, and varying weather conditions do not allow them to keep to a precise calendar. From their first spring onwards, however, birds may make use of additional means of orientation, because they are now returning to an already familiar goal. We have learned from the adult starlings in Perdeck's experiment (Fig. 7.2) that these other means are much more flexible than the genetic programme in coping with unforeseen conditions.

These homing capabilities will be the principal subject of this chapter. However, before dealing with the mechanisms on which homing is based, we will look at the performances which make a search for explanations necessary.

7.3 HOMING AFTER EXPERIMENTAL DISPLACEMENT

7.3.1 Experiments with wild birds

To investigate homing ability in detail, it is necessary to know the exact home site of the individual animal. This was not known in

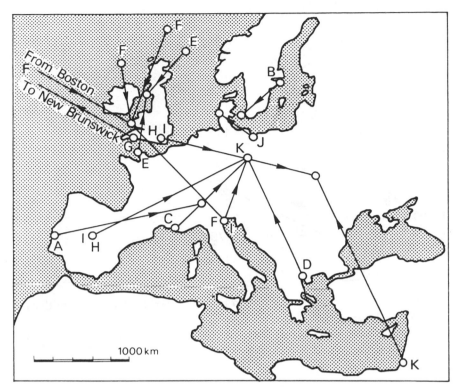

Figure 7.5 Examples of long homing flights in Europe. A, alpine swift (*Apus melba*); B, black-headed gull (*Larus ridibundus*); C, red-backed shrike (*Lanius collurio*); D, wryneck (*Jynx torquilla*); E, lesser black-backed gull (*Larus fuscus*); F, Manx shearwater (*Puffinus puffinus*); G, Leach's petrel (*Oceanodroma leucorhoa*); H, swallow (*Hirundo rustica*); I, starling (*Sturnus vulgaris*); J, Arctic tern (*Sterna paradisea*); K, white stork (*Ciconia ciconia*). (Modified from Matthews, 1968.)

the starling experiment shown in Fig. 7.2. Usually, therefore, birds are displaced from their nesting sites where it is possible to carry out an exact check on their return. Systematic homing experiments with birds of many species have been conducted since the beginning of this century (see tables published by Matthews 1955, 1968). Among these species were starlings, swallows, terns, gulls, shearwaters, petrels, penguins, and many others. Distances of displacement ranged from a few to several thousand kilometres. Some examples of long-distance homing in Europe are given in Fig. 7.5. The most spectacular homing experiments have been conducted with three species of Procellariiformes: shearwaters, petrels and

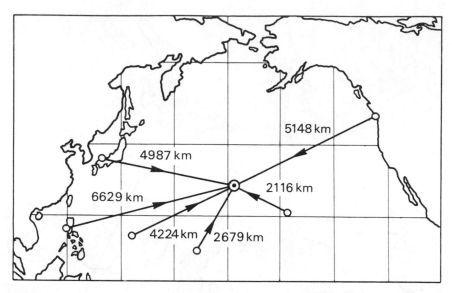

Figure 7.6 The top performances in homing after displacements to distant sites were those of the Laysan albatros (*Diomedea immutabilis*). Arrows join release sites with the nesting site on Sand Island, Midway Atoll. (After data from Kenyon and Rice, 1958.)

albatrosses. Fourteen out of 18 Laysan albatrosses (*Diomedea immutabilis*) displaced over distances of more than 2000 km (2000–6600 km) returned within 7–60 days (10 in 7–20 days; Fig. 7.6).

Another famous example of returns over distances of several thousand kilometres is shown in Fig. 7.7. Six *Zonotrichia* sparrows succeeded in returning twice to their wintering site in California after displacement first over 2900 km to Lousiana and then, after being recaptured the next winter, over 3860 km to Maryland. Between displacements and re-observation the following winter, these sparrows probably made a flight back to their breeding grounds in Canada or Alaska before again migrating south to California.

In experiments with other species, many birds homed over several hundred kilometres, sometimes remarkably fast, often more slowly, taking several days or weeks, and a considerable percentage of the displaced birds did not return at all. Theoretically, many low-level homing performances by birds could be explained as a possible result of random flights, or perhaps systematic searches, without the availability of any information on the direction of the goal. On the other hand, non-returns and low homing speed do not

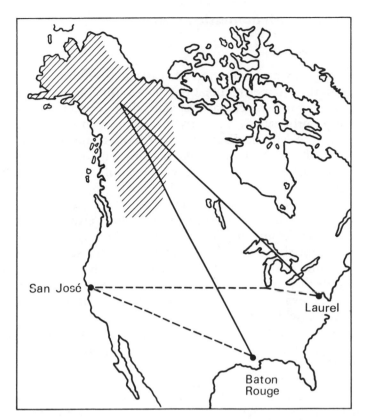

Figure 7.7 A further case of successful homing from very long distances. Golden-crowned sparrows (*Zonotrichia atricapilla*) and white-crowned sparrows (*Z. leucophrys gambeli*) wintering at San José, California, were displaced to Baton Rouge, Louisiana. The next winter some of them were recaptured at San José and taken to Laurel, Maryland, but they were again recorded at San José a year later. After each displacement, the birds very probably flew back to the breeding grounds in northwestern Canada or Alaska (hatched area). (Modified from Mewaldt, 1964.)

prove the absence of goal orientation. By telemetric studies it has been shown that even in such cases birds may well be home-oriented from the beginning, but travel slowly, with detours and in small stages, interrupted by feeding and resting, instead of returning in a fast non-stop flight (Able *et al.*, 1984). Homing performances do not solely depend on orientation, but at least as much on birds' motivation to return fast or return at all to the place where they had been caught.

7.3.2 Experiments with domestic pigeons

(a) Why homing pigeons?

Researchers on avian navigation are particularly lucky in being able to perform their experiments on a 'laboratory animal' which has been bred over many generations and selected precisely because of its homing properties. Pigeons have been used for carrying messages in the Middle East and the Mediterranean for thousands of years. For more than a century, homing pigeons have been bred as a special stock used for races involving millions of birds all over the world every year. Breeders rear offspring only from the fastest homers, so stocks have been selected that are clearly superior in homing performance to other breeds and also to their wild ancestor, the rock pigeon (*Columba livia*). This superiority probably derives to a much greater degree from an increased motivation to return without much delay (due to factors such as resting, feeding and joining other pigeons), than to an improvement in orientational capability (Alleva *et al.*, 1975). It is also this almost monomaniac homing drive which seldom allows these birds to make an early stop, as many wild birds do after release; it forces them to fly away immediately, so giving the investigator an opportunity to observe their initial orientation.

A second important advantage of working with homing pigeons is the fact that hundreds of them can be housed in a compact, man-made shelter, where they live all the year, are easily available for experimental purposes, and where their arrival after a release can be precisely recorded. It is because of these unique advantages, that most efforts to analyse the mechanisms that make avian homing possible have relied on this domesticated strain. As a result, most of the present chapter deals with homing pigeons.

(b) General features of pigeon homing

In races organized by pigeon fanciers, birds are released in large flocks comprising thousands of birds. During the summer season, pigeons housed in a given area are always transported in the same direction, with distances increasing stepwise. Unfortunately, this enormous mass of data is almost useless for research purposes, because data from pigeons flying together cannot be treated as independent units in statistical analyses, and because homing always performed from the same direction would be possible by the use of a compass only, and this makes it unsuitable for the investigation of true navigation. For research purposes pigeons are released

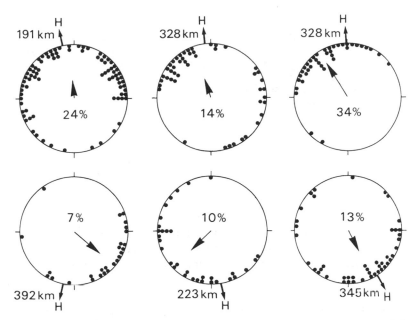

Figure 7.8 In releases from unfamiliar sites, homing pigeons, even if they are inexperienced at homing, mostly show homeward directedness, even if this varies in degree. However, bearing scattering, average deflection from the home bearing, and homing success vary widely. The diagrams show the initial orientation of pigeons from six different home sites, all inexperienced and all released from the same site. Each dot shows the vanishing bearing of one pigeon, while the inner arrow represents the mean vector, whose length is inversely proportional to the bearing scattering: its maximum possible length is equal to 1 (as represented by the radius of the circle). The outer arrow indicates the home direction. The percentage of birds which homed is given for each group. (Modified from Wallraff, 1970.)

individually; in most cases, each bird is followed up by field–glass observation until it vanishes from sight. Results are most suitable for analysis if release sites are arranged symmetrically around the loft.

When pigeons which have so far only made spontaneous flights from their home site (flight restricted to a radius of at most a few kilometres) are released farther from home, their initial orientation is usually characterized by three typical properties (Fig. 7.8): (1) the majority of the bearings are in a semicircle around home ±90°; (2) there is a considerable, and variable, degree of angular scatter of the individual bearings; (3) in many cases, even the mean direction

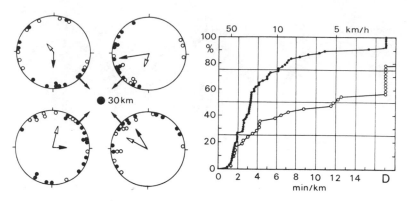

Figure 7.9 Initial orientation and homing performance are usually better in experienced than in inexperienced pigeons. The left-hand diagrams show the initial orientation from four sites symmetrically arranged around the home site; the release sites were unfamiliar both to experienced birds (filled circles) and inexperienced birds (open circles). On the right, the homing performances in the same experiment are shown by the cumulative percentages of returns on the release day (min/km) and on the following days (D). For other explanations see Fig. 7.8. (Modified from Wallraff, 1989a.)

does not point homewards but deviates significantly from home to the left or right.

When inexperienced pigeons are displaced over distances of 200–300 km, only a minority return to the loft. Most of the birds, however, approach home to some degree, as shown by the distribution of sites at which many of them are found on the days following release, mostly in pigeon lofts somewhere on the way (an example is shown in Fig. 7.20a). Thus, most of the non-homers do not fail to home because they are unable to find the way, but simply because they were insufficiently persistent in flying and searching for home.

Pigeons with some homing experience are much more reliable in homing from almost anywhere within a radius of several hundred kilometres. This is due to selection of the successful homers and to individual improvement. Figure 7.9 shows examples of initial orientation and homing performance in both inexperienced and experienced pigeons. It should, however, be stressed that no example can be fully representative of pigeons' performances in general. Initial orientation, homing speeds and return rates all vary to a great degree, depending on factors such as the geographical area, loft site, season and day of release, pigeon stock, experience and age of the birds, and familiarity with the release area. In most cases, the actual return times exceed the possible minimum achievable by a straight non-stop flight by a factor of 2, 3, 4, or more, indicating that flights include considerable detours or are interrupted by stops.

In conclusion, admirable as the goal-finding capacities of homing pigeons and other birds are, it should be borne in mind that the level of performances achieved does not indicate a powerful, always reliable, or perfectly efficient navigational system. On the other hand, we should also recall that passive displacement creates an artificial situation. Returns from spontaneous excursions, which allow the unimpeded usage of a multiplicity of environmental clues are certainly more successful.

The standard techniques for recording homing behaviour, which are based on simple visual observation at the release site, have sometimes been supplemented by visual tracking from a helicopter, by telemetric methods using radiotransmitters with or without additional airplane tracking, and by automatically operating flight recorders that memorize all bearings of a pigeon during its journey home (Fig. 1.1d). The latter method has just begun to yield valuable results (Fig. 7.26). A transmitter–airplane combination can be very useful in collecting data, but the method is extremely expensive and time-consuming. In reality, almost all the data on which our knowledge is based have been obtained using the classical 'primitive' techniques.

7.4 INSTRUMENTS OF NAVIGATION

Many of the terms applied in studies on bird navigation, such as compass, map or chart, and dead reckoning, have been borrowed from human navigation and are therefore liable to be somewhat misleading. Even so, the most practical policy is to use these terms as short-cut designations.

Homing to a goal that lies outside the direct sensory range requires both positional and directional information. It has been found empirically that in birds the two kinds of information, usually termed 'map' and 'compass', can be distinguished.

7.4.1 Use of directional information (compasses)

Keeping a constant course over very long distances requires external directional reference systems that are available in all the areas crossed. On a global scale, only two reference systems are known to exist, the geomagnetic field and celestial bodies

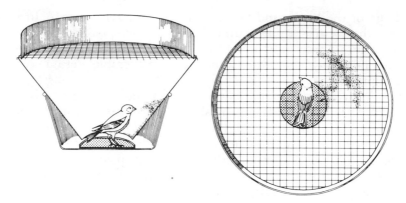

Figure 7.10 The Emlen funnels are small circular cages often used to record the directional preferences of small migratory birds. Here a funnel is shown in longitudinal section, and as seen from above. The bird stands on an ink pad and leaves footprints on the funnel at each escape attempt. (From Emlen, 1975a.)

(comprising a variety of visual clues connected with them). Both types are known to be used by birds.

(a) The geomagnetic field

The geomagnetic vector provides an immediate directional reference, indicating north and south, which is invariable in time and fairly constant over almost the entire globe. The evaluation of astronomical clues is much more complicated and requires additional measurements and knowledge, as the positions of the sun and stars vary greatly with the time of day, season and geographic latitude. Besides this, they are not available under overcast skies.

If birds were able to derive reliable compass information continually from the geomagnetic field, there would be little or no reason for them to utilize the less convenient astronomical clues as an alternative or an extra source of information. In reality, however, birds actually prefer to resort to astronomical references; if magnetic and celestial signals are at angular variance to each other, the birds' immediate response usually fits the astronomical conditions (see Wallraff, 1991, for references).

Without visual cues, directedness tends to be weaker; at first glance, birds often appear disoriented (Fig. 7.11c,d) when tested in the widely used 'Emlen funnels' (Fig. 7.10) or other circular cages. Second-order statistics, however, reveal that their weak directional

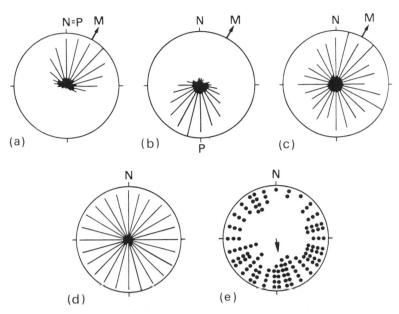

Figure 7.11 Passerine birds migrating at night orientate by means of the stars, apparently disregarding the magnetic field; when visual cues are excluded, it is only second-order statistics which sometimes reveal a weak orientation in the expected direction. (a–c) Record of the spring migratory activity of one indigo bunting (*Passerina cyanea*). Radii indicate the relative frequency of activity towards the respective directions in an Emlen funnel. In (a) the north of the planetarium sky (P) coincided with geographical north (N) and was close to magnetic north (M), in (b) celestial north was reversed, in (c) the stars were extinguished. (d, e) The pooled results of the autumn migratory orientation of 12 indigo buntings tested in the absence of meaningful visual cues in a normal magnetic field. The data refer to 122 bird-nights. (d) The frequency distribution of all activities per 15° sector, which does not reveal directional preferences; these are, however, shown in (e) by the distribution of the mean bearings calculated for each of the individual bird-nights. The mean of these means is indicated by the arrow. (Modified from Emlen, 1967, 1975b.)

preference is usually fairly consistent and reproducible in independent tests with either the same or different individual birds (Fig. 7.11e). It was first shown by Merkel and Wiltschko (1965) and Wiltschko (1968) that the direction preferred by European robins (*Erithacus rubecula*) can be predictably deflected by the application of artificial magnetic fields. More recently, magnetic compass

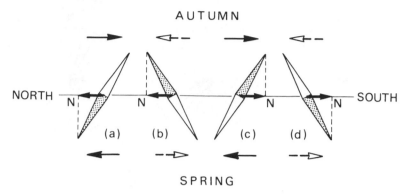

Figure 7.12 Schema of orientation by means of a magnetic inclination compass at median northern latitudes in a normal magnetic field (a) and in manipulated fields (b–d). Top and bottom arrows show flight directions of birds in autumn and spring. Compass needles (north-seeking pointer dotted) are thought to be free to rotate in a vertical North–South plane. The thick arrow indicates the horizontal component of the magnetic field vector. Birds fly north in spring and south in autumn in natural conditions (a) and with the whole field reversed (c); they fly in the opposite direction when either the vertical (b) or the horizontal field component (d) is reversed. (According to Wiltschko and Wiltschko, 1972; schema from Wallraff, 1984.)

orientation has been shown in several other migrant species as well. Wiltschko found two properties characterizing this mechanism (Wiltschko and Wiltschko, 1988, 1991): (1) its range of operation is restricted to a certain window of intensities but can be adapted to other levels if the birds are exposed to them for several days; (2) north and south are not distinguished on the basis of magnetic polarity but on the basis of the dip angle of the magnetic vector (Fig. 7.12). Thus, at the magnetic equator, where this vector is horizontal, the compass is ambiguous; once a bird has crossed the equator, it will take north for south (by changing from situation a to b in Fig. 7.12). The problem of how transequatorial migrants cope with this problem has not yet been investigated.

For a discussion on magnetic sensitivity, see section 7.6.

(b) The sun

G. Kramer (1951) showed that a starling which displayed migratory restlessness in a round cage changed its direction when the apparent sun position was changed by means of a mirror. It was then shown that birds take into account the sun's apparent motion in the course of the day and that for this purpose they use their endogenous so-called circadian clock (Hoffmann, 1954). As in other animals which rely on it, the sun compass mechanism of birds has been shown to consider only the sun's azimuth (p. 7), whereas its altitude is neglected. Except at the poles, the sun's azimuth rate of change varies in the course of a day. Moreover, the shape of the curve which represents the rate of change as a function of the time of day depends on latitude and season. Adaptations of the birds' sun compass system to varying geographical and seasonal conditions have only occasionally been investigated so far and have not yet been generally clarified (cf. Schmidt-Koenig et al., 1991). Even if little is known about applications of the sun compass in actual migration, it has been clearly demonstrated that it is an integral part of the pigeons', and probably other birds', homing system (section 7.4.4).

The sun's position is used less directly as a directional reference by night migrants. At dusk, birds of various species have been shown to orient their activities with respect to the glow of the setting sun above the horizon and/or to polarized light patterns caused by the sun (Moore, 1987; Able, 1989; Helbig, 1990).

(c) The stars

F. Sauer (1957) found that European warblers (*Sylvia* spp.) kept in a round cage orient their migratory restlessness (Zugunruhe) with respect to the starry sky or even to the simulated stellar sphere of a planetarium. More recently, this ability has also been demonstrated in other migrant species (cf. Fig. 7.11a,b). Owing to its two-dimensional extension, the invariable pattern of stars provides more information than that given by the sun. The pattern of fixed stars can, in principle, be used to determine north or south independently of the time. Viewed as a whole, however, the stellar pattern shifts its position, so that its appearance changes with time. Thus, an experimentally produced discrepancy between the visible stellar sphere and the bird's time scale may or may not induce directional shifts. Both of the possible results have been recorded in different cases, probably depending on the particular circumstances under which the unnatural condition was presented (see Wallraff, 1984, for references).

It has been shown that mallard (*Anas platyrhynchos*) and European teals (*Anas crecca*) possess great skill in learning, distinguishing and memorizing complex stellar patterns (Wallraff, 1972). At least some species are able to calibrate the stellar pattern to geographically meaningful directions by observation of the celestial rotation during the weeks before migration starts (Emlen, 1970; Wiltschko and Wiltschko, 1991).

(d) Interactions in the usage of different clues

Many studies on the ontogeny of compass systems used by young migrants have aimed to distinguish between genetically encoded and individually learned or imprinted components of an integrated compass (Able, 1991; Wiltschko and Wiltschko, 1991). In some species at least, an intended direction is genetically determined as an angle to the geomagnetic vector. This angle can then be transposed to the starry sky with reference to which it may be easier to keep a constant course. Several other kinds of calibration of one compass by means of another have been reported, yet a clear general picture of the interaction between different compasses and the underlying ontogenetic processes has not yet emerged. As different species and different experimental methods have been involved, it remains uncertain what levels of generalization will eventually be attained. Actual migratory flights are not only oriented with respect to the manifold 'compass clues' but also to landscape features and, of course, to the wind (Richardson, 1991).

Depending on its direction, speed and turbulence, the wind can be a favourable or a contrary factor. Migratory birds show they are able to select favourable winds, as departures mostly occur with more or less following winds and when meteorological conditions justify the prediction that they will continue to support the flight for its entire duration. On many occasions, compensation for drift occurs by day. This can be done by choosing a heading which makes it possible to maintain the intended direction relative to the ground. Over unfamiliar areas, at least, this requires the use of a compass (Alerstam, 1990; Richardson, 1991). In some cases, migrants which maintain a fixed compass direction reach their goal along a curved track caused by drift (Fig. 7.13).

7.4.2 Use of motion–dependent information (path integration)

In Chapter 3 it is shown that ants and other arthropods record and integrate all their movements while going on an excursion from their home. By referring to their integrator, they are permanently informed on their direction and distance relative to home. In this context, direction is determined with reference to the sun. Such a path-integration system, which requires an uninterrupted input flow during the animal's movements, has so far been found to

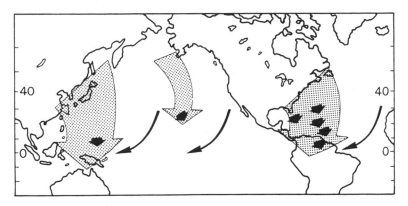

Figure 7.13 In migratory flights, a fixed compass direction pointing windward of the goal can compensate for drift and leads to that goal. In this diagram of three transoceanic migrations, the thin arrows indicate northeast trade winds, the big dotted arrows the tracks of the birds which migrate on a broad front, and the thick filled arrows their headings calculated from radar tracks and winds blowing at the altitude of the birds. (Modified from Williams and Williams, 1990.)

operate adequately only in cases where the animal moves independently. Passive displacements can be kept under control only if routes are particularly simple, excluding complicated turns and loops, and if the environment is visually accessible. Under such conditions, path integration seems to operate over distances of up to a few hundred metres, not only in wasps (Ugolini, 1987) and hamsters (Etienne *et al.*, 1988), but also in walking geese (*Anser anser, A. indicus, Branta leucopsis*) (Saint-Paul, 1982).

In homing experiments with pigeons, birds are usually transported inside a vehicle which allows no visual contact with the environment, on roads with many curves and over distances of up to several hundred kilometres. In some experiments, pigeons were irregularly rotated during the outward journey and all conceivably useful input channels were interrupted or disturbed. These birds were still able to orientate their courses in a goal-directed manner (Wallraff, 1980). It seems quite clear, therefore, that pigeons are able to home from unfamiliar areas without using a path–integration mechanism (for their collection of positional information *en route* see section 7.5.2).

This conclusion does not, of course, exclude the possibility that birds may use path integration, or more primitive forms of route-based orientation, when performing spontaneous flights away from home and back.

7.4.3 Use of positional information (maps)

It has been shown that pigeons are able to deduce positional information from external signals they receive at distant sites to which they were transported passively. This implies that birds possess means that are functionally equivalent to human maps. Two kinds of potential avian maps will now be considered.

First, the 'mosaic map', 'familiar area map' or 'topographical map' (see Baker, 1984, for details), like human maps, duplicates individual features making up a landscape, including their spatial relationships with each other and with the home site (Fig. 7.14a,b). Landmarks and spatial relationships have to be learned by individual exploration. Thus the range of this map depends on the range of a bird's previous experience. At sites outside a 'familiar area' the map is useless for homing. The borders of the familiar area are probably less distinct than Fig. 7.14b would indicate, and may exceed the range of a bird's earlier physical presence. Visual landmarks, for instance, can often be seen over quite long distances, and windborne odours may signal conditions prevailing far away. There are certainly different grades of familiarity, which usually fall with increasing distance from home. It can hardly be doubted that birds make use of such 'topographical maps'.

The 'gradient map' or 'grid map' is suitable for goal-finding from unfamiliar distant areas. In principle, this map is unlimited in extension (Fig. 7.14c,d). It is thought to be based on at least two gradients of any physical substrate which extend over a sufficiently large area. Assuming that the gradients extend monotonically beyond the familiar area, scalar values of the respective physical parameters at the current position of an animal, as compared with those remembered from the home site, could provide information on the animal's position with respect to home. Although the range of a gradient map is basically unlimited, its range of useful applicability may actually be limited by the sizes of gradient fields to which it refers. For a bird with a 'map' as given in Fig. 7.14d, for instance, the actual pattern of gradients shown in Fig. 7.14c would be misleading at some sites. If displaced to the northwestern or southeastern corner of Fig. 7.14c, the bird would fly away from home.

7.4.4 Use of positional plus directional information (map and compass)

Neither positional nor directional information alone is sufficient to determine the right bearing towards a distant goal with which there

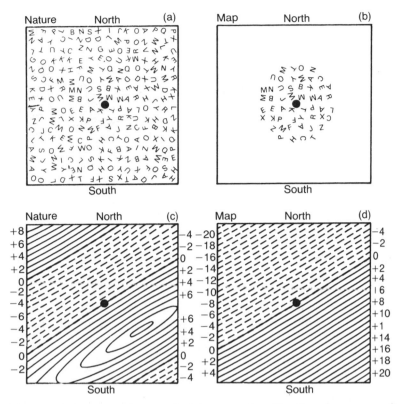

Figure 7.14 Schemata of maps. In (a) a mosaic of landmarks surrounding a bird's home site (central dot) is shown, and in (b) the corresponding mosaic (topographical) map, which is limited by the bird's range of experience. In (c) the isolines (arbitrary units) of a gradient field are represented, and in (d) the bird's corresponding gradient map as established by the extrapolation of home site conditions. For the sake of simplicity, gradients of only one variable are shown in (c) and (d), but, to complete site localization, at least two gradient fields intersecting at sufficiently large angles are required. The square sections shown are thought to be on different scales: the side of the square in (a) and (b) cannot be greater than a few hundred kilometres, those in (c) and (d) are at least 1000 kilometres. (From Wallraff, 1985.)

is no direct sensory contact. A grid map based only on scalar values of variables x and y (Fig. 7.15), for instance, would not tell an animal where to go from a point P in order to reach home H, unless it was also informed of the directions towards which x and y increase and decrease. This information could be obtained in one of

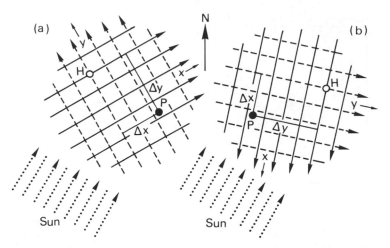

Figure 7.15 Scheme of how a grid map may be used for navigational purposes. The picture shows a grid of coordinates (gradients) x and y with home site H and a bird's current position, P. The grids themselves are identical in (a) and (b) but they have a different orientation geographically, and therefore, with respect to the sun at a given time of day. (From Wallraff, 1974.)

three ways: (1) the signals used for site localization are not simple scalar values but include a directional aspect. This condition would be met, for instance, if the birds deduced their position in relation to home from the sun's altitude and rate of change of altitude (Pennycuick, 1960), in which case the sun would itself act as a positional and a directional reference. In theory, the geomagnetic vector provides information on latitude as well as on the directions towards which it increases and decreases. (2) If the positional signals do not include directional information, the birds might scan the field on a trial-and-error basis to determine the directions of gradients empirically. This would require suitable relationships between the steepness of gradients, the sensory sensitivity of birds, and the spatial range of the scanning flights. Moreover, some *ad hoc* directional reference would be needed (e.g. landmarks or the sun, but not necessarily a complete compass system). (3) As in the previous case, the positional signals are scalar values only, but in addition the birds have some assumptions or knowledge about the compass alignment of the gradients of x and y. Thus, map and compass systems are independent, but are calibrated with each other. The bird knows, for instance, that variable x increases towards northeast and variable γ towards northwest (according to Fig. 7.15a). If this is true, a question arises as to where the proposed additional knowledge comes from.

Empirical findings show that in pigeon homing a compass mechanism is involved which can be manipulated separately. This supports the third hypothesis. Once it had been found that birds make use of the sun for direction-finding and that the birds' sun compass can be rotated by shifting their internal clock, it required only one more logical step to apply the clock-shifting procedure in homing experiments with pigeons (Schmidt-Koenig, 1958). Clock-shifts can easily be produced by exposing the birds for several days to an artificial light–dark regime which is out of phase with the natural day. Figure 7.16 shows and explains the orientation of displaced pigeons whose internal clock was phase-shifted forward by 6 h. The observed angular shift of initial orientation corresponds quite well with the shift predicted on the basis of the sun's orbit at the site and time at which the experiments were conducted (Neuss and Wallraff, 1988). This correspondence can reasonably be interpreted only by assuming that, in a first phase, by 'map reading', an intended compass direction is calculated, and, in a second phase, this direction is determined by means of a sun compass) (Kramer, 1953) map and compass concept. A result like this could not be expected if either of the above alternatives (1) or (2) were realized. In both these cases the birds would directly refer to the gradients themselves, recognition of which would not be affected by the clock-shift, at least in this way. For the birds, it would make no difference whether the grid was oriented according to Fig. 7.15a or b.

The appropriate linkage between map and compass, including knowledge of or assumption about the compass bearings in which the values of relevant physical variables increase and decrease (e.g. the decision that in Fig. 7.15 alignment (a) is correct rather than (b) or any other), must either have been created phylogenetically or by a learning process during a long-term stay in the home area. Empirical findings suggest that in pigeons the latter alternative occurs (Wallraff, 1974, and below).

So far we have considered the case of a grid map. If we adopt the hypothesis of a topographical map (Fig. 7.14a,b), the situation is different, and what matters is whether the landmarks used are visual or, for instance, olfactory. In the latter case, the compass direction with respect to home would have to be learned individually for each site or area either by direct overflight or indirectly, by means of wind-borne information (section 7.5.3). This knowledge would be necessary for goal-oriented homing, and

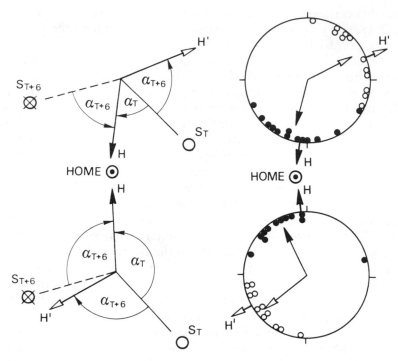

Figure 7.16 The effect of clock-shift on pigeon initial orientation shown by two real experiments. In two groups of pigeons the physiological clocks were shifted forward by keeping the birds in a light–dark regime which was brought forward by 6 h with respect to the natural day. The birds were released, in alternation with unshifted controls, at two sites about 30 km north and south of home and observed with binoculars until they vanished. The diagrams on the left show theoretically expected bearings at two release sites north (top) and south (bottom) of home on the assumption of perfect site localization and the use of a sun compass. If at time T (here 10.00 a.m.) the control birds select an angle α_T with regard to the sun, S_T, they fly in home direction H. At time T, the clock of the shifted birds shows $T + 6$ h. At that time (here 4.00 p.m.) the sun would be at position S_{T+6}; H would be reached by selecting an angle α_{T+6}. By keeping this angle with respect to the actually visible sun S_T, the birds would achieve course H'. Under the given conditions (summer morning in southern Germany), H' is approximately 120° left of H (like S_T from S_{T+6}). The diagrams on the right show actually observed initial bearings. Each peripheral circle symbolizes one pigeon; a filled circle shows a control and an open circle shows an experimental bird. Central arrows indicate mean direction and angular scatter. The experimental pigeons were as well oriented toward H' as the controls toward H. (Data from Neuss and Wallraff, unpublished.)

the effects of clock-shifts would be the same as those described above.

If the topographical map consists of visual landmarks, the involvement of a compass, and hence of the sun, might be helpful but would not be indispensable. A visual landscape provides an extended pattern which itself includes directional relationships between its constituent parts, so that pilotage without the involvement of a compass (as defined in Chapter 1) would be possible. If, however, the sun is used as a prominent component of the visual pattern, clock-shifts might produce confusion, because they make the sun appear dislocated with regard to the rest of the pattern. The particular circumstances, e.g. degree of familiarity with the area or conspicuousness of landscape features, may then determine whether the birds place greater reliance on the sun or the topography, or decide on a compromise, or else display disorientation. Empirically, the relationships between non-olfactory topographical maps and the sun compass have not yet been clarified definitively (Wallraff, 1991).

7.5 PIGEON HOMING: THE RESULTS OF EXPERIMENTAL RESEARCH

Modern research on pigeon navigation began in the early 1950s. The first important findings included homeward directedness at unfamiliar sites (Matthews, 1951), the homing capability of aviary pigeons (Kramer and Saint-Paul, 1954, see section 7.5.3) and the use of a sun compass.

The homing behaviour of pigeons is currently interpreted in the light of the map-and-compass concept, with most attention and research work focused on the nature of the map. Two approaches have been developed to solve this problem. The first consists of conceiving and testing navigation hypotheses based on physical parameters which may allow position fixing on the earth's surface. Following this line of research, Mattews (1953) proposed his sun-arc hypothesis, according to which birds compare the temporal and astronomical features of the sun arc at the release site with those at the loft site and deduce differences in longitude and latitude between the two localities. A second hypothesis, that of Yeagley (1947, 1951), supposes that birds may navigate by detecting the geographic variations both in vertical component of the magnetic field and Coriolis' force. A third hypothesis, which is independent of local physical parameters, puts forward the view that pigeons record the values of all linear and angular accelerations during the outward journey and integrate them to calculate home distance and direction through a mechanism of path integration (Barlow, 1964). After many experiments specifically devised to test these hypotheses, especially the first, they now commend little support.

Hypotheses on navigation based on the earth's magnetic field received impetus from the discovery of a magnetic compass in passerine birds (Merkel and Wiltschko, 1965) and still have some supporters. In theory, pigeons may make use of the earth's magnetic field by comparing its parameters at home and at their current position (and so obtaining at least an indication of difference in latitude), or by using the magnetic compass to determine the direction and distance of displacement during passive transportation. However, the attempts to test these ideas gave contradictory or negative results and it is even doubtful whether pigeons use the magnetic field for directional purposes (Moore, 1988).

The second approach to the map problem uses empirical, trial-and-error methods, which aim to ascertain the nature of the cues the birds rely on to navigate. They consist either in looking for so far untested sensory capabilities or in impairing the efficiency of sense organs and then testing homing behaviour. Using the former method it has been shown that pigeons are sensitive to polarized light, infrasound, and small differences in barometric pressure (section 7.6), but there is no evidence so far that these sensory capabilities are involved in position-fixing. The latter method has shown that pigeons released with vision drastically reduced by frosted lenses can fly homeward and orientate up to a few kilometres from the loft (Schmidt-Koenig and Schlichte, 1972). Conversely, pigeons deprived of olfaction are impaired in initial orientation and homing (Papi *et al.*, 1971, 1972). The successive investigation showed that pigeons rely on an olfactory mechanism to home. Some important aspects of the mechanism are still unknown or have been subjected to criticism by several authors. However, the results presently available already form a coherent picture of pigeon homing and demonstrate two main phenomena: (1) displaced pigeons rely on local odours (not originating from their loft areas) which they smell during transportation and at the release site to determine their position with respect to the loft; (2) pigeons acquire an olfactory map at the loft by smelling wind-borne odours. The above phenomena will be discussed in the next sections in agreement with recent reviews (Papi, 1986, 1991; Wallraff, 1990, 1991).

7.5.1 When and why anosmic pigeons are impaired in homing

Pigeons can be made anosmic by a number of different methods. The most radical is the bilateral sectioning of the olfactory nerve,

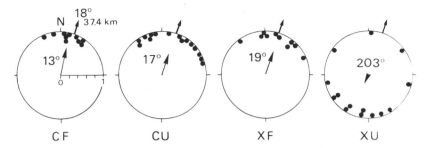

Figure 7.17 Anosmic pigeons are incapable of homeward orientation at unfamiliar sites but not at familiar ones. The diagrams show the results of a release of four groups of pigeons from the same site to test the effects of anosmia and familiarity. Pigeons with an unimpaired sense of smell oriented homewards regardless of whether they were familiar (CF) or unfamiliar to site (CU), but anosmic birds only did so if familiar with the site (XF). The orientation of anosmic birds unfamiliar with the site (XU) did not differ from random. Home direction and distance are given in the first diagram; the length of the mean vectors can be read with the scale. (Original: courtesy of A. Gagliardo.)

but if only one nerve is cut and the controlateral nostril is occluded, the experimental birds are still incapable of smelling, while the differences with respect to controls (one nerve cut, the ipsilateral nostril plugged) are minimized. Considering the uneasy breathing it provokes, the occlusion of both nostrils is not usually made on animals that have to fly, but very thin plastic pipes may be introduced through nasal passages, from nostrils to choanae, so that the air they breathe cannot reach the olfactory membranes. Other methods imply a successive recovery of smelling abilities: the treatment of olfactory membranes with local anaesthetics produces an almost complete anosmia lasting for a few hours at most, while with $ZnSO_4$ it lasts for some days.

Olfactory deprivation has little or no effect on pigeons released from familiar sites or from any site in an area with which they have become familiar: pigeons take off in the home direction and do home, their performance being equivalent, or only slightly inferior, to that of controls. Conversely, homing ability from unfamiliar sites is heavily impaired (Fig. 7.17) by all the anosmia-inducing methods mentioned above. Transient anosmia produces important delays in homing, and permanent anosmia even impairs homing success. The percentage of lost anosmic subjects increases in relation to their distance from home: with a distance of more than 50 km the homing success of inexperienced pigeons is negligible

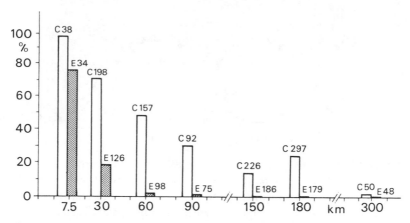

Figure 7.18 Homing success of inexperienced birds is proportional to distance of release, but in anosmic pigeons it decreases more steeply than in intact birds. Here the percentage of controls (C, open bars) and anosmic birds (E, dashed bars) which were successful in homing from different distances is shown. The number of birds released is given; they all belonged to the same loft. (Modified from Wallraff, 1989c.)

(Fig. 7.18). With distances less than about 50 km, some birds do manage to find home, probably by looking for familiar landmarks around their loft. When released far enough from home, anosmic pigeons are usually not homeward oriented if they have been carefully prevented from smelling during passive transportation and at the release site before their take-off. Even so, anosmic pigeons do not usually orientate at random, but tend to head in a preferred compass direction (PCD), which is loft-specific. Thus the capacity for homeward orientation must be tested by releasing pigeons from sites located symmetrically with respect to home. This method prevents a possible coincidence between PCD and home direction from simulating a non-existent homing ability. If pigeons are

Figure 7.19 Pigeons from different geographic areas turned out to be similarly disoriented when made anosmic (diagrams on the right), whatever the degree of homeward directedness in controls (diagrams on the left). In each diagram, the bearings recorded in 4−9 test releases were pooled by setting the home direction (H) to 0°. The smaller circles give the direction of the release sites with respect to home. The number (n) of birds released and length (a) of the mean vector are shown. (After different authors, from Papi, 1991.)

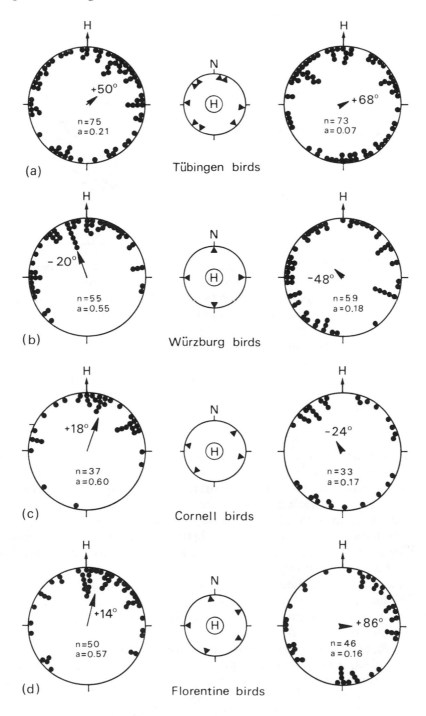

(a)   Tübingen birds

(b)   Würzburg birds

(c)   Cornell birds

(d)   Florentine birds

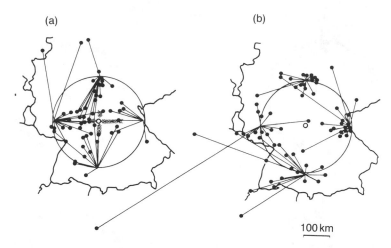

Figure 7.20 Most inexperienced pigeons fail to home when released from great distances, but approach home if not deprived of their sense of smell (a), whereas they spread in all directions if made anosmic (b). The birds belonged to a loft near Würzburg in Germany (open circle at the centre) and were released from four sites symmetrically located on the circumference of the circle, 180 km from the loft. The maps show part of the boundary of Germany and the recovery sites (filled circles connected with the respective release site). Pigeons that homed are represented by arrowheads pointing to the centre (32 controls, no anosmic birds). (Modified from Wallraff, 1990.)

incapable of homeward orientation, the bearings pooled with respect to the home direction are distributed more or less at random. After correct experimentation, the incapacity for homeward orientation shown by anosmic pigeons could be demonstrated in pigeons from several geographic areas (Fig. 7.19).

One issue of major interest is whether the disruption of homing ability in anosmic pigeons is a result of general behavioural troubles and/or a loss of homing motivation, or whether it depends on their incapacity to navigate in the absence of olfactory cues. The following facts support the second hypothesis: (1) anosmic pigeons released from unfamiliar sites cover large distances but head in directions not related to that of home (Fig. 7.20); (2) anosmic pigeons, as stated above, perform almost normally from familiar sites; and (3) if two groups of pigeons are transported in containers, one ventilated with unfiltered, and the other with filtered air, and then all are released after having been made anosmic by a local anaesthetic, only pigeons ventilated with filtered air are disoriented.

Experiment (1) proves that anosmic pigeons are not demotivated or disturbed in a non-specific way, and experiments (2) and (3) that anosmia in itself does not impair homing, while experiment (3) makes it clear that the cause of impairments in homing is to be found in the absence of any olfactory information before release.

7.5.2 To navigate pigeons use olfactory cues picked up
en route and at the release site

The results reported above lead to the conclusion that pigeons are able to fix their position in relation to home by smelling odorous substances dispersed in the atmosphere. This happens not only at the moment of release but also during passive transportation. This is supported by the results of detour experiments, in which two groups of pigeons are transported to the same release site by two different routes, which diverge strongly in their first leg. If birds can smell outside odours during transportation, the initial bearing of each group mostly deflects in a predictable way, since the batches, in selecting their course, take into account both the position of the release site and that of the sites crossed during the outward journey. Roughly, one can state that the take-off direction is a compromise between the home direction and the average home direction from the area crossed during the outward journey. Conversely, if birds are prevented from smelling atmospheric odours, no deflection between the two groups occurs (Fig. 7.21). In addition, homeward directedness decreases. The same phenomenon was observed in birds that had been transported directly but were also prevented from smelling outside odours. However, the contribution made by the olfactory information gathered *en route* varies widely.

In releases at very great distances, olfactory cues picked up *en route* can be essential for homeward orientation, as cues at the release site no longer convey useful information. Pigeons from Florence released at Würzburg (700 km north) are not homeward orientated when anosmically transported, but still steer home if they are able to smell outside odours *en route* (Ioalè et al., 1983). This indicates that the olfactory map has a limited range. It is possible that this varies seasonally and geographically.

These results led to experiments that investigated the results of a radical manipulation of olfactory experience before a test release. In what have been called simulation experiments, pigeons were only able to smell at one site (in one series also on the way to it); they were then transported without access to atmospheric odours to a release site located in the opposite direction with respect to the

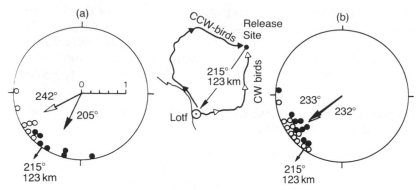

Figure 7.21 The initial orientation of two groups of pigeons can be significantly different if they are transported to the release site by two circuitous routes which are strongly divergent in their first leg. This detour effect disappears if the birds are prevented from smelling outside odours during transportation. The map shows the routes followed: birds transported by the clockwise route were expected to deflect counter-clockwise and vice versa. (a) The detour effect in normally transported birds. The same birds had been released from the same site 4 days before, following anosmic transportation: they did not deflect (b). Filled circles indicate birds expected to deflect counterclockwise (CCW), open circles indicate birds expected to deflect clockwise (CW). (Modified from Papi *et al.*, 1984.)

loft and released after anaesthesia of their olfactory membranes. The experiments, which were performed by two different teams with slightly different procedures (Benvenuti and Wallraff, 1985; Kiepenheuer, 1985), produced the same results: the pigeons oriented according to the position of the site where they could smell, so that they took off in the direction opposite to that of home (Fig. 7.22). Further experiments were performed by keeping pigeons in the loft, for 2 h, in a container (750 l) filled with air from the next release site or another site. There was a difference in orientation between controls and experimentals, which departed at random (Kiepenheuer, 1986), but it is uncertain whether pigeons can actually extract positional information from a small volume of air.

The natural tendency of pigeons to use olfactory cues picked up *en route* can be used to condition them to *en route* artificial odours. In a recent experiment, pigeons were treated with amyl acetate during journeys to and releases from a northerly site, while they were exposed to the smell of benzaldehyde during an equal number of releases from a release site in the south. In critical releases, birds

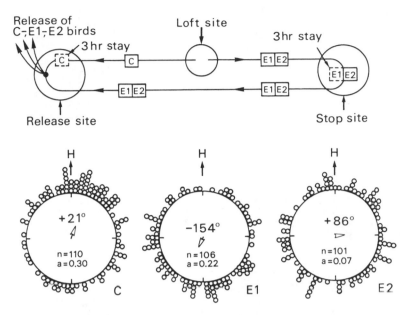

Figure 7.22 Pigeons can be deceived about home direction by manipulating their olfactory experience during transportation to the release site. As shown in the top schema, three groups of pigeons (C, E1, E2) were tossed from the same release site. They had had their olfactory mucosae anaesthetized just before release. Previously, C-birds had smelt outside odours for 3 h at the release site, and E1-birds at a site located in the opposite direction, while E2-birds had never been allowed to smell in the course of the experiment. C-birds turned out to be homeward oriented, and E1-birds oriented in a direction opposite to that of home, whereas E2-birds were randomly oriented (bottom diagrams). The number of birds released (*n*) and the length of the mean vector (*a*) are given. (After Benvenuti and Wallraff, 1985, from Papi, 1986.)

performed better in orientation and homing when exposed to the odour appropriate to the direction of displacement than in the opposite case (Ganzhorn, 1990).

7.5.3 The olfactory map: how pigeons obtain it

The first steps in the study of map ontogenesis were made by Kramer in the 1950s by restricting pigeons' experience from fledging time onwards (see Kramer, 1959, for references). After his death, others continued his pioneer work, bringing our knowledge

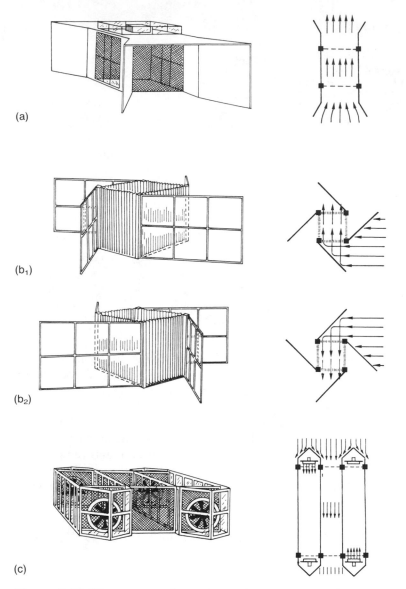

(a)

(b₁)

(b₂)

(c)

Figure 7.23 The pattern of association between wind direction and wind-borne odours that pigeons make at the loft can be altered by rearing them in special cages. Sketches and horizontal sections of some of the cages used are shown here. (a) Cage open to winds from two opposite quadrants. Deflector cages deflecting the wind direction clockwise (b₁) and counter-clockwise (b₂). (c) Corridors with fans. In the central corridor a group of control birds was exposed to natural winds blowing from two opposite quadrants (as in (a)); in the left corridor a second group of controls received an air current blowing in the same direction; in the right corridor experimentals faced an air current from the opposite direction. (Modified from Papi, 1986.)

to its present, relatively satisfactory state (see Papi, 1991, for references).

The first discovery made was that pigeons, kept from fledging time in aviaries whose capacity is about $200 \, m^3$, are able to home from distant sites. If the aviary walls are screened, the effects on homing ability depend on the type of screen utilized: materials that obstruct the view but do not block the passage of air do not impair performance; on the other hand, materials obstructing winds but not the view do impair pigeon homing ability. At a later stage, the use of partially screened aviaries has shown that pigeons are not homewards orientated if kept in an aviary open to winds blowing from one direction. Conversely, if two opposite sides of the aviary are open (Fig. 7.23a), pigeons only orientate homewards from those directions from which they were used to perceiving the wind. A further experiment clearly showed that pigeons take into account the directions from which winds blow. If the lofts are fitted with deflectors which produce a clockwise or counterclockwise deviation of the winds from whatever direction they come (Fig. 7.23b), the initial bearing of pigeons deflects correspondingly, independently of the direction in which the release site is located (Fig. 7.24).

It is hard to arrange an inversion of wind direction. Pigeons can, however, be exposed to an artificial air current from the direction opposite to that of the wind that is actually blowing, with the aim of giving them the same wind-borne information from the reversed cardinal point. As controls, one can use birds exposed to the natural winds or birds exposed to artificial winds blowing in the same direction as the natural ones. This was done by keeping pigeons in corridor lofts, some of which were equipped with fans (Fig. 7.23c). Pigeons were then tested in releases from the directions coinciding with the axis of the loft corridor; unlike control birds, birds exposed to reversed artificial winds were oriented in a direction opposite to that of home and their homing was impaired.

At this point, it was clear that each wind carries information useful for homing from the direction it blows from, and the question arises whether the wind-borne information is olfactory in nature. Positive answers were given by the following experiments. (1) If pigeons kept in aviaries screened against winds are allowed to fly outside with their nostrils plugged, they do not become capable of orientation, unlike birds prevented from smelling inside their aviary but allowed to smell outside it. (2) Still more enlightening results were obtained from an experiment with deflector loft pigeons which had been subjected to section of the brain's anterior commissure to prevent interhemispheric transfer of information. For periods of 3 days, they were alternately exposed to clockwise-

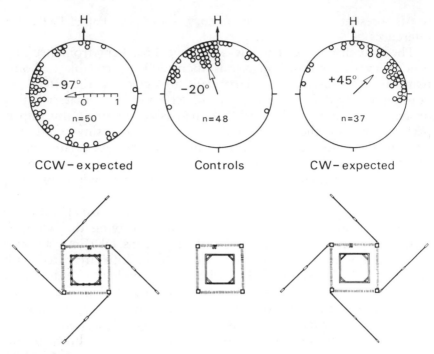

Figure 7.24 In pigeons reared in lofts which deflect wind, initial orientation deflects in a corresponding way. The circular diagrams show the initial bearings of controls (centre) and experimental birds (right and left). The data from three release sites located in different directions were pooled by setting the home direction (H) to 0°. The sketch of the corresponding cage is shown below. Note that in the two experimental groups the amount of deflection with respect to controls was roughly the same. (After data from Baldaccini *et al.*, 1975.)

deflected winds with one nostril plugged, and to counterclockwise-deflected winds with the opposite nostril plugged. Test releases were then made with one or other nostril plugged: birds deviated according to the treatment they underwent, i.e. in a clockwise direction when the free nostril was the one they had had free while exposed to clockwise winds, and in a counterclockwise direction in the opposite case.

Further experiments aimed to test more directly the hypothesis that had inspired most of this line of research, i.e. the idea that pigeons, when winds blow, receive information about the odours prevailing in the surrounding area. In addition, pigeons would associate the odours carried by each wind to their loft with its direction. When transported far from the loft, they could determine the home direction if odours perceived at the release site and/or *en route* had already been perceived at the loft as wind-borne and if they were associated with a specific direction. The home direction would, in fact, be the opposite one.

In a first experiment, pigeons kept in corridor lofts were screened against natural winds, but exposed to artificial air currents containing added odorants (olive oil, turpentine). They were expected to associate the artificial odours with their direction. In release tests, in fact, after exposure to one of the odorants, they flew in the direction predicted regardless of the true home bearing. In a second, more clear-cut experiment, a group of experimentals and a group of controls were reared in two separate aviaries which were exposed to winds from all directions. From time to time, the cage of the experimentals was exposed to an air flow from a specific direction (NNW) which carried a strong smell of benzaldehyde. Both experimentals and controls, which were untreated, were homeward oriented in normal releases. Conversely, when both groups were exposed to the odour of benzaldehyde *en route*/and or at the release site, the correct orientation of controls remained unaffected, whereas the experimentals oriented in a southerly direction approximately opposite that of the odorous treatment (Fig. 7.25).

In conclusion, these experiments show that pigeons build up their navigational map in their earliest months of life, even a few weeks can be enough, provided that they are exposed to natural winds without olfactory limitations (in fact, 3-month-old pigeons, or even younger ones, are already able to navigate). Clearly, they associate the olfactory patterns of winds with their direction and use this information to home from unfamiliar sites. There is no direct evidence that the improvement of homing ability in experienced homers is attributable to additional olfactory experience aloft and during transportation. We do, however, know that the map is open to further acquisitions and modifications at the loft. Pigeons transferred to a new site learn to find the new loft from unfamiliar sites, but do not lose the ability to fly to the old loft. Even pigeons that are kept at a single site are able to adjust their maps to new situations, as happens, for instance, when they are transferred from counterclockwise to clockwise deflectors, which invert the direction of their bias.

7.5.4 Two homing mechanisms

The interest aroused by olfactory navigation, a puzzling, still incompletely understood mechanism, runs the risk of over-

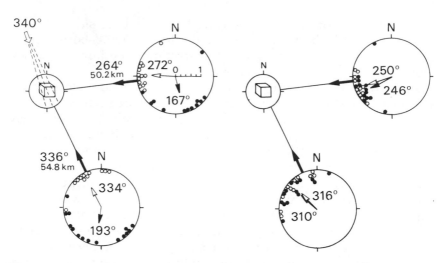

Figure 7.25 Artificial odours can direct pigeons in a specific direction following a treatment with odorous air currents at the home loft. Two batches of pigeons were reared in two cages exposed to winds from all directions. Experimental birds (left) were also treated at the loft with an artificial wind carrying an odour of benzaldehyde from NNW. They oriented towards the opposite direction in two releases from E and SSE if exposed to the benzaldehyde odour before release (filled circles), but flew homewards if not so exposed (open circles). Control birds (right), which had not been treated with benzaldehyde at the loft, flew home no matter whether they were exposed to it before release (filled circles), or not exposed (open circles). (After data from Ioalè *et al.*, 1990.)

shadowing the role of a second homing mechanism. As seen above, anosmic pigeons home from familiar sites with performances comparable to those of intact controls, and even home from un-familiar sites if these are not too far from the loft. This shows that there is a homing mechanism that is independent of olfaction and is based on familiar landmarks, which are probably mainly visual in nature. This is consistent with the fact that pigeons are capable of discriminating between projected images of a familiar location and those of an unfamiliar site (Wilkie *et al.*, 1989). When pigeons resort to this mechanism, they still take into account the sun's position, at least initially. In fact, clock-shifted pigeons deviate with respect to controls, even from sites of previous numerous take-offs and even when made anosmic in order to exclude reliance on the olfactory map-and-compass mechanism. However, one can suppose that, in later legs of their flight, they switch to a pilotage (as defined in

Chapter 1) disregarding the sun's position. There is no evidence that this happens, but the correction of the course of clock-shifted birds, as found by reconstructing their home flight (Fig. 7.26), might be attributed to an exclusive reliance on landmarks. The efficiency of the mechanism based on familiar landmarks decreases with the distance from home and from familiar areas, whereas that of the mechanism based on olfaction probably increases until a peak level is reached. The curves in Fig. 7.27 merely demonstrate a general principle. Real conditions may be much less regular, and even average shapes and levels would certainly differ greatly if we were able to draw realistic curves for pigeons with different levels of experience or different geographic areas, or different species of birds. Even in pigeons whose homing experience has been logged, we can, at best, roughly estimate the spatial range of the area within which they utilize familiar visual landmarks, or would be able to utilize them if necessary. In wild birds, whose individual experience is totally unknown, the range of effective piloting might be considered much larger than that drawn in Fig. 7.27, especially in directions of preceding migrations.

The idea of a dichotomy of the homing process is strongly supported by experiments showing that different parts of the brain appear to be involved in homing from familiar and unfamiliar areas (section 7.6).

7.5.5 What is the physical substrate of the olfactory map?

One of the major difficulties encountered in studies on olfactory orientation within nature is the determination of the chemical composition of orienting cues. Salmon are a typical example; despite the interest, which is partly economic, in the cues leading them to native streams, their nature is still unknown. In an experiment on salmon involving olfactory deceit, the use of artificial odorants, morpholine and phenethyl alcohol (p. 191), gave a positive result similar to that obtained on pigeons with benzaldehyde and other odorants. This only shows that the process of map building is flexible, but does not help in identifying the natural cues.

Speculating about the evolutionary process which led pigeons (and certainly many other bird species) to acquire their olfactory map, one may consider that many birds use the sense of smell to select or find food, often flying against the wind (Bang and Wenzel, 1985; Waldvogel, 1989), so showing a typical osmotactic behaviour. One can suppose that birds are alerted by winds carrying food odours and learn what direction they can be expected

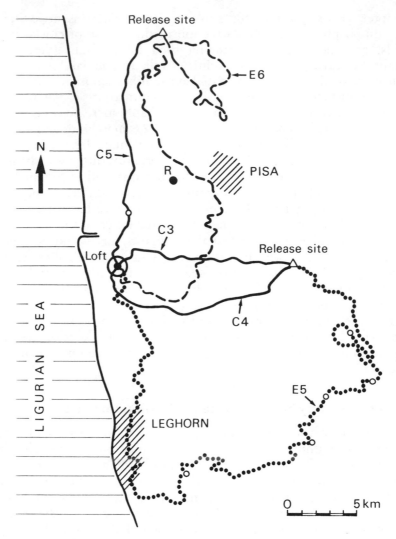

Figure 7.26 Clock-shifted pigeons released in a familiar territory usually take a deflected course but later correct it and eventually find home. The picture shows the tracks of two 6 h fast clock-shifted birds (E5, E6) released from two sites, north and east of the loft. As expected, they initially deflected clockwise with respect to the controls (C3, C4, C6) released from the same sites. Open circles indicate stops. The tracks were reconstructed by means of flight recorders. (Modified from Papi *et al.*, 1991.)

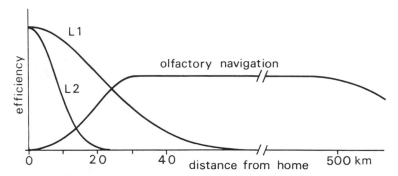

Figure 7.27 Schema showing the orientational efficiency of the mechanisms based on familiar landmarks and on the olfactory-based mechanism. The range of the former may be different depending on the range of individual spatial experience (L1, L2). The distance given and the shape of the curves are merely indicative. (From Wallraff, 1991.)

from, so acquiring an olfactory windrose. The wild ancestor of the homing pigeon, the rock pigeon, is known to cover long distances to drink and feed, and may have begun by building up a windrose-like prototype of the olfactory map. In successive steps, odours of different kinds may be associated with the paths along with pigeons usually fly until a true olfactory map is developed. Unfortunately, this will remain pure speculation so long as the physical substrate of the olfactory map remains unknown.

The only point that has been decided is that the olfactory cues used by pigeons consist of gaseous substances that are dispersed in the atmosphere. This is shown by an experiment on three batches of pigeons whose olfactory membranes were sprayed by an anaesthetic just before their release. During transportation while awaiting release, they could breathe unfiltered air, or ventilated air blown through a fibreglass filter which retains aerosol particles, or air blown through a filter of activated charcoal which retains substances dispersed in the atmosphere in a molecular state. Only the birds of the third group were not homeward oriented (Fig. 7.28).

Nothing certain is known about the origin and the distribution of the odorants that allow birds to navigate. F. Papi and his team have expressed the hypothesis 'that there exist odorant substances, which, for every area, give rise to a different pattern of olfactory stimulation, that is, to a characteristic odour'. These odours might

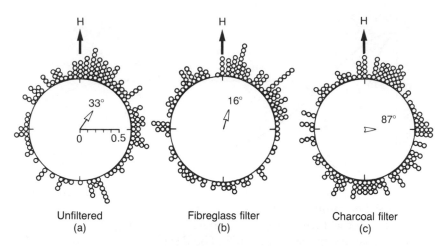

Figure 7.28 In searching for the physical substrate of olfactory navigation, it has been found that charcoal filters remove information from environmental air. The three diagrams show the initial orientation in birds which, both during transportation and at the release site, breathed (a) unfiltered air, (b) air passed through a filter made of fibreglass paper retaining large portions of the solid and liquid aerosol particles, or (c) air passed through an additional filter consisting of activated charcoal. All the three groups were released after local anaesthesia of their olfactory mucosae. The data of 15 test releases were pooled by setting home direction (H) to 0°. The first two groups are homeward oriented, the third is not. (Modified from Wallraff and Foà, 1981.)

originate in the soil and/or vegetation, and birds would become acquainted with the odours of different areas either by direct experience or because they are carried by winds to the loft. The map, therefore, would be a mosaic of irregularly distributed olfactory areas, partly built up by indirect acquisition. The related mechanism would be that of map-based navigation (see Table 1.1).

Reliance on an olfactory map presents theoretical difficulties if it has to work over hundreds of kilometres. H.G. Wallgraff therefore concluded that the cocktails of odours characterizing different areas should include components with a sufficient degree of long-range regularity to make extrapolations from one site (home) to other areas possible. This would require the existence of spatial gradients of olfactory stimuli which could, for instance, be a result of changes in the proportions between a number of relevant trace compounds.

The main difficulty in this context is the enormous instability of the atmosphere in many parts of the world which prevents (or seems to prevent) the conservation of fairly stable gradient configurations. A grid map based on atmospheric trace gases probably needs to be dynamic in order to be adaptable to varying meteorological conditions. A conceivable navigation system based on very much simplified fictitious atmospheric gradients has been developed by model calculations (Wallraff, 1989b, 1991). Suitable real trace compounds, however, have not yet been shown to exist. It should be emphasized that a mechanism with an efficiency as low as that exhibited in pigeon homing does not require very reliable or noise-free external conditions. A probabilistic solution, allowing chance a considerable role, would be sufficient to explain the observed phenomena (Wallraff, 1989a).

7.5.6 Criticism and discussions about olfactory navigation

Initially, a very cool reception was given to the hypotheses, ideas and findings that were favourable to olfactory navigation. Many critical comments, alternative explanations and claims that experimental results could not be replicated may be found in the literature. Even so, 20 years after the first findings, olfactory navigation is still the focus of attention (Schmidt-Koenig, 1987), as shown by a high proportion of the titles in the current literature. A complete history of all controversies cannot be given here, and the reader should consult the recent papers of critics (Schmidt-Koenig, 1987; Schmidt-Koenig and Ganzhorn, 1991; Waldvogel, 1989; Wiltschko and Wiltschko, 1989) and of supporters of olfactory navigation (Papi, 1989, 1991; Wallraff, 1990, 1991) for detailed information and bibliography. An attempt will now be made to present a balanced report on the controversy.

Researchers from several laboratories, working on pigeons housed in Italy, have confirmed the olfactory nature of their navigational mechanism, so that the soundness of the results obtained by the Italian team is hard to challenge. On this basis, the same confidence is given to results obtained elsewhere which are in accordance with those obtained in Italy. In this sense the series of experiments performed in Bavaria represent an equivalent repertory of evidence to that obtained on the other side of the Alps. Further support for olfactory navigation is given by results obtained in Switzerland, New York State, Utah, and most recently Ohio (Bingman and Mackie, 1992).

Opponents of the olfactory hypothesis tend to support two main views. First, that olfactory information may be necessary but is not sufficient to explain navigation. No known substance dispersed in the atmosphere meets the requirements of the hypothesis; it is therefore asserted that 'other factors' of an unknown nature are necessary as well. This argument remains weak until the nature of the 'other factors' is known, but it is difficult to reject until the physical basis of olfactory navigation has been discovered. This position could be labelled 'navigation not only by smell'.

The second assertion made by some critics is that, even if olfactory information is needed for navigation over some areas, it is not so over others, where non-olfactory cues might be used. On this hypothesis, pigeons use a redundant complex of cues of various kinds to navigate, tending to prefer those which offer the most reliable information for navigating over the region where they live (Keeton, 1980; Wiltschko et al., 1987). This 'not everywhere by smell' position is based on results which could not be replicated. Experiments specifically designed to decide the argument would determine whether pigeons made permanently anosmic and released at least 40 km beyond the familiar area are able to orientate and home. Until positive results of this kind are achieved, homing from unfamiliar areas without an olfactory capability remains a claim.

An idea parallel to that of geographical differences was that pigeons of different strains use different cues: 'not every strain navigates by smell'. Repeated experiments have, however, shown that different breeds of pigeons perform in a similar way when reared in the same loft and trained by the same methods. When deprived of the sense of smell, both Italian and American pigeons reared in Italy were impaired in homing (Benvenuti et al., 1990).

The olfactory approach to the pigeon navigation problem has turned out to be productive, and has led to a series of achievements, the most important of which is the discovery that olfactory cues are essential for navigation in an unfamiliar area. Despite this, the state of the art calls for further research, which would certainly benefit from fresh methods and ideas, and the presence of new researchers.

7.6 PHYSIOLOGICAL ASPECTS

A number of structures and functions are regarded as the result of adaptation to migration and long-distance flights (see Gwinner, 1990, for references). In conformity with the scope of the present book, the central nervous mechanisms involved in navigation, and some of the sensory functions required for it, will be reviewed here shortly.

The ability of pigeons to distinguish differences in altitude between a memorized sun position and the current one – an assumption of the sun's arc hypothesis (p. 287) – has been successfully tested with conditioning methods by Whiten (1978). Investigation of visual capacity has revealed a series of interesting facts (see Beason and Semm, 1991, for references). Evidence that birds can detect the plane of polarized light has been obtained both by manipulating the patterns of light from the sky reaching passerine migrants hopping in Emlen's funnels, and using heart-rate conditioning techniques in pigeons. Furthermore, pigeons can learn to orientate with respect to an overhead source of polarized light. It has been proposed that double cones may be involved in the detection of the plane of polarized light. Near-ultraviolet reception has been demonstrated in a number of species, including the pigeon and other non-migratory species: it is believed to be a capability of most, if not all, diurnal birds. In many species it is probably used for orientational and navigational purposes, as in arthropods. In fact, the greatest amount of polarization found in the vault of the sky is located where short wavelength radiation is richest. The mechanism of near-ultraviolet detection is unknown.

Besides pigeons, olfaction has been reported to be involved in the homing of swifts (*Apus apus*), starlings (*Sturnus vulgaris*) and Leach's petrels (*Oceanodroma leucorhoa*), but a number of other species use the sense of smell to select or find food (see Papi, 1991, for references). The development of olfactory bulbs is dramatically different in different species, but its adaptive significance is poorly understood. Sensitivity to a number of odorants has been demonstrated by electrophysiological methods and by recording visceral responses to odours, such as respiration and heart rate changes. The degree of success of different conditioning methods varies widely; very simple procedures are sometimes the most successful. Only a few species have been tested and these tests have used only a small number of compounds to determine olfactory sensitivity. The resulting threshold values were found to range between 10^{-5} and 10^{-7} M; they were rarely as low as 10^{-9} M.

Even if sound sources from the soil, such as frog choruses or even ground echoes of flight calls (Griffin, 1976), have been proposed as orientational cues assisting birds migrating in the dark, high and medium frequency sounds have usually been disregarded as useful cues in long-distance orientation. On the other hand, following a suggestion from William von Arx (Griffin, 1969), infrasounds have attracted the attention of researchers. Natural loud sources of infrasounds, such as thunderstorms, earthquakes and ocean waves can be detected by appropriate instruments at a distance of hundreds or thousands of kilometres. Pigeons are now known to be sensitive to sounds below 10 Hz at amplitudes within the range of those naturally occurring at these frequences. This was ascertained by using a conditioned cardiac response; surgical ablations showed that reception occurs in the inner ear, but the precise structure responsible for it has not yet been identified. With waves as long as these, binaural localization of the source is impossible and birds could only detect the directions of the source by means of the Doppler shift, while flying towards and away from it (Kreithen, 1979). Considering also that natural infrasound sources are seldom constant enough in time and space to be considered reliable vibrational beacons, it appears very improbable that birds resort to them.

Using cardiac conditioning, it was found that pigeons react to air pressure changes as small as 10 mm of water or even less. This sensitivity, if shared by migratory species, may be useful in controlling flight altitude in the dark and in forecasting weather changes (Kreithen, 1979).

Magnetic sensitivity in birds has been the subject of a great number of investigations (Wiltschko and Wiltschko, 1988, 1990; see Wiltschko and Wiltschko (1991) for references). Theoretical and experimental approaches are in progress, but neither the structure nor the mechanism of the magnetoreceptor have been discovered so far. At the behavioural level, it has been found extremely difficult to obtain irrefutable evidence showing that magnetic perception is at work at all. Many attempts to condition birds to artificial magnetic stimuli were unsuccessful. Noisy data produced within orientation cages, like those shown in Fig. 7.11c-e, still yield the best indication of a reliance on magnetic sense for orientational purposes.

There are many experiments in which a magnetic treatment affects orientation with unexpected or unpredictable results. Such experiments show responsiveness to magnetic treatment, but do not prove magnetic orientation. It has recently been shown that a magnetic treatment on pigeons, which disturbs initial orientation, simultaneously affects the concentration of opiate receptors. Thus the observed effect on orientation has been considered an indication of incomplete compensation for stress and not an interference in the navigational mechanism. The fact that a similar disturbance in orientation is produced by an opiate antagonist, naloxone, supports the same conclusion (Papi et al., 1992). Likewise, the responsiveness of single neurons in the basal optic root and in the tectum opticum (Semm *et al.*, 1984; Semm and Demaine, 1986) does not demonstrate the involvement of these structures in magnetic orientation. The pineal gland is also responsive to magnetic stimuli, but its extirpation does not affect navigation in pigeons (Papi *et al.*, 1985).

While research on the bird's sensory world is still needed to complete our knowledge of orientational capability and processes, the spatial navigation system of pigeons is now sufficiently well known to allow an experimental approach to the location and function of the nervous centres that underlie homing.

Homing behaviour in different situations has been investigated after ablation of four different telencephalic regions: (a) the Wulst in the anterior forebrain, (b) the dorsomedial hippocampal region (HI), consisting of medial hippocampus and dorsomedial para-hippocampus, (c) a ventrolateral area comparable with the mammalian pyriform cortex, PC, and (d) the postero-dorso-lateral neostriatum (PDLNS), which is believed to be equivalent to the prefrontal cortex in mammals (Fig. 7.29). Loss of the Wulst does not impair homing in any way; ablated birds are excellent controls in experiments with other lesions (Bingman *et al.*, 1988). On the other hand, HI turned out to be important for memory processes in the context of spatial behaviour. HI-ablated pigeons appear to be affected by retrograde memory loss and become unable to use familiar visual landmarks or to recognize their own loft. Despite this, their ability to use the olfactory mechanism to home remains unimpaired and they resort to it at both familiar and unfamiliar sites. Thus they always behave as if they were in a unknown environment; when made anosmic they become incapable of homeward orientation even at sites from which they have homed several times before the operation. After HI-ablation, pigeons can relearn to recognize their loft and landmarks, but always show a decrease in homing speed, probably as a result of some residual impairment in the use of spatial landmarks (Bingman *et al.*, 1989). Up to this point, HI only appeared to be involved in the familiar landmark-based mechanism, but more recent experiments have shown that HI-ablation makes it impossible to learn the first navigational map or a new one. Both young pigeons subjected to HI-ablation before having acquired the olfactory map, and adult ones transferred to a new distant loft remain unable to orient homeward from unfamiliar sites (Bingman *et al.*, 1990; Bingman and Yates, in press).

The working of the olfactory map is completely disrupted in PC-ablated pigeons. Unlike HI-ablated birds, PC-operated ones

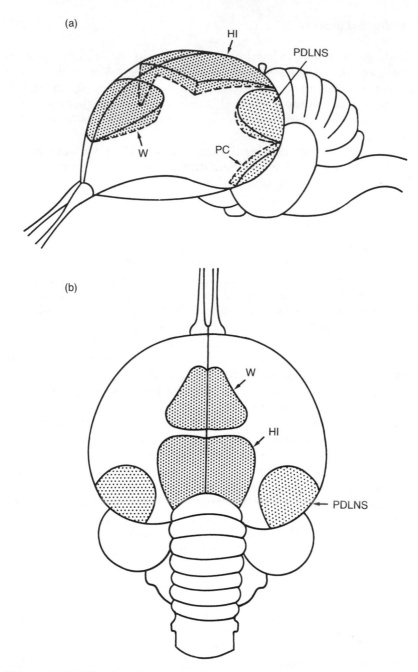

Figure 7.29 The involvement of specific brain areas in navigation has been tested by the lesion method. These areas are shown in the pigeon brain in (a) lateral and (b) dorsal view. W, Wulst; HI, hippocampal region; PDLNS, postero-dorso-lateral neostriatum; PC, pyriform cortex. (original: courtesy of A. Gagliardo.)

recognize familiar landmarks and their loft, and reach home without difficulty from familiar sites, but are impaired in homeward orientation and in homing from unfamiliar sites. In other words, they behave as if they were anosmic, though still showing cardiac responses to odorous stimuli. Since extensive projections reach the PC from the olfactory bulb, it may be inferred that the navigational impairment of PC-ablated birds is due to their inability to process olfactory information for spatial tasks (Papi and Casini, 1990).

The ablation of the fourth region, PDLNS, has an effect similar to that of PC-ablation as regards initial orientation and homing incapacity from unfamiliar sites. The one major difference with respect to PC-ablated birds was that the homing performances of PDLNS-ablated ones from familiar sites were significantly worse than those of controls, even if they were not as disastrous as in the case of unfamiliar sites (Gagliardo and Divac, 1992). Further experiments are needed to allow a full understanding of these results.

7.7 SUMMARY AND CONCLUSIONS

Observations and experiments on homing have been far more numerous on birds than on any other animal group, as migratory behaviour, site fidelity, frequent foraging flights, and diurnal habits render many species an ideal object for homing studies.

The young of some species face their first migratory flight on the basis of their ability to select and keep a specific compass direction for a given time. In later migratory flights birds rely on an additional ability, that of true navigation, which makes position fixing possible, and, as a result, allows compensation for any displacements off the right course.

A great number of experimental displacements to distant, unfamiliar areas have demonstrated a capacity for true navigation in many migratory and sedentary species, but the underlying mechanisms are still poorly understood. Almost everything we know about them derives from research on homing pigeons, which, among various advantages, have that of displaying homeward directedness soon after release, even if with a widely varying degree of accuracy and concentration of bearings. Only a few inexperienced birds home from distances as great as 180–300 km, but the distribution of the unsuccessful birds' recovery sites does show a tendency for them to approach home. Experienced birds perform better, both in initial orientation and in homing.

True navigation requires both directional and positional

information. In the literature, the two kinds of information are metaphorically compared to the knowledge gained by consulting a compass and a map, respectively. Birds are known to rely on three kinds of compasses: magnetic, solar, and stellar. It appears that the compass information that derives from the earth's magnetic field is 'noisy', and birds usually prefer to resort to astronomical cues. The birds' sun compass mechanism has the same features as in other animal groups: reference to the sun's azimuth, and compensation for the sun's apparent movement by means of a circadian clock. Migrating birds use the solar glow after sunset and related patterns of polarization in the sky for navigational purposes. The birds' star compass may work without compensation for the time of day, but such compensation has also been reported. Compasses are assumed to play an important role in choosing favourable winds and compensating for drift during migration. It is not yet possible to formulate conclusions about ontogenetic development of the compasses and their interactions that are valid for all bird species.

Phenomena of path integration so far appear to be of little importance, while position fixing mainly takes place through a geocentric system, for which two kinds of maps have been proposed. The 'mosaic' map (also called 'topographical' or 'familiar area' map) has a range depending on the extension of the overflown areas (direct acquisition) but may be enlarged using distant visual landmarks or windborne information about odours prevailing far away (indirect acquisition). The 'grid' or 'gradient' map is supposed to be made up of two or more physical factors and to allow comparisons between the local physical values and those remembered from the goal site. Experimental manipulation of the sun compass by shifting the underlying circadian clock shows that a compass mechanism is involved in pigeon homing, independently of the kind of map used.

The problem of the mysterious power of pigeons to orient homewards at distant, unfamiliar sites has been approached either by testing hypotheses based on known physical parameters which may allow position fixing or by manipulating the sensory capacities of birds. The latter approach has led to the discovery that olfaction plays an essential role in pigeon navigation. Pigeons prevented from smelling are impaired in initial orientation and homing from unfamiliar, but not from familiar sites. This shows that pigeons rely on two homing mechanisms, one olfaction-dependent and one based on non-olfactory familiar landmarks. For navigation from unfamiliar sites pigeons do not resort to odorous cues originating from the home site, but to local cues. Those smelt at the release site often turn out to be sufficiently informative, while the influence of those perceived during passive transportation is demonstrable by means of the effect of circuitous outward journeys, the 'detour' effect, or of anosmic transportation. In releases from very distant sites, olfactory information gathered *en route* can be essential for homing, so showing that the olfactory map is spatially limited. By manipulating olfactory experience before release, one can show that pigeons orient according to the position of the last site where they smelt environmental odours, rather than the real position of the release site.

The idea that the olfactory map of pigeons is built up at the loft by associating wind-borne odours with the wind direction is supported by the results of experiments that aim to prevent or alter the association between odours and wind direction. The method used has been that of subjecting pigeons to odorous air currents blowing from a specific direction or in rearing birds in special cages, which were completely or partly screened against the wind. In other lofts, wind was deflected in a specific, regular way, or was inverted. Results suggest that the olfactory map is able to change indefinitely in response to the acquisition of new data. Birds' navigational information is extracted from substances dispersed in the atmosphere, probably in a molecular state, but anything else – their nature, origin, or distribution – is unknown and a matter of speculation.

Both the picture of pigeon navigation reported in this chapter and the results and interpretation of single experiments supporting it have been challenged and criticized by some authors, so giving rise to one of the most lively discussions in the field of animal navigation. These critics often argue either that additional cues of an unknown nature may be necessary besides the olfactory ones, or that pigeons from some areas or strains may rely on non-olfactory mechanism(s). If alternative cues and/or mechanisms do exist, they will become clear in the process of being discovered and described, not simply as a result of being claimed or invoked.

Every review on bird navigation ends with the obvious state-ment that birds rely on refined, complex machanisms of spatial orientation which are still poorly understood. What perhaps is most useful in a final comment is to indicate the gaps whose bridging would provide the greatest stimulus to further advances. In the sensory field, the main aim is to discover how the earth's magnetic field is perceived and how the related information is processed. The only mechanism of true navigation that has been demonstrated so far, that of the homing pigeon, needs to be underpinned by ascertaining which odorants convey navigational information, and discovering their nature, origin, and distribution. A neurobiological approach to the central mechanism of homing, even if only undertaken recently, has already proved to be rewarding. Investigation on compass orientation and genetic programmes of navigation in migratory species have yielded brilliant results, but the concentration of so many research efforts on pigeons has led to an unbalanced situation with respect to wild species. One goal of future research is to determine if and how the mechanisms found to operate in pigeons are incorporated in the navigational machinery of other species, and how they have been adapted to various ecological and functional requirements.

REFERENCES

Able, K.P. (1980) Mechanisms of orientation, navigation, and homing, in *Animal Migration, Orientation and Navigation* (ed. S.A. Gautreaux, Jr), Academic Press, New York, pp. 283–373.

Able, K.P. (1989) Skylight polarization patterns and the orientation of migratory birds. *J. Exp. Biol.*, **141**, 241–56.

Able, K.P. (1991) The development of migratory orientation mechanisms, in *Orientation in Birds* (ed. P. Berthold), Birkhäuser Verlag, Basel, pp. 166–79.

Able, K.P., Gergits, W.F., Cherry, J.D. and Terrill, S.B. (1984) Homing behavior of wood thrushes (*Hylocychla mustelina*). *Behav. Ecol. Sociobiol.*, **15**, 39–43.

Alerstam, T. (1990) *Bird Migration*, Cambridge University Press, Cambridge.

Alleva, E., Baldaccini, N.E., Foà, A. and Visalberghi, E. (1975) Homing behaviour of the rock pigeon. *Monitore Zool. Ital., N.S.*, **9**, 213–24.

Baker, R.R. (1978) *The Evolutionary Ecology of Animal Migration*, Hodder and Stoughton, London.

Baker, R.R. (1984) *Bird Navigation: the solution of a mystery?*, Hodder and Stoughton, London.

Baldaccini, N.E., Benvenuti, S., Fiaschi, V. and Papi, F. (1975) Pigeon navigation: effects of wind deflection at home cage on homing behaviour. *J. Comp. Physiol.*, **99**, 177–86.

Bang, B.G. and Wenzel, B.M. (1985) Nasal cavity and olfactory system, in *Form and Function in Birds*, Vol. 3 (eds A.S. King and J. McLelland), Academic Press, New York, pp. 165–225.

Barlow, J.S. (1964) Inertial navigation as a basis for animal navigation. *J. Theor. Biol.*, **6**, 76–117.

Beason, R.C. and Semm, P. (1991) Sensory basis of bird orientation, in *Orientation in Birds*, (ed. P. Berthold), Birkhäuser Verlag, Basel, pp. 106–25.

Benvenuti, S., Brown, A.I., Gagliardo, A. and Nozzolini, M. (1990) Are American homing pigeons genetically different from Italian ones? *J. Exp. Biol.*, **148**, 235–43.

Benvenuti, S. and Wallraff, H.G. (1985) Pigeon navigation: site simulation by means of atmospheric odours. *J. Comp. Physiol.*, **156A**, 737–46.

Berthold, P. (ed.) (1991a) *Orientation in Birds*, Birkhäuser, Basel.

Berthold, P. (1991b) Spatiotemporal programmes and genetics of orientation, in *Orientation in Birds* (ed. P. Berthold), Birkhäuser Verlag, Basel, pp. 86–105.

Berthold, P. and Querner, U. (1981) Genetic basis of migratory behavior in European warblers. *Science*, **212**, 77–9.

Bingman, V.P., Bagnoli, P., Ioalè, P. and Casini, G. (1989) Behavioral and anatomical studies of the avian hippocampus, in *The Hippocampus – New Vistas, Neurology and Neurobiology Series* (eds V. Chan-Palay and S.L. Palay), Alan Liss, New York, pp. 377–92.

Bingman, V.P., Ioalè, P., Casini, G. and Bagnoli, P. (1988) Hippocampal ablated homing pigeons show a persistent impairment in the time taken to return home. *J. Comp. Physiol.*, **163A**, 559–63.

Bingman, V.P., Ioalè, P., Casini, G. and Bagnoli, P. (1990) The avian hippocampus: evidence for a role in the development of the homing pigeon navigational map. *Behavl. Neurosci.*, **104**, 906–11.

Bingman, V.P. and Mackie, A. (1992) Importance of olfaction for homing pigeon navigation in Ohio, USA. *Ethol. Ecol. Evol.*, in press.

Bingman, V.P. and Yates, G. (1992) Hippocampal lesions impair navigational learning in experienced homing pigeons. *Behavl. Neurosci.*, in press.

Emlen, S.T. (1967) Migratory orientation in the Indigo bunting, *Passerina cyanea*. *Auk*, **84**, 309–42, 463–82.

Emlen, S.T. (1970) Celestial rotation: its importance in the development of migratory orientation. *Science*, **170**, 1198–201.

Emlen, S.T. (1975a) The stellar-orientation system of a migratory bird. *Scient. Am.*, **233**, 102–11.

Emlen, S.T. (1975b) Migration: orientation and navigation, in *Avian Biology*, Vol. 5 (eds D.S. Farner and J.R. King), Academic Press, New York, pp. 129–219.

Etienne, A.S., Maurer, R. and Saucy, F. (1988) Limitations in the assessment of path dependent information. *Behaviour*, **106**, 81–111.

Gagliardo, A. and Divac, I. (1992) Effects of ablation of the equivalent of the mammalian prefrontal cortex on pigeon homing. *Behav. Brain Res.*, (in press).

Ganzhorn, J.U. (1990) Konditionierung verfrachteter Brieftauben: eine 'neue' Methode zur Analyse der Karte von Brieftauben. *J. Orn.*, **131**, 21–31.

Griffin, D.R. (1969) The physiology and geophysics of bird navigation. *Q. Rev. Biol.*, **44**, 255–76.

Griffin, D.R. (1976) The audibility of frog choruses to migrating birds. *Anim. Behav.*, **24**, 421–7.

Gwinner, E. (ed.) (1990) *Bird Migration: physiology and ecophysiology*, Springer Verlag, Berlin.

Helbig, A.J. (1990) Depolarization of natural skylight disrupts orientation of an avian nocturnal migrant. *Experientia*, **46**, 755–8.

Helbig, A., Berthold, P. and Wiltschko, W. (1989) Migratory orientation of blackcaps (*Sylvia atricapilla*): population-specific shifts of direction during the autumn. *Ethology*, **82**, 307–15.

Hoffmann, K. (1954) Versuche zu der im Richtungsfinden der Vögel enthaltenen Zeitschätzung. *Z. Tiepsychol.*, **11**, 453–75.

Ioalè, P., Nozzolini, M. and Papi, F. (1990) Homing pigeons do extract directional information from olfactory stimuli. *Behav. Ecol. Sociobiol.*, **26**, 301–5.

Ioalè, P., Wallraff, H.G., Papi, F. and Foà, A. (1983) Long-distance releases to determine the spatial range of pigeon navigation. *Comp. Biochem. Physiol.*, **76A**, 733–42.

Keeton, W.T. (1980) Avian orientation and navigation: new developments in an old mystery, in *Acta XVII Congressus Internationalis Ornithologici (1978)* (ed. R. Nöhring), Dt. Ornithol. Ges., Berlin, pp. 137–57.

Kenyon, K.W. and Rice, D.W. (1958) Homing in Laysan albatrosses. *Condor*, **60**, 3–6.

Kiepenheuer, J. (1985) Can pigeons be fooled about the actual release site position by presenting them information from another site? *Behav. Ecol. Sociobiol.*, **18**, 75–82.

Kiepenheuer, J. (1986) Are site-specific airborne stimuli relevant for pigeon navigation only when matched by other release-site information? *Naturwissenschaften*, **73**, 42–3.

Koskimies, J. (1950) The life of the swift, *Micropus apus* L. in relation to the weather. *Ann. Acad. Sci. Fenn, 4A Biol.*, **15**, 1–151.

Kramer, G. (1951) Eine neue Methode zur Erforschung der Zugorientierung und die bisher damit erzielten Ergebnisse. *Proc. X Ornithol. Congr. Uppsala 1951*, pp. 269–80.

Kramer, G. (1953) Wird die Sonnenhöhe bei der Heimfindeorientierung verwertet? *J. Orn.*, **94**, 201–19.

Kramer, G. (1959) Uber die Heimfindeleistung unter Sichtbegrenzung aufgewachsener Brieftauben. *Verh. Dt. Zool. Ges.*, **52**, 168–76.

Kramer, G. and Saint-Paul, U. (1954) Das Heimkehrvermögen gekäfigter Brieftauben. *Orn. Beob.*, **51**, 3–12.

Kreithen, M.L. (1979) The sensory world of the homing pigeon, in *Neural Mechanisms of Behavior in the Pigeon* (eds A.M. Granda and J.H. Maxwell), Plenum, New York, pp. 21–33.

Lack, D. (1958) Swifts over the sea at night. Weather movement of swifts 1955–1957. *Bird Study*, **5**, 126–42.

Matthews, G.V.T. (1951) The experimental investigation of navigation in homing pigeons. *J. Exp. Biol.*, **28**, 508–36.

Matthews, G.V.T. (1953) Sun navigation in homing pigeons. *J. Exp. Biol.*, **30**, 243–67.

Matthews, G.V.T. (1955) *Bird Navigation*, 1st edn, Cambridge University Press, Cambridge.

Matthews, G.V.T. (1968) *Bird Navigation*, 2nd edn, Cambridge University Press, Cambridge.

Merkel, F.W. and Wiltschko, W. (1965) Magnetismus und Richtungsfinden zugunruhiger Rotkehlchen (*Erithacus rubecula*). *Vogelwarte*, **23**, 71–7.

Mewaldt, L.R. (1964) California sparrows return from displacement to Maryland. *Science*, **146**, 941–2.

Moore, B.R. (1988) Magnetic field and orientation in homing pigeons: experiments of the late W.T. Keeton. *Proc. Natl. Acad. Sci., U.S.A.*, **85**, 4907–9.

Moore, F.R. (1987) Sunset and the orientation behaviour of migrating birds. *Biol. Rev.*, **62**, 65–86.

Neuss, M. and Wallraff, H.G. (1988) Orientation of displaced homing pigeons with shifted circadian clocks: prediction vs observation. *Naturwissenschaften*, **75**, 363–5.

Papi, F. (1986) Pigeon navigation: solved problems and open questions. *Monitore Zool. Ital.*, N.S., **20**, 471–517.

Papi, F. (1989) Pigeons use olfactory cues to navigate. *Ethol. Ecol. Evol.*, **1**, 219–31.

Papi, F. (1991) Olfactory navigation in birds, in *Orientation in Birds* (ed. P. Berthold), Birkhäuser Verlag, Basel, pp. 52–85.

Papi, F. and Casini, G. (1990) Pigeons with ablated pyriform cortex home from familiar but not from unfamiliar sites. *Proc. Natl. Acad. Sci., U.S.A.*, **87**, 3783–7.

Papi, F., Fiore, L., Fiaschi, V. and Benvenuti, S. (1971) The influence of olfactory nerve section on the homing capacity of carrier pigeons. *Monitore Zool. Ital.*, N.S., **5**, 265–7.

Papi, F., Fiore, L., Fiaschi, V. and Benvenuti, S. (1972) Olfaction and homing in pigeons. *Monitore Zool. Ital.*, N.S., **6**, 85–95.

Papi, F., Ioalè, P., Fiaschi, V., Benvenuti, S. and Baldaccini, N.E. (1984) Pigeon homing: the effect of outward-journey detours on orientation. *Monitore Zool. Ital.*, N.S., **18**, 53–87.

Papi, F., Ioalè, P., Dall'Antonia, P. and Benvenuti, S. (1991) Homing strategies of pigeons investigated by clock-shift and flight path reconstruction. *Naturwissenschaften*, **78**, 370–3.

Papi, F., Luschi, P. and Limonta, P. (1992) Orientation-disturbing magnetic treatment affects the pigeon opioid system. *J. Exp. Biol.*, **166**, 169–79.

Papi, F., Maffei, L. and Giongo, F. (1985) Pineal body and bird navigation: new experiments on pinealectomized pigeons. *Z. Tierpsychol.*, **67**, 257–68.

Pennycuick, C.J. (1960) The physical basis of astronavigation in birds: theoretical considerations. *J. Exp. Biol.*, **37**, 573–93.

Perdeck, A.C. (1958) Two types of orientation in migrating starlings, *Sturnus vulgaris* L., and chaffinches, *Fringilla coelebs* L., as revealed by displacement experiments. *Ardea*, **46**, 1–37.

Richardson, W.J. (1991) Wind and orientation of migrating birds: a review, in *Orientation in Birds* (ed. P. Berthold), Birkäuser Verlag, Basel, pp. 226–49.

Saint-Paul, U. von (1982) Do geese use path integration for walking home?, in *Avian Navigation* (eds F. Papi and H.G. Wallraff), Springer, Berlin, pp. 298–307.

Sauer, F. (1957) Die Sternenorientierung nächtlich ziehender Grasmücken.

(*Sylvia atricapilla, borin* and *curruca*). *Z. Tierpsychol.*, **14**, 29–70.

Schmidt-Koenig, K. (1958) Experimentelle Einflussnahme auf die 24-Stunden-Periodik bei Brieftauben und deren Auswirkungen unter besonderer Berücksichtigung des Heimfindevermögens. *Z. Tierpsychol.*, **15**, 301–31.

Schmidt-Koenig, K. (1987) Bird navigation: has olfactory orientation solved the problem? *Q. Rev. Biol.*, **62**, 31–47.

Schmidt-Koenig, K. and Ganzhorn, J.U. (1991) On the problem of bird navigation. *Perspect. Ethol.*, **9**, 261–83.

Schmidt-Koenig, K., Ganzhorn, J.U. and Ranvaud, R. (1991) The sun compass, in *Orientation in Birds* (ed. P. Berthold), Birkhäuser Verlag, Basel, pp. 1–15.

Schmidt-Koenig, K. and Schlichte, H.J. (1972) Homing of pigeons with impaired vision. *Proc. Natl. Acad. Sci., U.S.A.*, **69**, 2446–7.

Semm, P. and Demaine, C. (1986) Neurophysiological properties of magnetic cells in the visual system of the pigeon. *J. Comp. Physiol.*, **159**, 619–25.

Semm, P., Nohr, D., Demaine, C. and Wiltschko, W. (1984) Neural basis of the magnetic compass: interactions of visual, magnetic and vestibular inputs in the pigeon's brain. *J. Comp. Physiol.*, **155**, 283–8.

Ugolini, A. (1987) Visual information acquired during displacement and initial orientation in *Polistes gallicus* L. (Hymenoptera, Vespidae). *Anim. Behav.*, **35**, 590–5.

Waldvogel, J.A. (1989) Olfactory orientation by birds, in *Current Ornithology*, Vol. 6 (ed. D.M. Power), Plenum Press, New York, pp. 269–321.

Wallraff, H.G. (1970) Über die Flugrichtungen verfrachteter Brieftauben in Abhängigkeit vom Heimatort und vom Ort der Freilassung. *Z. Tierpsychol.*, **27**, 303–51.

Wallraff, H.G. (1972) An approach toward an analysis of the pattern recognition involved in the stellar orientation of birds, in *Animal Orientation and Navigation* (eds J.R. Galler *et al.*), NASA SP-262, Washington, DC, pp. 211–22.

Wallraff, H.G. (1974) *Das Navigationssystem der Vögel*, R. Oldembourg Verlag, München.

Wallraff, H.G. (1980) Does pigeon homing depend on stimuli perceived during displacement? I. Experiment in Germany. *J. Comp. Physiol.*, **139**, 193–201.

Wallraff, H.G. (1984) Migration and navigation in birds: a present-state survey with some digressions to related fish behaviour, in *Mechanisms of Migration in Fishes* (eds J.D. McCleave *et al.*), Plenum Press, New York, pp. 509–44.

Wallraff, H.G. (1985) Theoretical aspects of avian navigation, in *Acta XVIII Congr. Int. Ornith.* (eds V.D. Ilyichev and V.M. Gavrilov), Nauka, Moscow, pp. 284–92.

Wallraff, H.G. (1989a) Simulated navigation based on unreliable sources of information (models on pigeon homing, Part 1). *J. Theor. Biol.*, **137**, 1–19.

Wallraff, H.G. (1989b) Simulated navigation based on assumed gradients of atmospheric trace gases (models on pigeon homing, Part 2). *J. Theor. Biol.*, **138**, 511–28.

Wallraff, H.G. (1989c) The whereabouts of non-homing homing pigeons: recoveries of normal and anosmic birds, in *Orientation and Navigation: birds, humans and other animals*, Royal Institute of Navigation, London, paper no. 10.

Wallraff, H.G. (1990) Navigation by homing pigeons. *Ethol. Ecol. Evol.*, **2**, 81–115.

Wallraff, H.G. (1991) Conceptual approaches to avian navigation systems, in *Orientation in Birds* (ed. P. Berthold), Birkhäuser Verlag, Basel, pp. 128–65.

Wallraff, H.G. and Foà, A. (1981) Pigeon navigation: charcoal filter removes relevant information from environmental air. *Behav. Ecol. Sociobiol.*, **9**, 67–77.

Whiten, A. (1978) Operant studies on pigeon orientation and navigation. *Anim. Behav.*, **26**, 571–610.

Wilkie, D.M., Wilson, R.J. and Kardal, S. (1989) Pigeons discriminate pictures of a geographic location. *Anim. Learn. Behav.*, **17**, 163–71.

Williams, T.C. and Williams, J.M. (1990) The orientation of transoceanic migrants, in *Bird Migration: physiology and ecophysiology* (ed. E. Gwinner), Springer Verlag, Berlin, pp. 7–21.

Wiltschko, R. and Wiltschko, W. (1989) Pigeon homing: olfactory orientation – a paradox. *Behav. Ecol. Sociobiol.*, **24**, 163–73.

Wiltschko, W. (1968) Über den Einfluss statischer Magnetfelder auf die Zugorientierung der Rotkehlchen (*Erithacus rubecula*). *Z. Tierpsychol.*, **25**, 537–58.

Wiltschko, W. and Wiltschko, R. (1972) Magnetic compass of European robins. *Science*, **176**, 62–4.

Wiltschko, W. and Wiltschko, R. (1988) Magnetic orientation in birds. *Curr. Orn.*, **5**, 67–121.

Wiltschko, W. and Wiltschko, R. (1991) Magnetic orientation and celestial cues in migratory orientation, in *Orientation in Birds* (ed. P. Berthold), Birkhäuser Verlag, Basel, pp. 16–37.

Wiltschko, W., Wiltschko, R., Grüter, M. and Kowalski, U. (1987) Pigeon homing: early experience determines what factors are used for navigation. *Naturwissenschaften*, **74**, 196–7.

Yeagley, H.L. (1947) A preliminary study of a physical basis of bird navigation. *J. Appl. Phys.*, **18**, 1035–63.

Yeagley, H.L. (1951) A preliminary study of a physical basis of bird navigation. Part II. *J. Appl. Phys.*, **22**, 746–60.

Chapter 8
Mammals

J. Bovet

8.1 HOME RANGES AND HOMING

In comparison with what has been seen in previous chapters, the study of homing behaviour in mammals is marked by a steady reference to the concept of **home range** (see Chapter 1), i.e. 'the area over which an animal normally travels in pursuit of its routine activities' (Jewell, 1966) during a stated period of time (e.g. a few weeks, a season, etc.). The current trend is to describe home range size and shape in a probabilistic sense (Worton, 1987). For instance, home range can be viewed as the smallest area that accounts for a meaningful percentage of the animal's space utilization (Anderson, 1982). Such probabilistic descriptions illustrate the normal, routine character of travels within a home range, and also its necessary complement, namely that travels outside the home range do occur but are infrequent and non-routine. The evidence available for many mammals is that the smallest area in which they spend, say, 90% of their time is one or two orders of magnitude smaller than the surrounding area where they spend the remaining 10%. It is thus reasonable to postulate that, as a result of this difference in 'utilization density', a mammal has much more familiarity with its small home range than with any part of the huge surrounding area where it travels occasionally; and that the strategies and mechanisms it uses to find its way within its home range are not necessarily the ones it can use outside. In the mammalian literature, homing usually refers to travelling back to one's home range, and I shall focus here on this kind of travel. Studies of movements within one's home range will be referred to only in as much as they contribute to the understanding of how mammals find their way back to their home range.

Animal Homing. Edited by Floriano Papi. Published in 1992 by Chapman & Hall, 2–6 Boundary Row, London SEI 8HN. ISBN 0 412 36390 9.

Mammals perform homing in two types of natural circumstances. Homing occurs when an individual makes a **migration** from one seasonal home range to another that it occupied the previous year. It also occurs when an animal returns home from a short-term foray off home range limits, thus completing an **excursion**. Homing to one's home range can also be induced experimentally in mammals, using displacement procedures as discussed in Chapter 1.

8.2 METHODOLOGICAL PROBLEMS

The direct study of mammalian travel is marred by technical difficulties. Mammals are commonly 'classified' as small or large. Because of their size, small mammals escape easily from their observer's field of perception, even though they might still be close by. Large mammals, on the other hand, are likely to quickly outrun or attack their observer. A majority of mammals are nocturnal, which makes observation difficult. As a result, the direct study of their movements is often limited to the visual observation of very short stretches of travel (e.g. a few seconds immediately after release in a displacement experiment). In some cases, the efficiency of this technique can be improved by fitting animals with incandescent, fluorescent or chemiluminescent markers, which increases the range over which they can be seen moving. A somewhat equivalent but strongly 'condition-dependent' technique is to survey the tracks left by individually known animals on an appropriate substratum, e.g. snow.

Much of what is known of mammalian movements rests, therefore, on the application of the capture–marking–recapture (CMR) method, or of radiotracking techniques. As discussed in Chapter 1, the CMR method is not appropriate to reveal the 'fine grain' of movements. When used in connection with a displacement experiment, however, it usually provides a valid measure of the proportion of animals displaced that actually return home, the so-called **homing success**. But it provides little more than that. Radiotracking allows for a better assessment of movements, particularly when it can be combined with direct visual observation, what Harris *et al.* (1990) called **radio-assisted surveillance**. Prior to a tracking session, the observer uses radio-signals to locate the animal scheduled for study, and to approach it until visual contact is made; the observer then follows the animal, keeping visual contact as long as possible; if contact is lost, the observer re-establishes it using radio-signals, and so on. The recent proliferation of tracking studies on mammals has produced first-rate data on actual routes followed in natural or experimentally induced homing movements. As shown subsequently, this sheds light on the strategies which can be at work in homing behaviour.

8.3 NATURAL INSTANCES OF HOMING IN MIGRATIONS

8.3.1 Site fidelity

Many species of bats, cetaceans, marine carnivores, and ungulates perform seasonal migrations, the gross patterns and functions of which have been reviewed recently by Fenton and Thomas (1985), Wuersig (1989), Riedmann (1990) and McCullough (1985), respectively. The length of one migratory move varies from tens (many bats, many ungulates), to hundreds (several bats, caribou), to thousands of kilometres (many baleen whales and seals). These distances are at least five times the home range diameter in the species concerned, and usually much more. The movements are therefore between two discrete spatial units, and are not mere home range shifts.

To qualify as homing behaviour, a migratory movement must bring the animal back to a previously occupied home range. Evidence for such **site fidelity** does not require that home range sizes and shapes remain the same from year to year, or are known with precision. The return of a bat to a previously occupied summer roost, of a seal to a previously used breeding ground, or of a deer to a previously exploited winter yard are considered as evidence of site fidelity, and hence of homing. Site fidelity is more difficult to establish in whales, due to the nature and behaviour of these animals. In their case, it is usually (and reasonably) considered that resightings of identifiable individuals at yearly intervals at the same place are sufficient evidence of site fidelity.

Repeated fidelity to both summer and winter home ranges has been well documented in several species of ungulates. Thus in mule deer (*Odocoileus hemionus*), among 30 radiotagged individuals that Thomas and Irby (1990) could track over more than 1 year in Idaho (USA), 29 did not shift summer or winter home range between years, bee-line distances between individual seasonal ranges being between 11 and 115 km. This confirmed a conclusion reached for the same species, using different methods, by Garrott et al. (1987) for two Colorado (USA) populations totalling nearly 100 individuals. The same is true for wapiti (*Cervus elaphus*): Morgantini and Hudson (1988) observed the same herds during four consecutive winters and three consecutive summers in Alberta (Canada). Among 18 individually tagged animals, 16 used the same winter home ranges and 11 the same summer home ranges throughout the study. Bee-line distances between individual winter and summer ranges were between 26 and 68 km. This kind of double-site fidelity has also been documented in other ungulates, e.g. white-tailed deer (*O. virginianus*; Nelson and Mech, 1981), moose (*Alces alces*; Andersen, 1991) and mountain sheep (*Ovis canadensis*; Geist, 1971). Caribou (*Rangifer tarandus*), however, appear to show seasonal fidelity to population ranges rather than to individual home ranges (e.g. Fancy *et al.*, 1989).

Individual fidelity to a summer home range, a winter home range, or both, has been demonstrated, in the grey bat (*Myotis grisescens*) by Tuttle (1976), using the CMR method. Over several years, he tagged 40 182 individuals in 53 caves and made 19 691 recoveries at 120 places within a 750 × 500 km area in the south-eastern USA. Straight lines connecting individual winter and summer sites ranged from 17 to 437 km in all possible directions. The bats that shared the same cave in summer did not all hibernate in the same winter cave, and vice versa, which resulted in individual rather than population migration lines. Even though they do not hibernate, some species of tropical bats have a similar pattern of site fidelity to 'seasonal' home ranges several hundred kilometres apart (Gopalakrishna, 1985/86).

In migratory marine mammals, current evidence of site fidelity is available for one seasonal range only. Working on humpback whales (*Megaptera novaeangliae*) in Massachusetts Bay (USA), which is the southernmost feeding (summer) area for that species in the northwest Atlantic, Clapham and Mayo (1987) reported that 18 of 21 adult females spotted there in 1979 were resighted each of the six following summers, the remaining three being 'missed' for 1 or 2 years only; and that 11 of 13 individuals recorded there in 1979, 1980 or 1981 as first-year calves were resighted each year in 1982–85, the remaining two being seen only 1 year each. Site fidelity to feeding areas around the southwest part of Nova Scotia (Canada) was also displayed by several right whales (*Balaena glacialis*) studied by Kraus *et al.* (1986). On the other hand, repeated resightings of known individuals in the same breeding area winter after winter have been reported in the grey whale (*Eschrichtius robustus*) along the coast of Baja California (Mexico) (Swartz, 1986).

Among marine carnivores, some species of seals and sea-lions are well known for their spectacular site fidelity to small and crowded breeding grounds (see Riedmann, 1990, for references). It is usually assumed that outside the breeding season these animals do not hold home ranges and have a rather nomadic way of life, driven by the availability of food. But a case of site fidelity to a non-reproductive home range has been recently reported for a group of male New

Zealand sea-lions (*Phocarctos hookeri*; Beentjes, 1989). Furthermore, the male sea otters (*Enhydra lutris*) studied along the Californian coast by Jameson (1989) returned year after year to the same breeding home ranges, of 40 ha on average, from relatively well-defined non-breeding areas up to more than 100 km away.

8.3.2 Individual migration routes

The most detailed and accurate records of actual routes of travel during migration are those provided by satellite-tracking for the barren-ground caribou in northwest North America (Craighead and Craighead, 1987; Fancy *et al.*, 1989). Figure 8.1 gives an example of a female's track in Alaska, that compares well with other tracks obtained in other areas, using the same technique. Common features of these tracks are that (1) they are made of a series of essentially 'straight' segments, the length of which is at least 100 km (and often more); (2) the orientation of these segments does not systematically correspond to that of nearby linear landscape features (e.g. rivers). Of course, the word 'straight' should not be taken in this context as 'laser beam straight', but rather as when travelling by road 'straight' from Montréal to Vancouver over the Trans-Canadian Highway: there is a clear, general orientation of travel, even though there are many short-term or short-range deviations for trivial reasons such as finding appropriate food or sleeping sites, or avoiding small-scale obstacles.

The migratory tracks of other ungulates appear to be as straight as those of caribou. There is, however, some ambiguity in the data available concerning the relationship between landscape features and the orientation of their routes. Thus, Gruell and Papez (1963) claimed that mule deer in northeast Nevada (USA) travel along direct routes crossing mountain ranges during migrations as shown in Fig. 8.2, which is a good illustration of inter-individual diversity in length and orientation of migration routes. The mule deer studied in nearby southeast Idaho by Thomas and Irby (1990) also migrated in fairly straight routes over tens of kilometres, but did so initially at least along natural corridors formed by valleys oriented northwest–southeast. Similarly, the wapiti of the Jackson herd in Wyoming (USA) migrate along direct routes that are along as well as across valleys (Boyce, 1991). However, in the wapiti herds studied by Morgantini and Hudson (1988) in Alberta (Canada), the direct migratory routes were along valleys oriented ENE–WSW. Straightness of travel has also been documented in migrating white-tailed deer by Rongstad and Tester (1969) and Verme (1973).

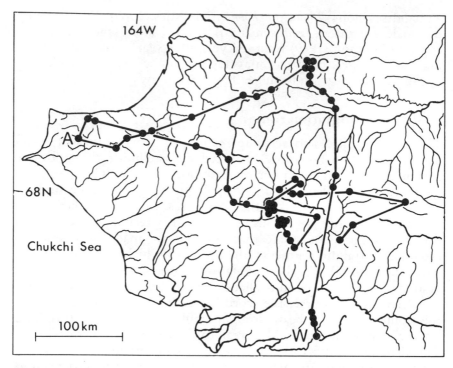

Figure 8.1 Migration route of an adult, female caribou (*Rangifer tarandus*) in northwest Alaska, USA, in 1984. The heavy line connects successive satellite-fixes (filled circles). The caribou left the winter range (W) on 15 May and arrived on the calving ground (C) on 30 May, where she calved on 5 June, and stayed until 16 June. She then moved to the herd's aggregation area (A), where she stayed on 4 and 5 July. She spent the summer travelling east, with occasional 1–2 week stays in localized areas. The last fix was obtained on 7 October, while she was moving toward the winter range. Due to the hydrographic features of the area (thin lines = rivers), the caribou's route was probably as often across, as it was parallel to, valleys. [After Craighead and Craighead (1987), and sheet 'Fairbanks' of the World Map (1:2 500 000) of the USSR Main Administration of Geodesy and Cartography, Moscow, 1973.]

   In northeast Manitoba (Canada), which is the southernmost part of the species' range in North America, polar bears (*Thalarctos maritimus*) are forced onto land during summer, due to the total absence of ice in this area of Hudson Bay at this time of year. Most bears return to the sea ice in November or early December. However, pregnant females stay inland for the winter where they

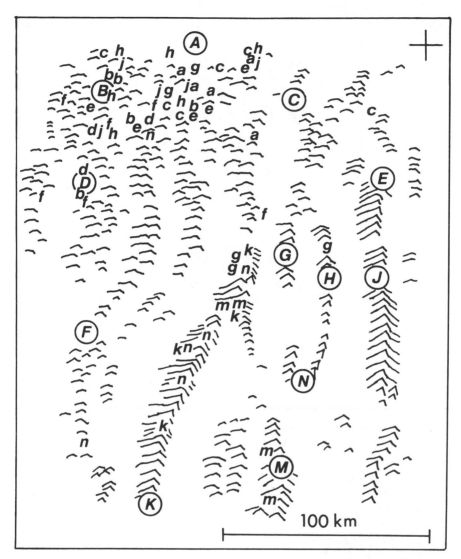

Figure 8.2 Winter and summer ranges of individual mule deer (*Odocoileus hemionus*) in northeast Nevada, USA. Each circled capital letter identifies a 'communal' winter range, and each lower-case letter identifies the place where a deer that had spent a previous winter in a range with the corresponding capital letter was seen or killed during the summer. Mountain ranges in the area are coarsely oriented north–south, as shown symbolically; mountains in the north and west are about 600 m higher than the valley floors, in the southeast about 1400 m. The cross in the upper right corner (northeast) is for the common borderstone between Nevada, Utah and Idaho. (After Gruell and Papez, 1963.)

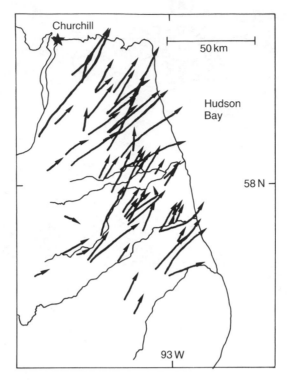

Figure 8.3 Spring migration routes of polar bears (*Thalarctos maritimus*) in northeast Manitoba, Canada. Each arrow represents a segment of route left as a track in the snow by a mother polar bear and her cub(s), recorded at any one spring between 1980 and 1984. Major rivers are shown. In the area crossed by bears, no location is more than 100 m in elevation. (After Ramsay and Andriashek, 1986.)

give birth to cubs, and return to the sea ice only between mid-February and early April (Derocher and Stirling, 1990). Between 1980 and 1984, Ramsay and Andriashek (1986) made aerial records of long stretches of footprint tracks left in the snow by 63 different females migrating back to sea with their newly born cubs. These tracks were straight over tens of kilometres (Fig. 8.3). Their headings were significantly clustered around a northeast mean bearing (39°, $P < 0.001$), which led these families direct toward the area where Derocher and Stirling (1990) located most of their bears on ice in spring. With respect to possible topographical corridors, Ramsay and Andriashek (1986) stated that tracks of travelling bears rarely followed river courses for more than 2–3 km but, instead, passed over windswept tundra and only crossed rivers *en route*. It should also be stressed that these routes did not represent the shortest way to the sea.

Such detailed mappings of actual, individual migration routes are lacking for flying or swimming mammals. Circumstantial evidence often suggests, however, that migration routes are relatively straight. Considering various elements such as times of multiple recoveries of single individuals, places of recoveries between summer and winter ranges during migration periods, or weight loss during migration, Tuttle (1976) concluded that the migration routes of grey bats were usually direct. Some followed topographical corridors, others did not. On fairly similar grounds, baleen whales are usually believed to follow 'straight' migration routes. In some cases, this is clearly related to a major topographical feature, such as in grey whales that migrate close to and along the North American Pacific coast (Swartz, 1986). However, the relationship with topography is lacking in pelagic forms like the humpback whale (e.g. Stone et al., 1987, 1990; Clapham and Mattila, 1990).

8.4 NATURAL INSTANCES OF HOMING IN EXCURSIONS

8.4.1 Occurrence and function of excursions

The occurrence of excursions is documented in a number of species of terrestrial mammals by the occasional sightings, captures or radio-locations of 'excursionists' outside their home range. As a result of technical difficulties, similar evidence is still largely lacking in flying or swimming mammals. But I mention here, for the record, the established occurrence of excursions in the grey bat (*Myotis grisescens*; Tuttle, 1976), the dugong (*Dugong dugon*; Marsh and Rathbun, 1990) and Commerson's dolphin (*Cephalorhynchus commersoni*; Buffrenil et al., 1989).

In a few studies, the occurrence of excursions could be easily related causally to some catastrophic events and to the need for the animals to escape their short-term adverse effects: natural flooding − Townsend's moles (*Scapanus townsendii*; Giger, 1973); military manoeuvres − coyotes (*Canis latrans*; Gese et al., 1989); release of hunting dogs − white-tailed deer (*Odocoileus virginianus*; Sweeney et al., 1971); withdrawal of local food resources − brown rats (*Rattus norvegicus*; Taylor and Quy, 1978); logging operations − deer mice (*Peromyscus maniculatus*; Gashwiler, 1959). In other cases, excursions were likely to be caused by reproductive needs, e.g. when males searched for reproductive females [e.g. brush-tailed possums (*Trichosurus vulpecula*; Ward, in Cowan, 1983); lesser bushbabies (*Galago senegalensis*; Bearder and Martin, 1979); red squirrels (*Tamiasciurus hudsonicus*; Lair, in Bovet, 1984, 1990)]; or by alimentary needs, e.g. when packs of wolves (*Canis lupus*) made extraterritorial raids to deer wintering areas (Messier, 1985).

However, in most instances, their cause or function cannot be inferred so simply from context. Still, there is a definite trend for excursions to become more frequent in potential dispersers, shortly before or at about the time when dispersal normally occurs (dispersal being an emigration, usually performed at a subadult age, from a natal home range to a new, reproductive home range). This link with dispersal has been documented in wolves (Ballenberghe, 1983; Messier, 1985; Potvin, 1988; Fuller, 1989), coyotes (Andelt and Gipson, 1979), red foxes (*Vulpes vulpes*; Woollard and Harris, 1990), black bears (*Ursus americanus*; Rogers, 1987a), roe deer (*Capreolus capreolus*; Bideau *et al.*, 1987), white-tailed deer (Nelson and Mech, 1981), Columbian ground squirrels (*Spermophilus columbianus*; Wiggett *et al.*, 1989) and Merriam's kangaroo rats (*Dipodomys merriami*; Jones, 1989). Whether these predispersal excursions could be viewed as tests of potential future home ranges is a far-reaching question because of its importance in current discussions about the adaptive value of dispersal (e.g. Anderson, 1989; Jones, 1989). The scant evidence available suggests that dispersers do not settle in places previously visited during excursions, as shown for wolves by Potvin (1988) and for Columbian ground squirrels by Wiggett *et al.* (1989). This suggests that dispersers test several potential places until they find the right one, returning to the safety of the natal range after each 'negative' test. Alternatively, predispersal excursions could be viewed as a 'practice' for the great 'move' to come: the animals learn how to survive and travel alone in unfamiliar areas. The fact that the ground squirrels began actual dispersal from their natal home range along routes taken during previous excursions is also suggestive of a kind of 'testing or learning the exits'.

On the other hand, excursionist behaviour is in no way restricted to those sex or age classes that are prone to disperse, or to reactions to catastrophic events. In many species, it is displayed by all classes 'without obvious reason'. Baker (e.g. 1982) has repeatedly stressed that the urge to explore is widespread in the animal kingdom and is satisfied by these excursions; and Cowan (e.g. 1983) that the function of exploration (thus of excursions) may simply be the acquisition of information about the individual's environment.

8.4.2 Excursion routes

While the existence of excursions is fairly well documented, hard data on distances travelled, duration and actual routes are still scant.

In his study of wolves' extraterritorial movements in southwest Québec (Canada), Messier (1985) recorded 56 instances of solitary excursions by 23 different individuals, based on aerial radio-fixes taken every 1 or 2 days. For pups ($n = 9$ excursions), yearlings ($n = 21$) and adults ($n = 26$), the average durations ($\pm$ SE) of these excursions were 9.1 ($\pm$ 2.6), 13.3 ($\pm$ 3.2) and 10.3 ($\pm$ 2.1) days, respectively; and their average one-way lengths, i.e. bee-line distances from territorial border to farthest place reached were 22.2 ($\pm$ 5.3), 21.7 ($\pm$ 2.6) and 25.5 ($\pm$ 4.2) km, respectively. The narrowness of standard error ranges suggests fairly stereotyped patterns of distance and duration. But bee-line orientation of these excursions (from original home range to farthest place reached) could be in any direction. When specifically asked about actual routes, Messier (personal communication) commented that all fixes for a given excursion were within a 25° arc.

Other authors reported detailed data on excursions by producing actual maps of outbound and inbound (homing) routes followed. Figure 8.4 gives examples for eight species. With two exceptions, these examples can be viewed as typical excursions away from and back to one's home range. The first exception is that shown for a bear, the example given here (200 km) being much longer than the typical 30 km or so. The second exception is that shown for a stoat, which was so often absent from its 'home range' that the latter should be considered as a 'base camp' at most. The polecat, the stoat, the bushbaby and the squirrel made their excursions to visit potential mates. The fox made this and other similar excursions immediately after settling down as a disperser in its new home range. The bear and the mouse probably made theirs for alimentary reasons; the bear stayed more than 1 month, and the mouse 30 min in the distant area before moving back home. The deer moved away from its home range to escape hunting dogs; the homing part of its excursion was not mapped in the original publication, but inferred from a statement by Sweeney *at al.* (1971) to the effect that while returning, the deer took the most direct routes to their ranges rather than backtracking along the chase routes.

The main features of excursion routes are (1) that both the outbound and the homing routes are typically 'fairly straight', and (2) that the homing routes are not necessarily backtracks of the outbound routes (Fig. 8.4). Two further points not shown in Fig. 8.4 but documented in the original publications are (3) that there is usually no evidence that the routes are linked with topographical corridors other than accidentally, and (4) that for a given species, excursions tend to be characterized by a standard, species-specific length (except for the bear, the examples in Fig. 8.4 correspond to

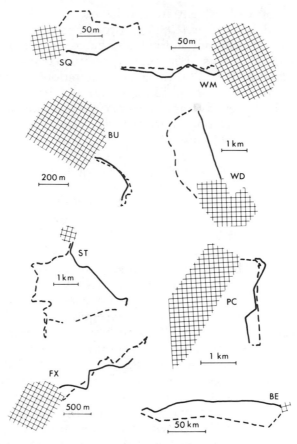

Figure 8.4 Examples of excursion routes. Cross-hatched areas are home ranges, dotted areas are excursions' goals. Dashed lines are outbound trips, full lines are homing trips. SQ: red squirrel (*Tamiasciurus hudsonicus*), radio-assisted surveillance (after Lair, in Bovet, 1984). WM: wood mouse (*Apodemus sylvaticus*), direct visual observation with chemiluminescent marker (after Benhamou, 1990 and personal communication). BU: lesser bushbaby (*Galago senegalensis*), radio-fixes from mobile ground stations (after Bearder and Martin, 1979). WD: white-tailed deer (*Odocoileus virginianus*), radio-fixes from mobile ground stations and direct visual observation (after Sweeney *et al.*, 1971). ST: stoat (*Mustela erminea*), radio-assisted surveillance and radio-fixes from mobile ground stations (after Sandell, 1986). PC: polecat (*Mustela putorius*), radio-assisted surveillance, tracking discontinued before polecat actually re-entered home range (after Weber, 1989 and personal communication). FX: red fox (*Vulpes vulpes*), radio-fixes from stationary ground stations (after Fabrigoule and Maurel, 1982 and personal communication). BE: black bear (*Ursus americanus*), aerial radio-fixes (after Rogers, 1987a). See text for further details.

that standard). There is, however, a well-documented partial exception to this set of 'rules'. The excursionist Columbian ground squirrels studied by Wiggett *et al.* (1989) followed topographical features, such as an old road or the top of a steep river bank (that were admittedly relatively straight), and they backtracked along the outbound routes while homing.

8.5 EXPERIMENTAL INDUCTION OF HOMING

8.5.1 Objectives of homing experiments, past and current

Like in other taxonomic groups, there are two broad categories of experimental studies of homing in mammals. Experiments of the first category are based on the displacement of free-ranging animals and their release off home range limits. They thus induce actual homing attempts on the part of the animals. Because of the 'real world' distances involved, they are made at least partly in the field, with all the problems of control of conditions in natural settings. They are the topic of this section. Experiments of the second category are usually carried out in controlled laboratory conditions, with captive animals. They do not imply actual homing in the sense of returning back to one's home range, and often involve some form of training. Their aim is to check specific predictions on mechanisms potentially implied in homing behaviour. These experiments are discussed in section 8.6.

The simplest kind of displacement experiment consists of releasing a mammal outside its home range and then seeing whether or not it can be resighted or recaptured at home. This has been done, often casually, with a number of mammalian species, and has revealed homing abilities across the Class. It is so easy to perform, and it looks so 'naturalistic' that the kind of homing it induces has long been considered as the phenomenon to understand and explain, rather than an experimental artefact for understanding and explaining natural instances of homing. In a seminal paper, Murie and Murie (1931) reported that after deer mice (*Peromyscus maniculatus*) had been displaced in opaque containers and released at distances well beyond their assumed normal range of movements, a small yet sizeable number managed to home, as evidenced by recaptures in the original home ranges. They stated three potential explanations: (1) the mice were sufficiently familiar with the terrain to find their way back; (2) the mice returned by chance; (3) the mice travelled back by means of a 'homing instinct' or 'sense of direction', whatever that may be. The substance of these statements has been the universal theoretical background in the study of homing in mammals for the next 50 years. This has had important conceptual and methodological consequences. Explanations (1) and (3) have been assumed to imply that, in order to be able to home, mammals must use a system of reference acquired or inherited prior to experimental displacement. Fearing interferences from information that the animals may acquire during displacement and eventually use for homing by **trail following** or **route reversal** (see Chapter 1), most authors have purposely tried to suppress this unwanted source of information by displacing their subjects in opaque, little ventilated containers and/or along circuitous routes. This amounts to investigating only the possibility of **true navigation** by preventing the use of **route-based orientation**. On the positive side, however, authors have usually been careful to discriminate between naive individuals, used for the first time in a homing experiment, and experienced ones, used previously, and assumed to have a system of reference improved by previous experimental homing trips.

It is to Baker's (1981, 1982) merit that he stated explicitly that a displacement experiment is a mimic of a natural round-trip excursion or migration, in which the experimenter controls (or manipulates) the features of the outward journey, e.g. its length, its sinuosity, or the type of sensory information available to the animals, etc. It is only since this kind of statement was made that authors (including myself) began looking earnestly at the relationships between natural and experimental homing in mammals, as well as at a possible role for route-based orientation processes other than route reversal.

8.5.2 Homing success

The basic experiment is to displace several individuals in similar conditions (distance, direction, etc.), and see how many (if any) can be eventually located back home, thus yielding a measure of **homing success** (see p. 6). It has been performed systematically and repeatedly with several species of bats, rodents, and large terrestrial carnivores. The use of bats and rodents is easy to understand, given their small size and large numbers. The use of bears or wolves is more surprising. It is actually the by-product of management problems, e.g. how to get rid of so-called 'nuisance' individuals without killing them. Earlier work of this type on bats and rodents has been thoroughly reviewed by Davis (1966) and Joslin (1977), respectively; work on black bears (*Ursus americanus*) has been reviewed recently by Rogers (1986b).

The many basic experiments performed with naive individuals point to a clear decrease in homing success with increasing displace-

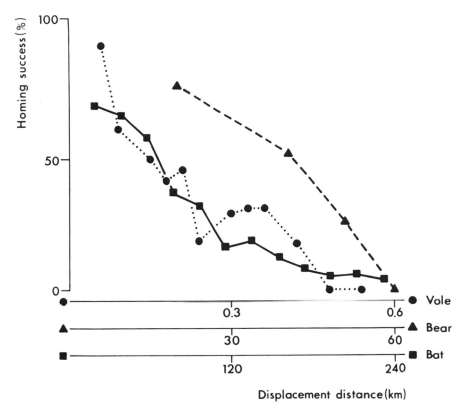

Figure 8.5 Relationship between homing success and displacement distance in naive meadow voles (*Microtus pennsylvanicus*), *n* = 460 (after Robinson and Falls, 1965); black bears (*Ursus americanus*), *n* = 112 (after McArthur, 1981); and Indiana bats (*Myotis sodalis*), *n* = 700 (after Hassell, 1963).

ment distances. Three examples are presented in Fig. 8.5. These experiments also show a clear trend in species with larger home ranges to keep sustained homing success levels over larger distances. So, for instance, muskrats (*Ondatra zibethicus*) have relatively large home ranges and a 50% homing success over distances of the order of 2000 m (Mallach, 1972). Deer mice, cotton rats (*Sigmodon hispidus*), and Gapper's red-backed voles (*Clethrionomys gapperi*) have smaller home ranges, and a 50% homing success over about 350, 350 and 250 m, respectively (Murie, 1963; DeBusk and Kennerly, 1975; Bovet, 1980). Meadow voles (*Microtus pennsylvanicus*) and California voles (*M. californicus*) have still smaller home ranges, and a 50% homing success over about 140 m each (Robinson and Falls, 1965;

Fisler, 1962). Furthermore, house mice (*Mus domesticus*) with large home ranges homed over much larger distances than conspecifics with small home ranges (Anderson *et al.*, 1977).

Displacing rodents along a direct route and/or in sight of the landscape does not seem to improve their homing success (e.g. Robinson and Falls, 1965; Sims and Wolfe, 1976; Bovet, 1991; but see Newsome *et al.*, 1982, for a conclusion to the contrary). Displaced wolves (*Canis lupus*) or bears are usually anaesthetized prior to displacement. There is no indication that this treatment affects their homing success (Fritts *et al.*, 1984; Rogers, 1986b). There is also general agreement that homing times (between release and first resighting at home) are paradoxically independent of displacement distance. This is considered to result from inappropriate means for assessing the exact time when an animal re-enters its home range.

In other aspects, the basic experiment provides less clear-cut evidence. Juvenile rodents and bears have low homing success figures compared to adults (e.g. Seidel, 1961; Murie, 1963; Robinson and Falls, 1965; and Rogers, 1986b, on bears). However, Mueller (1966) did not find differences in homing success between 'older' and 'younger' little brown bats (*Myotis lucifugus*). In some studies females, in others males, have better homing success than the other sex, which likely results from random variations among small samples. The effects that crossing 'unfavourable' habitats might have on homing success of rodents are unclear, ranging from negative (Griffo, 1961) to neutral (Jamon *et al.*, 1986) to positive (DeBusk and Kennerly, 1975).

Several authors have used the basic experiment to evaluate the role of gross sensory functions in the homing process, e.g. whether an animal deprived of vision is still able to home. Answers to such questions are inconsistent, and hard to interpret. Cooke and Terman (1977) and Parsons and Terman (1978) compared the homing success of white-footed mice (*Peromyscus leucopus*) that were either blinded, anosmic or intact, and found no effect for anosmia, and occasional negative effects for blindness. This indicates that vision is useful, but not necessary, and that olfaction is not necessary, but not necessarily useless. Compared to controls, the little brown bats blindfolded by Mueller (1966) had a normal homing success (but increased homing times) over short distances, and a decreased homing success over distances more than 32 km. Williams *et al.* (1966) reported comparable results for the greater spear-nosed bat (*Phyllostomus hastatus*). The difficulty in interpreting this type of result is best exemplified by the five little brown bats that Mueller (1966) deafened by stuffing their ears with cotton: none homed, but also none flew more than 80 m. Their lack of homing success was due primarily to their inability to use their sonar, which is essential for short-range control of flight; but whether deafening also impaired a long-distance orientation mechanism remains unknown.

The basic experiment has also been used to evaluate the effect of experience on homing success, with successful naive homers released a second time either at the same, or at another release site. Homing success and speeds of mouse-like rodents increased in both conditions, compared to first releases (e.g. Robinson and Falls, 1965; Furrer, 1973). Here again, the result is difficult to interpret. It could mean that the mice improved their system of reference during their first homing trip (e.g. they learned the landmarks), or that the first release made a selection for 'good' homers which became the only subjects available for second release. Mueller (1966) performed a similar experiment on little brown bats. Here, homing success was dramatically smaller after second than after first releases, particularly so when second releases were made at the same place as the first!

8.5.3 Initial orientation

As seen in Chapter 7, the analysis of 'vanishing bearings' of experimental subjects when they leave the release site is a powerful tool in the study of bird navigation. Several students of mammals have also tried to record and analyse vanishing bearings. In those situations where releases occurred in a field setting and in conjunction with a basic experiment, most authors stated that the vanishing bearings of brush rabbits (*Sylvilagus bachmani*) and all sorts of rodents were distributed randomly with respect to the home direction, irrespective of displacement distance and of whether the animals would eventually home or not (e.g. Griffo, 1961; Sims and Wolfe, 1976; Durup, 1982): the behaviour of the animals was clearly oriented toward potential shelters or familiar habitat types (e.g. Lehmann, 1956; Gentry, 1964; Jamon and Bovet, 1987; and Chapman, 1971, on rabbits). Similarly, little brown bats vanished in all directions, by day or night, and irrespective of displacement distance (Mueller, 1966). There is general agreement that when just released the animals are not motivated by homing but by seeking shelter. However, Bovet's (1971) deer mice vanished in the home direction, which could have resulted from a uniform distribution of potential shelters around the release sites (on a flat snow field), and/or from the fact that most of them were experienced homers. On the other hand, Jamon and Benhamou (1989) performed an unusual experiment in which they examined the vanishing bearings of wood mice (*Apodemus sylvaticus*) that had been attracted to traps set several tens of metres outside their home range. When released there following these active outbound forays with outward journey information available, the mice vanished in the home direction.

Another type of experiment has been performed with several kinds of rodents and with domestic cats (*Felis catus*). It consists of releasing displaced animals in the centre of a radially symmetrical arena (diameter 1–2 m) and recording their bearings when they reach the arena's periphery, or their relative amounts of activity in any radial sector. The same animals are used repeatedly, with or without being displaced back to their home range between tests. With respect to a possible homeward bias of these intra-arena forays or activity distribution patterns, the results are elusive. They range from clearly negative (Fisler, 1967; Freye and Pontus, 1973; Jamon *et al.*, 1986) to 'inconsistent' (Bovet, 1960; Fluharty *et al.*, 1976) to clearly positive (Lindenlaub, 1960; Mather and Baker, 1980; Karlsson, 1984; and Precht and Lindenlaub, 1954, on cats). There are obvious problems of replication among studies using similar methods and species (e.g. Lindenlaub, 1960 vs Bovet, 1960; Mather and Baker, 1980 vs Jamon *et al.*, 1986), or even within the same study (August *et al.*, 1989).

Initial orientation data thus appear to be marred with so many problems that one wonders whether they can be considered as a reliable tool for the study of homing in mammals.

8.5.4 Intermediate orientation

To alleviate the shortcomings of initial orientation studies, several authors have tried to evaluate the 'intermediate orientation' of their animals, i.e. their position at a short distance from the release site, at a time when they are assumed to be on their homing trip rather than seeking shelter. For small terrestrial mammals, a common technique is to set lines of traps along circles (or other closed figures) centred on the release site, with radii of a few tens of metres. Radio-fixes can be taken at a few kilometres from the release site for larger animals. Generally speaking, the results of such studies on rodents [or on a marsupial, the quokka (*Setonix brachyurus*)] suggest a random distribution of places of capture with respect to the home direction (Murie, 1963; Robinson and Falls, 1965; and Packer, 1963, on quokkas). However, when the data are analysed to eliminate the individuals which will eventually remain in the release site area (see p. 339) and/or to take into account that the shelter used by an animal is the real starting place of its homing trip (and not the release site), then a certain trend for the clustering of places of capture around the home direction may become apparent (Bovet, 1972, 1980). On the other hand, radio-fixes of larger mammals taken shortly after the animals have left the release site area tend to be clustered in the home direction [brown bears (*Ursus arctos*; Miller and Ballard, 1982); wolves (Fritts *et al.*, 1984)]. But in none of the cases discussed here is there any evidence that eventual homers are more home-oriented than non-homers.

8.5.5 Final fate of non-homers

The basic experiment says nothing about the final whereabouts of 'lost' animals, except that they were not relocated at home. This type of information is therefore found only in displacement experiments where radio-fixes, trapping or other procedures (e.g. records by hunters, wildlife officers, etc.) were used systematically and long enough at places other than previous home ranges. In such studies, a number of non-homers are found to establish a new home range in the release site area. This is documented for many mouse-like (e.g. Fisler, 1962, 1966; Robinson and Falls, 1965) or larger rodents [muskrat (Mallach, 1972); red squirrel (*Tamiasciurus hudsonicus*; Bovet, 1984, 1991); nutria (*Myocastor coypus*; Wolfe and Bradshaw, 1986)], for lagomorphs [brush rabbit (Chapman, 1971); brown hare (*Lepus europaeus*; Broekhuizen and Maaskamp, 1982)], and for the marten (*Martes americana*; Slough, 1989), with no simple relationship, if any, between displacement distance and the proportion of subjects which settle in the release site area (Bovet, 1972, 1980). By trapping at appropriate locations, Bovet (1978) recaptured 51 of 159 'lost' wood mice 250 m or more from the release site. A large majority (84%) of these recaptures were made at 250 m, and only 10 and 6% at 500 and 750 m, respectively. Fritts *et al.* (1984) recorded an 'end-point' for 25 of their 100 'lost' wolves. The range of distances between the release site and end-point was 23–302 km, but 60% were less than 70 km.

In black bears (Rogers, 1987b), the distribution of end-points around the release sites suggests that most non-homers did travel in the general direction of home (Fig. 8.6a). The matter is ambiguous for the wolves studied by Fritts *et al.* (1984; Fig. 8.6b). Bovet's (1978) non-homer wood mice were found randomly in all directions with respect to home (Fig. 8.6c).

8.5.6 Routes travelled while homing

The review that follows is based on a set of studies featuring a vast array of recording techniques: direct visual observation with or

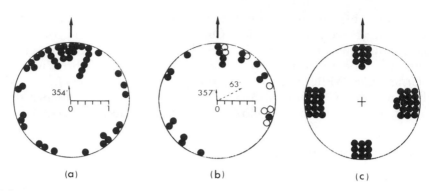

(a)   (b)   (c)

Figure 8.6 Bearings of end-points of non-homers after experimental displacement in three species of mammals, as seen from the release site (centre of circular diagrams) with respect to home direction (thick arrow). The thin arrows emanating from the centre of (a) and (b) represent mean vectors, whose lengths can be read with the scale. All subjects were naive. (a) Black bears (*Ursus americanus*), $n = 43$; all subjects displaced 64 km or more, end-points determined by radio-fixes, recaptures or kills: distances from release site to end-points not specified; the bearings are significantly clustered about a mean 354° ($P < 0.002$) (after Rogers, 1987b). (b) Wolves (*Canis lupus*), $n = 25$; subjects displaced any distance between 55 and 321 km, end-points determined as for bears; filled circles are end-points less than 100 km from release site, their average bearing is 357°, but their distribution is not significantly different from uniformity ($P \approx 0.2$); open circles are end-points more than 100 km from the release site, they are significantly clustered about a mean 63° ($P < 0.05$), which might be related to topographical features (after Fritts *et al.*, 1984). (c) Wood mice (genus *Apodemus*), $n = 51$; subjects displaced 250, 500 or 750 m, with opportunities to be recaptured as non-homers in the home direction, or 90°, 180° or 270° from it; distances from release site to end-points are 250, 500 or 750 m; the distribution of recaptures in the four directions is not significantly different from uniformity ($P \approx 0.3$) (after Bovet, 1978).

without luminescent markers; tracks in the snow; radio-locations taken virtually continuously or at intervals from receiving stations that were on the ground (fixed or mobile) or airborne; and radio-assisted surveillance. The inter-study consistency of results suggests that they are not methodological artefacts.

After release, most animals stay for minutes or hours in the first sheltered location encountered, grooming, dozing and/or watching [e.g. brush rabbits (Chapman, 1971); Brazilian spiny rats (*Proechimys roberti*; Alho, 1980); wood mice (Jamon *et al.*, 1986)], or engage for minutes or hours in an often frenzied, apparently aimless search, roaming close to and about the release site, with occasional breaks [e.g. domestic dogs (*Canis familiaris*; Müller, 1965); greater spear-nosed bats (Williams and Williams, 1970); boars (*Sus scrofa*; Lozan, 1980); bank voles (*Clethrionomys glareolus*; Mironov and Kozhevnikov, 1982); red squirrels (Bovet, 1984)]. Some stay several days near the release site before moving away [e.g. brush rabbits (Chapman, 1971); wolves (Fritts *et al.*, 1984)]. After some time, however, virtually all the animals move away along a long initial straight course (ISC). Some will keep to the same general direction. Others will make an obvious 'U'-turn after a while, return along a straight course closer to the release site without backtracking the ISC other than accidentally, move off in another direction, possibly return again, and so on. Figure 8.7 gives examples of these two patterns, which can co-exist within the same species or population.

The evidence about whether or not these ISCs are homeward oriented is inconsistent as yet. Investigators of dogs (Müller, 1965), wolves (Fritts *et al.*, 1984), black bears (Alt *et al.*, 1977) and brown bears (Miller and Ballard, 1982) say that they are, but do not provide many details. According to Lozan's (1980) schematic description, boars proceed from the start toward home, making long zigzags that are perpendicular to the home direction. Bovet (1984) found a definite, yet complex relationship between home direction and orientation in his red squirrels' ISCs, irrespective of displacement distance. Mironov and Kozhevnikov (1982) reported none for their bank voles. When released in an unusual habitat, several rodents oriented toward familiar vegetation type rather than toward home (e.g. Alho, 1980; Jamon and Bovet, 1987). However, the tracks made in the snow by Bovet's (1968) deer mice were definitely homeward oriented (most were admittedly from experienced mice, but the two examples in Fig. 8.7 are from naive individuals). In the bats studied by Williams and Williams (1970), the ISCs were clearly home-oriented at 20 km from home, less so at 30 km, and not at all at 60 km. It should be stressed here that the standard displacement procedure in all these experiments aimed at preventing the use of route-based orientation. In a recent experiment, Bovet (1991) found an improvement in the homeward orientation of red squirrels that could see the landscape during the straight-line off home range displacement, compared with squirrels that could not; but this treatment did not improve homing success or speed, or the length of ISCs.

U-turns back to the release site area are found in reports on greater spear-nosed bats (Williams and Williams, 1970), marten (Slough, 1989), red squirrels (Bovet, 1984, 1990, 1991), deer mice (Bovet, 1968), bank voles (Mironov and Kozhevnikov, 1982) and wood mice (Jamon and Benhamou, 1989). They also appear in

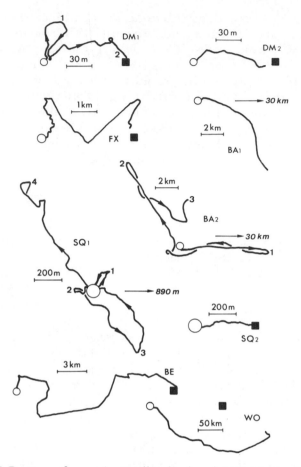

Figure 8.7 Routes of experimentally displaced individuals. Release site areas are identified by circles, original capture sites by black squares, except for BA₁, BA₂ and SQ₁, where their position must be inferred from arrows and distances given. Arrowheads and digits along the tracks of DM₁, BA₂ and SQ₁ indicate the succession of events. DM₁ and DM₂: deer mice (*Peromyscus maniculatus*), tracks in the snow (after Bovet, 1968). FX: red fox (*Vulpes vulpes*), radio-fixes from stationary ground stations (after Fabrigoule and Maurel, 1982). BA₁ and BA₂: greater spear-nosed bats (*Phyllostomus hastatus*), radio-fixes from stationary and mobile ground stations (after Williams, 1968). SQ₁ and SQ₂: red squirrels (*Tamiasciurus hudsonicus*), radio-assisted surveillance (after Bovet, 1990 and unpublished). BE: black bear (*Ursus americanus*), aerial radio-fixes (after Alt *et al.*, 1977). WO: wolf (*Canis lupus*), aerial radio-fixes (after Fritts *et al.*, 1984). All animals were successful in homing, except BA₂ and WO, but including BA₁ and SQ₁.

wolves (Fritts *et al.*, 1982) and black bears (Rogers, 1986a), but less typically and less frequently, which, together with their apparent absence in dogs (Müller, 1965), might reflect a difference between 'larger' and 'smaller' mammals. As a provisional evaluation, the distances travelled from the release site to a first U-turn correspond roughly to the standard one-way length of spontaneous excursions (e.g. compare the foxes or the squirrels in Figs 8.4 and 8.7). In the greater spear-nosed bats and the red squirrels, where their occurrence has been documented for various displacement distances, U-turns are frequent when displacement distance is clearly larger than the standard foraging (bats) or excursion (squirrels) range of the species, irrespective of direction of the first foray away from the release site. With shorter displacement distances (<20 km for bats, <400 m for squirrels), U-turns are common only in those individuals that started in a wrong direction. Squirrels that do not manage to find their way back home after several forays from the release site often quit moving, and establish a new home range not far from the release site (Bovet, 1984, 1990, 1991).

Little information is available concerning the influence of landscape corridors on the orientation of experimentally induced homing attempts. Williams and Williams' (1970) greater spear-nosed bats were unaffected. Müller's (1965) dogs often followed relatively straight topographical features (e.g. roads) for some distance. Chapman's (1971) brush rabbits moved to their home ranges using the available cover, which resulted in fairly zigzagging patterns.

8.6 STRATEGIES AND MECHANISMS

8.6.1 Operational definitions

For the purpose of this section, a homing strategy is a course of action taken by an animal in order to return to its home range. The concept of strategy implies a set of alternative rules of conduct, the choice of which is decided upon depending on circumstances, as and when they appear. On the other hand, mechanisms are means or tools by which the rules can be enforced.

8.6.2 A classical strategy based on random search and memory

The view most commonly held about homing in mammals is that it implies random search and/or the use of previously learned routes (i.e. pilotage or route reversal: see Chapter 1). This view is explained in detail in Davis (1966) and Joslin (1977), and has been

modelled formally by Wilson and Findley (1972) and Furrer (1973).
It states that animals are familiar with an off home range area
visited during previous dispersal, migrations or excursions, and that
they remember landmarks along the routes that lead from and to
the home range. If they are released within this familiar area, the
animals locate such landmarks, and then follow the homing route
they have memorized during previous trips. If they are released
outside their familiar area, known landmarks are missing. The
assumed rule of conduct is then to move straight away from the
release site, hopefully toward the familiar area. But since the animals
have no clue as to where it is, the directions are chosen randomly
with respect to the home direction.

This random–memory strategy accounts reasonably well for
many experimental results, such as the decrease in homing success
when displacement distances increase, or the usual lack of evidence
for initial or intermediate homeward orientation. Juveniles do not
home as well as adults do, because they have not yet acquired an off
home range familiar area. If one reasonably assumes that there is a
standard relationship between the size of the home range and that of
the familiar area around it, it explains why animals with smaller
home ranges do not home as well as animals with larger home
ranges. Furthermore, the models proposed by Wilson and Findley
(1972) and Furrer (1973) predict sizes of familiar areas that, if they
were circular, would have radii congruent with known ranges of
excursions. Without rejecting absolutely any idea of a 'homing
instinct' or 'sense of direction' (*sensu* Murie and Murie, 1931),
Wilson and Findley (1972) consider that the random–memory
strategy is a more parsimonious explanation than 'sophisticated'
navigational abilities.

The random–memory model has weaknesses, however. Its
explanatory powers relative to the decrease in homing success with
increasing displacement distances or to the size of the familiar area
are linked to the necessary condition that all the animals scatter
away from an unfamiliar release site, and that they do so uniformly
(i.e. individual azimuths are chosen randomly from a uniform
circular distribution) and radially (i.e. these azimuths are kept once
moving). This very condition, however, is incompatible with the
distances distribution of final locations of non-homers in experiments on wood mice (genus *Apodemus*; Bovet, 1982) or with the
directional distribution of these final locations in experiments on
black bears (*Ursus americanus*; Rogers, 1987b) (see p. 339). Moreover, the settlements near the release site, the U-turns, the lack of
relationship between individual patterns of initial or intermediate
orientation and eventual status as homer or non-homer that have

been described in previous sections, all indicate that many individuals do not scatter or do not keep their first chosen azimuth. On the other hand, the model implies that animals memorize all places and routes during their spontaneous excursions, and store them into what psychologists call the reference (long-term) memory, by contrast to the working (short-term) memory. It can be questioned whether it is adaptive to keep the details of infrequently used routes in the reference memory.

8.6.3 An alternative strategy based on directions and distances

The model described here stems from the observation that off home range excursions are typically made along straight routes, and over 'standardized' distances; that the return (homing) routes are not necessarily backtracks of the outbound routes; and that the patterns of travel displayed after experimental displacements bear much resemblance to patterns of natural travel, irrespective of displacement distance. Logically then, we may speculate that all a mammal need do for a successful round-trip excursion is to evaluate the general direction of the outbound trip, and to reverse it for the homing trip. Assuming that the similarity of routes in spontaneous and induced travel results from similar rules of conduct, we may further speculate that all a mammal need do for genuine (not accidental) homeward orientation after experimental displacement is to evaluate the direction of the passive outbound trip and reverse it for homing. It would thus home by **course reversal**, as defined in Chapter 1.

In excursions, given their standard range, the angular size of the 'homing target' (home range) is rather large, often several tens of degrees (see Fig. 8.4). The system allows, therefore, for a fair range of errors in the determination of directions, and does not require the accounting of short-range temporary deviations along the route, or of short-range trivial movements around the goal of the excursions. However, it could benefit from a safety device to counter the effects of major orientation mistakes. The U-turns described in the previous section provide a clue as to what the safety device could be: initially at least, they occur after a distance of travel which corresponds to the standard distance of homing routes in an excursion. I speculate, therefore, that a condition where home range is not reached after the standard excursion distance has been travelled is an indication to the animal that it made an orientation mistake. It then goes back to the origin of the homing trip, and tries another direction. This **critical distance strategy** (Bovet, 1984, 1987, 1990) is viewed as adaptive in natural excursions because it keeps a lost animal within short range of its home, which can then be reached by a pattern of systematic search similar to that of some arthropods (see Chapter 3). But it is a handicap to experimental long-distance homing because it concentrates the initial search efforts on too short a range.

In evaluating whether this model accounts for the experimental evidence of the previous section, it should be kept in mind that virtually all of this evidence was obtained after purposeful attempts to eliminate outward journey information, thus the very kind of information the model claims to be necessary for orientation. The part of the model that applies to these results is, therefore, the systematic search only, that explains the decrease in homing success with distance and the lack of homeward orientation at least as well as the random–memory model. Provisionally, it might even be speculated that the attempts to eliminate outward journey information were not altogether successful in the few experiments where some homeward orientation was actually found [e.g. in Bovet (1968, 1972, 1980, 1984), where animals were displaced in opaque containers, but along direct or fairly direct routes, and released without delay, as discussed by Jamon and Benhamou (1989)]. The improvement of initial homeward orientation but not of homing success in red squirrels (*Tamiasciurus hudsonicus*) that had access to visual outward journey information (Bovet, 1991) certainly fulfils a prediction of the directions-and-distances model.

Except for a different distance scale, both models would explain migrations as they explain excursions. There are good indications that in ungulates (McCullough, 1985; Andersen, 1991) and whales (Clapham and Mayo, 1987) at least, migratory routes are 'acquired' by social transmission, which explains the amount of inter-population or inter-individual variability mentioned earlier: young animals make at least a one-way migratory move, if not a complete round-trip with their mother, and once independent they establish seasonal home ranges in the same general areas as their mother. Thus, something is learnt about migratory routes; with the random–memory strategy, it is a series of landmarks; with the alternative strategy, it is a direction, and perhaps also a distance for more safety.

8.6.4 Mechanisms for determining and keeping directions

An important question in this debate on strategies is whether or not mammals are able to evaluate, determine, use, keep or reverse directions and if so by what means: if they are not, the strategy based on directions and distance has no grounds; but if they are, it would be amazing that they do not use this ability for long distance travel. I review here some lines of research which are relevant to the problem.

(a) Idiothetic path integration

In theory, an animal is able to complete round-trip excursions by integrating the proprioceptive information self-generated by its rotations and translations on the way out, and by bringing the integrator 'back to zero' on the way back. This mechanism of **idiothetic path integration** (see Chapter 1) has been modelled for Mongolian gerbils (*Meriones unguiculatus*) by Mittelstaedt and Mittelstaedt (1982), and extensively studied in golden hamsters (*Mesocricetus auratus*) by Etienne and co-workers (e.g. Etienne *et al.*, 1988, and references therein). It is a route-based orientation mechanism which does not require backtracking of the outbound path. Hamsters were actually able to relocate their peripheral nest-site from the centre of a circular arena (diameter about 2 m), using only this type of information collected on their way out from the periphery to the centre. But on theoretical grounds, this mechanism appears to have a complexity far above the needs of the directions-and-distances model: it requires the accurate integration of all small or large turns made and short or long partial distances covered during all phases of the round-trip, and implies the constant up-dating of information stored in the working memory. Furthermore, on the practical side, the system seems to be limited in the amount of cues it can register and compute during any one trip, which, together with its high error-sensitivity, does not make it a likely candidate as a long-distance homing mechanism (Potegal, 1987; Etienne *et al.*, 1988; Benhamou *et al.*, 1990).

(b) Distant landmarks

Several studies using various methods and species have shown that in the laboratory, rodents are able to accurately locate the position of a hidden shelter or food source on the basis of its geometrical relationships with an array of distant visual cues (e.g. Morris, 1981; Collett *et al.*, 1986; Schenk, 1987; Teroni *et al.*, 1987). This mechanism has an obvious value for locating places within a home range [e.g. a grey squirrel's (*Sciurus carolinensis*) food cache (McQuade *et al.*, 1986)]. But, basically, it is a mechanism of pilotage (as defined in Chapter 1), that requires training before an animal can use it to travel straight from a new starting place to the target (Sutherland *et al.*, 1987). It is therefore not a likely candidate to explain orientation from places visited infrequently or for the first time. It is, however, the best explanation for the better homeward orientation of greater spear-nosed bats (*Phyllostomus hastatus*) after shorter displacements (Williams and Williams, 1970; see p. 341).

On the other hand, there is no *a priori* reason why a mammal could not use one or several distant landmarks as route-based orientation cues to assess the direction of the outbound part of an excursion, and reverse it for homing, e.g. to go out 'toward the mountain' and go back 'away from the mountain', or to proceed to a kind of path integration similar to that of the ant, *Cataglyphis fortis* (see Chapter 3). The laboratory results of Teroni *et al.* (1987) and Etienne *et al.* (1990) are not encouraging: when confronted with conflicting information from different sources, their hamsters trusted first location-specific information from distant cues, then idiothetic outward journey information, and last only outward journey information from a distant cue. As I learned from a 30-year-long experience of solving personal homing problems as a field biologist, distant landmarks are usable for determining or keeping a direction when they are available; but they are not always available to the eyes of a standing human, and one can only wonder how often they are available to a mouse whose eyes are only 2–3 cm above ground level.

(c) Sun compass

The possible use of a sun compass has been much less studied in mammals than in arthropods, fish, reptiles or birds, probably on the reasoning that many mammals, being nocturnal, cannot use a sun compass. The ability to use the sun as a compass with time compensation has been demonstrated using standard procedures in the striped field mouse (*Apodemus agrarius*; Lüters and Birukow, 1963), the meadow vole (*Microtus pennsylvanicus*; Fluharty *et al.*, 1976) and the thirteen-lined ground squirrel (*Spermophilus tridecemlineatus*; Haigh, 1979). Big brown bats (*Eptesicus fuscus*) use the post-sunset glow as a directional cue to travel straight from their roost to their habitual foraging area several kilometres away (Buchler and Childs, 1982). If displaced to an unfamiliar area and released at glow time, they leave the release site toward the usual azimuth; but if released later, in the dark, they fly in all possible directions. If released in a planetarium with artificial glow, they fly in the 'correct' direction relative to the glow. The roost and the foraging area being steady, the bats must gradually adjust their flight azimuth relative to the glow as the year progresses. On the other hand, circumstantial evidence led Pilleri and Knuckey (1969) to the conclusion that common dolphins (*Delphinus delphis*) use the sun as a compass when moving along a west–east axis between various parts of their home range.

The major interest of these findings is not so much that mammals can use the sun as a compass, but rather that they can use a 'worldwide' compass, independent of local cues, and that their capability to determine, use and keep real azimuths is more than a mystical possibility. The finding is particularly enlightening in the case of the meadow vole, which is a classical example of a small mammal with a very small range of action (home range size <0.1 ha; Gaulin and FitzGerald, 1988).

(d) Magnetic compass

The question of whether mammals are able to use geomagnetic cues as directional information is still the object of scepticism, if not controversy. There is circumstantial and experimental evidence in support of magnetic orientation abilities. On the circumstantial side, it is claimed that the occurrence of cetacean live strandings is correlated temporally with so-called magnetic storms (Klinowska, 1986) and spatially with magnetic 'topography': these strandings occur at places where isopleths of magnetic intensity are perpendicular to coasts (Klinowska, 1985; Kirschvink *et al.*, 1986). Experimentally, Mather and Baker (1981) have measured the directional preferences of wood mice (*Apodemus sylvaticus*) released in an arena after being displaced 40 m from their home range along a direct route in a normal magnetic field (controls), or in a field with its horizontal component inverse from normal (experimentals). While the controls tended to prefer the homeward sector, the experimentals concentrated their activity in the sector away from home. These results were replicated by August *et al.* (1989), testing white-footed mice (*Peromyscus leucopus*). Using a different paradigm, Burda *et al.* (1990) induced mole-rats (*Cryptomys hottentotus*) that had the habit of building peripheral nests in a certain azimuth with respect to the centre of an experimental arena, to change this azimuth in proportion to changes in the orientation of the horizontal component of the magnetic field around the arena.

The scepticism stems in part from Mather and Baker's (1981) paradigm because, as seen earlier (p. 338), there are difficulties in replicating their control results. Actually, when the diagrams of August *et al.* (1989) are worked out so as to separate their two sets of experiments, it appears that the paradigm 'worked' at one of two study sites only. On the other hand, attempts to modify magnetically the trajectory of golden hamsters in a circular arena have failed (Etienne *et al.*, 1986), as have attempts to train bush opossums (*Monodelphis domestica*) and Djungarian hamsters (*Phodopus sungorus*) to search for food in a magnetically identifiable sector of a four-arm arena similar to that used by Mather and Baker (Madden and Phillips, 1987). The fact that Haigh's (1979) ground squirrels readily used a sun compass rather than a potential magnetic compass in a digging task somewhat analogous to the nest-building of mole-rats in the experiments by Burda *et al.* (1990), suggests either species–specific differences, or a hierarchy of mechanisms where the magnetic compass would rank low.

In their quest for a 'magnetic organ', authors studying swimming, flying or terrestrial mammals have repeatedly found deposits of particulate magnetite that they assume to be involved in magnetic bio-detection (e.g. Kirschvink *et al.*, 1985). If, as suggested by these authors, the appropriate directional cue is the orientation that these particles take during a natural or experimental outward journey, the question of why this mechanism would not be jammed by the magnetic effects of radio-collars, or of automobiles or airplanes used for experimental displacement remains open.

(e) Corridors

As seen in previous sections, there are instances where off home range travel is along landscape corridors, and one can ask whether these corridors are the primary directional cue in such cases. It should first be remembered that travel routes seem to be as often across as along corridors. Furthermore, examination of routes often suggests that mammals quit a corridor when and where it deviates from its original orientation (e.g. homing dogs, *Canis familiaris*; Müller, 1965) or choose the one that is aligned with the original orientation when and where there is a choice between two diverging corridors (e.g. migrating mule deer, *Odocoileus hemionus*; Thomas and Irby, 1990). It seems, therefore, that corridors are followed when and as long as their orientation approximates the general direction of travel, as is the case in migratory birds (Bruderer, 1982). Cetaceans appear to follow straight 'magnetic' corridors, which may cause stranding problems, because there is no relationship between coastal and magnetic topographies (Klinowska, 1985; Kirschvink *et al.*, 1986).

8.7 SUMMARY

Migratory mammals often display site fidelity to at least one seasonal home range, and thus perform homing behaviour when migrating back to that home range. When they are known, migration routes (thus homing routes) appear to be essentially straight (examples in Figs 8.1 and 8.3). The findings suggest, still somewhat inconsistently, that topographical corridors (e.g. valleys) are not a primary factor for the overall orientation of migrations. On the other hand, there is good evidence for much inter-individual variability in the length and orientation of migration routes, in bats and ungulates at least (example in Fig. 8.2).

Short-term excursions away from and back to one's home range are probably common in both migratory and non-migratory mammals. The little that is known about actual excursion routes suggests much similarity with migration routes. Both the outbound and the homing parts of an excursion are fairly straight, without the homing path being necessarily the backtrack of the outbound path (examples in Fig. 8.4), and topographical corridors do not seem to be a primary factor for the overall orientation of travel. But in contrast to migrations, excursions tend to be characterized by standard, species-specific lengths.

Displacement experiments with flying or terrestrial mammals show that homing success decreases with increasing displacement distances (examples in Fig. 8.5), and that individuals with larger home ranges are likely to home over greater distances. Homing success is negatively affected by some forms of gross sensory deprivation, e.g. of vision. On the other hand, homing success does not appear to be related to the sex of the animals, the availability or non-availability of outward journey information, or the orientation taken when leaving the release site area. The effects that age, previous homing experience, or types of habitats to be traversed might have on homing success remain unclear. Many non-homers settle down in new home ranges not far from the release sites. The routes followed by animals when they leave a release site area bear much analogy with routes followed in spontaneous excursions: they are made of straight stretches, the length of which often seems to correspond to the standard length of excursions (examples in Fig. 8.7). In squirrels at least, these routes reveal a pattern of systematic search by radial forays about the release site that is limited in time. It is doubtful that the first forays of small mammals are homeward oriented when attempts are made to prevent the animals from getting outward journey information during displacement. However, the few deliberate attempts made at providing such

information have resulted in a clear homeward orientation of initial
forays without, paradoxically, improving homing success. On the
other hand, the reports available on large terrestrial mammals
indicate homeward orientation in both homers and non-homers.
The lack of systematic displacement experiments with aquatic
mammals is noteworthy.

The literature discusses two types of strategies to explain homing
in mammals. One view, based primarily on large sets of indirect
data (i.e. CMR data) on experimentally induced homing trips, is
that chance and long-term memory of landmarks along routes
travelled previously are the key factors. The other view, based
primarily on considerably smaller sets of direct records of natural
and experimental homing routes, is that determination and short-
term memory of the gross direction of the outward journey is
the key factor, long-term memory of the magnitude of distances
usually travelled being an accessory safety factor, so to speak.
Laboratory studies show that, similar to arthropods or other
vertebrates, mammals can use several mechanisms of direction
determination that do not rely on close, local landmarks. Whether
and to what extent they use them in real world migrations or
excursions remains to be demonstrated.

8.8 APPENDIX: HOMING IN HUMANS

According to Jaccard (1932), the interest of the 'learned world' for
homing behaviour in humans goes back to reports by modern
(post-fifteenth century) European explorers on the amazing abilities
of their native guides in America, Asia, Africa or Australia to show
and/or trace the direct way back home from remote and (for the
explorers at least) cueless areas. Attempts to relate these human
feats to spectacular cases of homing in animals (mostly dogs, but
also horses, cats, insects and crabs) made an important part of a
series of articles, letters and comments on the general theme of
'Perception and instinct in the lower animals', fuelled by no less
than Charles Darwin, A.R. Wallace and G.J. Romanes (among
others) and published in the journal Nature in 1873 (Vols 7 and 8).
Later on, Jaccard (1932) made a critical review of the explorers'
or mountaineers' reports on human homing or direction finding
abilities that were available at the time. Essentially, his conclusions
were to the effect that these abilities must be acquired by individual
experience and training, and require good powers of attention,
memory and observation. Vision is the only sensory function
involved. Using current terminology, he might have said that
pilotage based on a cognitive map is used for orientation in a

familiar area; and that route-based orientation (route reversal, course reversal, or path integration, depending on circumstances) based on visual outward journey information is used for homing from an unfamiliar area. Jaccard also emphasized that explorers, mountaineers and their guides could get lost and that these cases were much more common than publicized. This comment conveys the same message as a remark made by Lewes (1873) on dogs: 'for one dog who finds his way home, hundreds are helpless when lost'.

Interest for human homing abilities was sparked again in the early 1980s by experiments by Baker (1980, 1981) in which, among other things, he displaced naive, untrained, blindfolded individuals by bus along circuitous routes, and then asked them to indicate the direction of the distant starting place. When analysed the same way as vanishing bearings in experiments with other animals, the distribution of individual directional choices was non-uniform, with a significant component in the home direction. Moreover, this 'correct' orientation could be significantly altered by the appropriate placement of magnets or activated Helmholtz coils against or around the head. Baker concluded that his subjects displayed route-based orientation abilities using magnetic outward journey information. Replicates of these experiments by other authors, including the control situation with no magnetic interference, have yielded inconsistent conclusions (Gould and Able, 1981; Westby and Partridge, 1986; and several papers by various authors in Kirschvink *et al.*, 1985). More recently, Baker (1987, 1989) has argued that when taken together for statistical analysis, the other authors' findings confirm his own original conclusions. The argument, however, is based on a procedure that might be unreliable (Bovet, 1992).

On the other hand, Bovet (1990) displaced humans by bus in a pilot experiment designed in the conceptual framework of the directions-and-distances model (p. 345). His results suggest that straightness of displacement route and access to visual outward journey information are essential for accurate homeward orientation.

REFERENCES

Alho, C.J.R. (1980) Homing ability in the wild rodent *Proechimys roberti* determined by radio-telemetry. *Rev. Brasil. Biol.*, **40**, 91–4.

Alt, G.L., Matula, G.J. Jr, Alt, F.W. and Lindzey, J.S. (1977) Movements of translocated nuisance black bears of Northeastern Pennsylvania. *Trans NE Fish Wildl. Conf.*, **34**, 119–26.

Andelt, W.F. and Gipson, P.S. (1979) Home range, activity and daily movements of coyotes. *J. Wildl. Mgt*, **43**, 944–51.

Andersen, R. (1991) Habitat distribution and the migratory behaviour of moose (*Alces alces*) in Norway. *J. Appl. Ecol.*, **28**, 102–8.

Anderson, D.J. (1982) The home range: a new nonparametric estimation technique. *Ecology*, **63**, 103–12.

Anderson, P.K. (1989) *Dispersal in Rodents: a resident fitness hypothesis*, American Society of Mammalogists, Provo.

Anderson, P.K., Heinsohn, G.E., Whitney, P.H. and Huang, J.P. (1977) *Mus musculus* and *Peromyscus maniculatus*: homing ability in relation to habitat utilization. *Can. J. Zool.*, **55**, 169–82.

August, P.V., Ayvazian, S.G. and Anderson, J.G.T. (1989) Magnetic orientation in a small mammal, *Peromyscus leucopus*. *J. Mammal.*, **70**, 1–9.

Baker, R.R. (1980) Goal orientation by blindfolded humans after long-distance displacement: possible involvement of a magnetic sense. *Science*, **210**, 555–7.

Baker, R.R. (1981) *Human Navigation and the Sixth Sense*, Hodder and Stoughton, London.

Baker, R.R. (1982) *Migration: paths through time and space*, Hodder and Stoughton, London.

Baker, R.R. (1987) Human navigation and magnetoreception: the Manchester experiments do replicate. *Anim. Behav.*, **35**, 691–704.

Baker, R.R. (1989) *Human Navigation and Magnetoreception*, Manchester University Press, Manchester.

Ballenberghe, V. van (1983) Extraterritorial movements and dispersal of wolves in Southcentral Alaska. *J. Mammal.*, **64**, 168–71.

Bearder, S.K. and Martin, R.D. (1979) The social organization of a nocturnal primate revealed by radio-tracking, in *Handbook on Biotelemetry and Radio-Tracking* (eds C.J. Amlaner and D.W. Macdonald), Pergamon Press, Oxford, pp. 633–48.

Beentjes, M.P. (1989) Haul-out patterns, site fidelity and activity budgets of male Hooker's sea lions (*Phocarctos hookeri*) on the New Zealand mainland. *Mar. Mamm. Sci.*, **5**, 281–97.

Benhamou, S. (1990) An analysis of movements of the wood mouse (*Apodemus sylvaticus*) in its home range. *Behav. Proc.*, **22**, 235–50.

Benhamou, S., Sauvé, J.P. and Bovet, P. (1990) Spatial memory in large scale movements: efficiency and limitation of the egocentric coding process. *J. Theor. Biol.*, **145**, 1–12.

Bideau, E., Vincent, J.P., Maublanc, M.L. and Gonzalez, R. (1987) Dispersion chez le jeune chevreuil (*Capreolus capreolus* L.): étude sur une population en milieu forestier. *Acta Oecol. (Oecol. Appl.)*, **8**, 135–48.

Bovet, J. (1960) Experimentelle Untersuchungen über das Heimfindevermögen von Mäusen. *Z. Tierpsychol.*, **17**, 728–55.

Bovet, J. (1968) Trails of deer mice (*Peromyscus maniculatus*) traveling on the snow while homing. *J. Mammal.*, **49**, 713–25.

Bovet, J. (1971) Initial orientation of deer mice (*Peromyscus maniculatus*) released on snow in homing experiments. *Z. Tiepsychol.*, **28**, 211–16.

Bovet, J. (1972) Displacement distance and quality of orientation in a homing experiment with deer mice (*Peromyscus maniculatus*). *Can. J. Zool.*, **50**, 845–53.

Bovet, J. (1978) Homing in wild myomorph rodents: current problems, in *Animal Migration, Navigation and Homing* (eds K. Schmidt-Koenig and W.T. Keeton), Springer, Berlin, pp. 405–12.

Bovet, J. (1980) Homing behavior and orientation in the red-backed vole, *Clethrionomys gapperi*. *Can. J. Zool.*, **58**, 754–60.

Bovet, J. (1982) Homing behavior of mice: test of a 'randomness'-model. *Z. Tierpsychol.*, **58**, 301–10.

Bovet, J. (1984) Strategies of homing behavior in the red squirrel, *Tamiasciurus hudsonicus*. *Behav. Ecol. Sociobiol.*, **16**, 81–8.

Bovet, J. (1987) Cognitive map size and homing behavior, in *Cognitive Processes and Spatial Orientation in Animal and Man*, Vol. 1 (eds P. Ellen and C. Blanc-Thinus), Nijhoff, Dordrecht, pp. 252–65.

Bovet, J. (1990) Orientation strategies for long distance travel in terrestrial mammals, including humans. *Ethol. Ecol. Evol.*, **2**, 117–28.

Bovet, J. (1991) Route-based visual information has limited effect on the homing performance of red squirrels, *Tamiasciurus hudsonicus*. *Ethology*, **87**, 59–65.

Bovet, J. (1992) Combining *V*-test probabilities in orientation studies: a word of caution. *Anim. Behav.*, **44** (in press).

Boyce, M.C. (1991) Migratory behavior and management of elk (*Cervus elaphus*). *Appl. Anim. Behav. Sci.*, **29**, 239–50.

Broekhuizen, S. and Maaskamp, F. (1982) Movement, home range and clustering in the European hare (*Lepus europaeus* Pallas) in the Netherlands. *Z. Säugetierkd.*, **47**, 22–32.

Bruderer, B. (1982) Do migrating birds fly along straight lines?, in *Avian Navigation* (eds F. Papi and H.G. Wallraff), Springer, Berlin, pp. 3–14.

Buchler, E.R. and Childs, S.B. (1982) Use of post-sunset glow as an orientation cue by the big brown bat (*Eptesicus fuscus*). *J. Mammal.*, **63**, 243–7.

Buffrenil, V. de, Dziedzic, A. and Robineau, D. (1989) Répartition et déplacements des dauphins de Commerson (*Cephalorhynchus commersonii* (Lacépède, 1804)) dans un golfe des îles Kerguelen; données du marquage individuel. *Can. J. Zool.*, **67**, 516–21.

Burda, H., Marhold, S., Westenberger, T., Wiltschko, R. and Wiltschko, W. (1990) Magnetic compass orientation in the subterranean rodent *Cryptomys hottentotus* (Bathyergidae). *Experientia*, **46**, 528–30.

Chapman, J.A. (1971) Orientation and homing of the brush rabbit (*Sylvilagus bachmani*). *J. Mammal.*, **52**, 686–9.

Clapham, P.J. and Mattila, D.K. (1990) Humpback whale songs as indicators of migration routes. *Mar. Mamm. Sci.*, **6**, 155–60.

Clapham, P.J. and Mayo, C.A. (1987) Reproduction and recruitment of individually identified humpback whales, *Megaptera novaeangliae*, observed in Massachusetts Bay, 1979–1985. *Can. J. Zool.*, **65**, 2853–63.

Collett, T.S., Cartwright, B.A. and Smith, B.A. (1986) Landmark learning and visuo-spatial memories in gerbils. *J. Comp. Physiol. A*, **158**, 835–51.

Cooke, J.A. and Terman, C.R. (1977) Influence of displacement distance and vision on homing behavior of the white-footed mouse (*Peromyscus leucopus noveboracensis*). *J. Mammal.*, **58**, 58–66.

Cowan, P.E. (1983) Exploration in small mammals: ethology and ecology, in *Exploration in Animals and Humans* (eds J. Archer and L.I.A. Birke), Van Nostrand Reinhold, Wokingham, pp. 147–75.

Craighead, D.J. and Craighead, J.J. (1987) Tracking caribou using satellite telemetry. *Nat. Geogr. Res.*, **3**, 462–79.

Davis, R. (1966) Homing performance and homing ability in bats. *Ecol. Monogr.*, **36**, 201–37.

DeBusk, J. and Kennerly, T.E. Jr (1975) Homing in the cotton rat, *Sigmodon hispidus* Say and Ord. *Am. Midl. Nat.*, **93**, 149–57.

Derocher, A.E. and Stirling, I. (1990) Distribution of polar bears (*Ursus maritimus*) during the ice-free period in Western Hudson Bay. *Can. J. Zool.*, **68**, 1395–403.

Durup, M. (1982) Etude du retour au gîte du mulot sylvestre (*Apodemus sylvaticus* L.). *Biol. Behav.*, **7**, 277–91.

Etienne, A.S., Maurer, R. and Saucy, F. (1988) Limitations in the assessment of path dependent information. *Behaviour*, **106**, 81–111.

Etienne, A.S., Maurer, R., Saucy, F. and Teroni, E. (1986) Short-distance homing in the golden hamster after a passive outward journey. *Anim. Behav.*, **34**, 696–715.

Etienne, A.S., Teroni, E., Hurni, C. and Portenier, V. (1990) The effect of a single light cue on homing behaviour of the golden hamster. *Anim. Behav.*, **39**, 17–41.

Fabrigoule, C. and Maurel, D. (1982) Radio-tracking study of foxes' movements related to their home range. A cognitive map hypothesis. *Q. J. Exp. Psychol.*, **34B**, 195–208.

Fancy, S.G., Pank, L.F., Whitten, K.R. and Regelin, W.L. (1989) Seasonal movements of caribou in arctic Alaska as determined by satellite. *Can. J. Zool.*, **67**, 644–50.

Fenton, B.M. and Thomas, D.W. (1985) Migrations and dispersal of bats (Chiroptera), in *Migration: mechanisms and adaptive significance* (ed. M.A. Rankin), Marine Science Institute, Port Aransas, pp. 409–24.

Fisler, G.F. (1962) Homing in the California vole, *Microtus californicus*. *Am. Midl. Nat.*, **68**, 357–68.

Fisler, G.F. (1966) Homing in the Western harvest mouse, *Reithrodontomys megalotis*. *J. Mammal.*, **47**, 53–8.

Fisler, G.F. (1967) An experimental analysis of orientation to the homesite in two rodent species. *Can. J. Zool.*, **45**, 261–8.

Fluharty, S.L., Taylor, D.H. and Barrett, G.W. (1976) Sun-compass orientation in the meadow vole *Microtus pennsylvanicus*. *J. Mammal.*, **57**, 1–9.

Freye, H.A. and Pontus, H. (1973) Die Heimfindeleistungen von Hausmäusen (*Mus musculus* Linné 1758). *Biol. Rundschau*, **11**, 84–94.

Fritts, S.H., Paul, W.J. and Mech, L.D. (1984) Movements of translocated wolves in Minnesota. *J. Wildl. Mgt*, **48**, 709–21.

Fuller, T.K. (1989) Population dynamics of wolves in North-Central Minnesota. *Wildl. Monogr.*, **105**, 1–41.

Furrer, R.K. (1973) Homing of *Peromyscus maniculatus* in the channelled scablands of East-Central Washington. *J. Mammal.*, **54**, 466–82.

Garrott, R.A., White, G.C., Bartmann, R.M., Carpenter, L.H. and Alldredge, A.W. (1987) Movements of female mule deer in Northwest Colorado. *J. Wildl. Mgt*, **51**, 634–43.

Gashwiler, J.S. (1959) Small mammal study in West-Central Oregon. *J. Mammal.*, **40**, 128–39.

Gaulin, S.J.C. and FitzGerald, R.W. (1988) Home-range size as a predictor of mating systems in *Microtus*. *J. Mammal.*, **69**, 311–19.

Geist, V. (1971) *Mountain Sheep*, University of Chicago Press, Chicago.

Gentry, J.B. (1964) Homing in the old-field mouse. *J. Mammal.*, **45**, 276–83.

Gese, E.M., Rongstad, O.J. and Mytton, W.R. (1989) Changes in coyote movements due to military activity. *J. Wildl. Mgt*, **53**, 334–9.

Giger, R.D. (1973) Movements and homing in Townsend's mole near Tillamook, Oregon. *J. Mammal.*, **54**, 648–59.

Gopalakrishna, A. (1985/86) Migratory patterns of some Indian bats. *Myotis*, **23/4**, 223–7.

Gould, J.L. and Able, K.P. (1981) Human homing: an elusive phenomenon. *Science*, **212**, 1061–3.

Griffo, J.V. Jr (1961) A study of homing in the cotton mouse, *Peromyscus gossypinus*. *Am. Midl. Nat.*, **65**, 257–89.

Gruell, G.E. and Papez, N.J. (1963) Movements of mule deer in Northeastern Nevada. *J. Wildl. Mgt*, **27**, 414–22.

Haigh, G.R. (1979) Sun-compass orientation in the thirteen-lined ground squirrel, *Spermophilus tridecemlineatus*. *J. Mammal.*, **60**, 629–32.

Harris, S., Cresswell, W.J., Forde, P.G., *et al.* (1990) Home range analysis using radio-tracking data – a review of problems and techniques particularly as applied to the study of mammals. *Mammal Rev.*, **20**, 97–123.

Hassell, M.D. (1963) A study of homing in the Indiana bat, *Myotis sodalis*. *Trans. Kentucky Acad. Sci.*, **24**, 1–4.

Jaccard, P. (1932) *Le Sens de la Direction et l'Orientation Lointaine chez l'Homme*. Payot, Paris.

Jameson, R.J. (1989) Movements, home range, and territories of male sea otters off Central California. *Mar. Mamm. Sci.*, **5**, 159–72.

Jamon, M. and Benhamou, S. (1989) Orientation and movement patterns of wood mice (*Apodemus sylvaticus*) released inside and outside a familiar area. *J. Comp. Psychol.*, **103**, 54–61.

Jamon, M., Benhamou, S. and Sauvé, J.P. (1986) Initial orientation and navigation in homing rodents, in *Orientation in Space* (ed. G. Beugnon), Privat, Toulouse, pp. 45–55.

Jamon, M. and Bovet, P. (1987) Possible use of environmental gradients in orientation by 'homing' wood mice. *Behav. Proc.*, **15**, 93–107.

Jewell, P.A. (1966) The concept of home range in mammals, in *Play, Exploration and Territory in Mammals* (eds P.A. Jewell and C. Loizos), *Symp. Zool. Soc. Lond.*, **18**, 85–109.

Jones, W.T. (1989) Dispersal distances and the range of nightly movements in Merriam's kangaroo rats. *J. Mammal.*, **70**, 27–34.

Joslin, J.K. (1977) Rodent long distance orientation ('homing'). *Adv. Ecol. Res.*, **10**, 63–89.

Karlsson, A.F. (1984) Age-differential homing tendencies in displaced bank voles, *Clethrionomys glareolus. Anim. Behav.*, **32**, 515–19.

Kirschvink, J.L., Dizon, A.E. and Westphal, J.A. (1986) Evidence from strandings for geomagnetic sensitivity in Cetaceans. *J. Exp. Biol.*, **120**, 1–24.

Kirschvink, J.L., Jones, D.S. and MacFadden, B.J. (eds) (1985) *Magnetite Biomineralization and Magnetoreception in Organisms*, Plenum, New York.

Klinowska, M. (1985) Cetacean live stranding sites relate to geomagnetic topography. *Aquat. Mamm.*, **11**, 27–32.

Klinowska, M. (1986) Cetacean live stranding dates relate to geomagnetic disturbances. *Aquat. Mamm.*, **11**, 109–19.

Kraus, S.D., Prescott, J.H., Knowlton, A.R. and Stone, G.S. (1986) Migration and calving of right whales (*Eubalaena glacialis*) in the Western North Atlantic. *Rep. Int. Whal. Comm, Special issue*, **10**, 139–44.

Lehmann, E. von (1956) Heimfindeversuche mit kleinen Nagern. *Z. Tierpsychol.*, **13**, 485–91.

Lewes, G.H. (1873) Instinct. *Nature*, **7**, 437–8.

Lindenlaub, E. (1960) Neue Befunde über die Anfangsorientierung von Mäusen. *Z. Tierpsychol.*, **17**, 555–78.

Lozan, A.M. (1980) Orientation of the wild boar toward positive stimuli at medium and great distances. *Vest. Zool.*, **1980**, 56–60 (in Russian).

Lüters, W. and Birukow, G. (1963) Sonnenkompaßorientierung der Brandmaus (*Apodemus agrarius* Pall.). *Naturwissenschaften*, **50**, 737–8.

Madden, R.C. and Phillips, J.B. (1987) An attempt to demonstrate magnetic compass orientation in two species of mammals. *Anim. Learn. Behav.*, **15**, 130–4.

Mallach, N. (1972) Translokationsversuche mit Bisamratten (*Ondatra zibethica* L.). *Anz. Schädlingskd. Pflanzensch.*, **45**, 40–4.

Marsh, H. and Rathbun, G.B. (1990) Development and application of conventional and satellite radio tracking techniques for studying dugong movements and habitat use. *Austr. Wildl. Res.*, **17**, 83–100.

Mather, J.G. and Baker, R.R. (1980) A demonstration of navigation by small rodents using an orientation cage. *Nature*, **284**, 259–62.

Mather, J.G. and Baker, R.R. (1981) Magnetic sense of direction in woodmice for route-based navigation. *Nature*, **291**, 152–5.

McArthur, K.L. (1981) Factors contributing to effectiveness of black bear transplants. *J. Wildl. Mgt*, **45**, 102–10.

McCullough, D.R. (1985) Long range movements of large terrestrial mammals, in *Migration: mechanisms and adaptive significance* (ed. M.A. Rankin), Marine Science Institute, Port Aransas, pp. 444–65.

McQuade, D.B., Williams, E.H. and Eichenbaum, H.B. (1986) Cues used for localizing food by the gray squirrel (*Sciurus carolinensis*). *Ethology*, **72**, 22–30.

Messier, F. (1985) Solitary living and extraterritorial movements of wolves in relation to social status and prey abundance. *Can. J. Zool.*, **63**, 239–45.

Miller, S.D. and Ballard, W.B. (1982) Homing of transplanted Alaskan brown bears. *J. Wildl. Mgt*, **46**, 869–76.

Mironov, A.D. and Kozhevnikov, V.S. (1982) Character of migrations of *Clethrionomys glareolus* within the home range and outside it. *Zool. Zh.*, **61**, 1413–18 (in Russian, with English summary).

Mittelstaedt, H. and Mittelstaedt, M.L. (1982) Homing by path integration, in *Avian Navigation* (eds F. Papi and H.G. Wallraff), Springer, Berlin, pp. 290–7.

Morgantini, L.E. and Hudson, R.J. (1988) Migratory patterns of the wapiti, *Cervus elaphus*, in Banff National Park, Alberta. *Can. Field Nat.*, **102**, 12–19.

Morris, R.G.M. (1981) Spatial localization does not require the presence of local cues. *Learn. Motiv.*, **12**, 239–60.

Mueller, H.C. (1966) Homing and distance-orientation in bats. *Z. Tierpsychol.*, **23**, 403–21.

Murie, M. (1963) Homing and orientation of deermice. *J. Mammal.*, **44**, 338–49.

Murie, O.J. and Murie, A. (1931) Travels of *Peromyscus. J. Mammal.*, **12**, 200–9.

Müller, B. (1965) Experimentelle Untersuchungen über das Heimfinden beim Hund. PhD Dissertation, University of Basle.

Nelson, M.E. and Mech, L.D. (1981) Deer social organization and wolf predation in Northeastern Minnesota. *Wildl. Monogr.*, **77**, 1–53.

Newsome, A.E., Cowan, P.E. and Ives, P.M. (1982) Homing by wild house-mice displaced with or without the opportunity to see. *Austr. Wildl. Res.*, **9**, 421–6.

Packer, W.C. (1963) Homing behaviour in the quokka, *Setonix brachyurus* (Quay and Gaimard) (Marsupialia). *J. R. Soc. West. Austr.*, **46**, 28–32.

Parsons, L.M. and Terman, C.R. (1978) Influence of vision and olfaction on the homing ability of the white-footed mouse (*Peromyscus leucopus noveboracensis*). *J. Mammal.*, **59**, 761–71.

Pilleri, G. and Knuckey, J. (1969) Behaviour patterns of some Delphinidae observed in the Western Mediterranean. *Z. Tierpsychol.*, **26**, 48–72.

Potegal, M. (1987) The vestibular navigation hypothesis: a progress report, in *Cognitive Processes and Spatial Orientation in Animals and Man*, vol. 2 (eds P. Ellen and C. Thinus-Blanc), Nijhoff, Dordrecht, pp. 28–34.

Potvin, F. (1988) Wolf movements and population dynamics in Papineau-Labelle Reserve, Quebec. *Can. J. Zool.*, **66**, 1266–73.

Precht, H. and Lindenlaub, E. (1954) Über das Heimfindevermögen von

Säugetieren I: Versuche an Katzen. *Z. Tierpsychol.*, **11**, 485–94.

Ramsay, M.A. and Andriashek, D.S. (1986) Long distance route orientation of female polar bears (*Ursus maritimus*) in spring. *J. Zool.*, **208**, 63–72.

Riedmann, M. (1990) *The Pinnipeds: seals, sea lions and walruses*, University of California Press, Berkeley.

Robinson, W.L. and Falls, J.B. (1965) A study of homing of meadow mice. *Am. Midl. Nat.*, **73**, 188–224.

Rogers, L.L. (1986a) Homing by radio-collared black bears, *Ursus americanus*, in Minnesota. *Can. Field Nat.*, **100**, 350–3.

Rogers, L.L. (1986b) Effects of translocation distances on frequency of return by adult black bears. *Wildl. Soc. Bull.*, **14**, 76–80.

Rogers, L.L. (1987a) Effects of food supply and kinship on social behavior, movements, and population growth of black bears in Northeastern Minnesota. *Wildl. Monogr.*, **97**, 1–72.

Rogers, L.L. (1987b) Navigation by adult black bears. *J. Mammal.*, **68**, 185–8.

Rongstad, O.J. and Tester, J.R. (1969) Movements and habitat use of white-tailed deer in Minnesota. *J. Wildl. Mgt*, **33**, 366–79.

Sandell, M. (1986) Movement patterns of male stoats *Mustela erminea* during the mating season: differences in relation to social status. *Oikos*, **47**, 63–70.

Schenk, F. (1987) Comparison of spatial learning in woodmice (*Apodemus sylvaticus*) and hooded rats (*Rattus norvegicus*). *J. Comp. Psychol.*, **101**, 150–8.

Seidel, D.R. (1961) Homing in Eastern chipmunk. *J. Mammal.*, **42**, 256–7.

Sims, R.A. and Wolfe, J.L. (1976) Homing behavior of the house mouse (*Mus musculus* L.). *J. Miss. Acad. Sci.*, **21**, 89–96.

Slough, B.G. (1989) Movements and habitat use by transplanted marten in the Yukon Territory. *J. Wildl. Mgt*, **53**, 991–7.

Stone, G.S., Flórez-Gonzalez, L. and Katona, S. (1990) Whale migration record. *Nature*, **346**, 705.

Stone, G.S., Katona, S.K. and Tucker, E.B. (1987) History, migration and present status of humpback whales *Megaptera novaeangliae* at Bermuda. *Biol. Conserv.*, **42**, 133–45.

Sutherland, R.J., Chew, G.L., Baker, J.C. and Linggard, R.C. (1987) Some limitations of the use of distal cues in place navigation by rats. *Psychobiology*, **15**, 48–57.

Swartz, S.L. (1986) Gray whale migratory, social and breeding behavior. *Rep. Int. Whal. Comm., Special issue*, **10**, 207–29.

Sweeney, J.R., Marchinton, R.L. and Sweeney, J.M. (1971) Responses of radio-monitored white-tailed deer chased by hunting dogs. *J. Wildl. Mgt*, **35**, 707–16.

Taylor, K.D. and Quy, R.J. (1978) Long distance movements of a common rat (*Rattus norvegicus*) revealed by radio-tracking. *Mammalia*, **42**, 63–71.

Teroni, E., Portenier, V. and Etienne, A.S. (1987) Spatial orientation of the golden hamster in conditions of conflicting location-based and route-based information. *Behav. Ecol. Sociobiol.*, **20**, 389–97.

Thomas, T.R. and Irby, L.R. (1990) Habitat use and movement patterns by migrating mule deer in Southeastern Idaho. *Northw. Sci.*, **64**, 19–27.

Tuttle, M.D. (1976) Population ecology of the gray bat (*Myotis grisescens*): philopatry, timing and patterns of movement, weight loss during migration, and seasonal adaptive strategies. *Occas. Papers Mus. Nat. Hist. Univ. of Kansas*, **54**, 1–38.

Verme, L.J. (1973) Movements of white-tailed deer in Upper Michigan. *J. Wildl. Mgt*, **37**, 545–52.

Weber, D. (1989) Beobachtungen zu Aktivität und Raumnutzung beim Iltis (*Mustela putorius* L.). *Rev. Suisse Zool.*, **96**, 841–62.

Westby, G.W.M. and Partridge, K.J. (1986) Human homing: still no evidence despite geomagnetic controls. *J. Exp. Biol.*, **120**, 325–31.

Wiggett, D.R., Boag, D.A. and Wiggett, A.D.R. (1989) Movements of intercolony natal dispersers in the Columbian ground squirrel. *Can. J. Zool.*, **67**, 1447–52.

Williams, T.C. (1968) *Nocturnal Orientation Techniques of a Neotropical Bat.* PhD Dissertation, The Rockefeller University.

Williams, T.C. and Williams, J.M. (1970) Radio tracking of homing and feeding flights of a neotropical bat, *Phyllostomus hastatus*. *Anim. Behav.*, **18**, 302–9.

Williams, T.C., Williams, J.M. and Griffin, D.R. (1966) The homing ability of the neotropical bat *Phyllostomus hastatus*, with evidence for visual orientation. *Anim. Behav.*, **14**, 468–73.

Wilson, D.E. and Findley, J.S. (1972) Randomness in bat homing. *Am. Nat.*, **106**, 418–24.

Wolfe, J.L. and Bradshaw, D.K. (1986) Homing behavior of the nutria. *J. Miss. Acad. Sci.*, **31**, 1–4.

Woollard, T. and Harris, S. (1990) A behavioral comparison of dispersing and non-dispersing foxes (*Vulpes vulpes*) and an evaluation of some dispersal hypotheses. *J. Anim. Ecol.*, **59**, 709–22.

Worton, B.J. (1987) A review of models of home range for animal movement. *Ecol. Model.*, **38**, 277–98.

Wuersig, B. (1989) Cetaceans. *Science*, **244**, 1550–7.

Animal index

The groups are listed in the same order as they appear in the book. For practical reasons the arrangement of the taxa varies within the single groups. As a rule, English names are given if they are also mentioned in the text.

Author index

Numbers in *italics* indicate pages on which the complete references are listed

Subject index